Principles of Pavement Engineering

Principles of Pavement Engineering

Third edition

Nick Thom

University of Nottingham, UK

Published by Emerald Publishing Limited, Floor 5,
Northspring, 21-23 Wellington Street, Leeds LS1 4DL.

ICE Publishing is an imprint of Emerald Publishing Limited

Other ICE Publishing titles:
Deterioration and Maintenance of Pavements, Second edition
David Cudworth and Mujib Rahman. ISBN 978-0-7277-6508-6
Highways: The Location, Design, Construction and Maintenance of Road Pavements, Fifth edition
Edited by Coleman O'Flaherty with David Hughes.
ISBN 978-0-7277-5993-1
The Shell Bitumen Handbook, Sixth edition
Robert N. Hunter (ed.). ISBN 978-0-7277-5837-8

A catalogue record for this book is available from the British Library

ISBN 978-1-83549-711-1

Cover photo: Jamesteohart/Shutterstock.com

Commissioning Editor: Michael Fenton
Content Development Editor: Cathy Sellars
Books Production Lead: Emma Sudderick

Typeset by KnowledgeWorks Global Ltd.
Index created by Lyn Nesbitt-Smith
Printed and bound by CPI Group (UK) Ltd, Croydon, CR0 4YY

Contents

Foreword

Pavements are built as ground-bearing structures to protect the subgrade from excessive deformation, so that serviceability under vehicular loading is maintained throughout their operational life. Pavement construction has evolved over the past 2000 years since the Romans built the first roads network. Improving ride quality, pavement durability, safety, sustainability and optimising whole-life cost and user expectations are the main factors that have driven the evolution of pavement engineering.

Roman roads provided durable all-weather infrastructure until the introduction of high-speed motor vehicles in the twentieth century. Asphalt surfacing over granular material was introduced to improve high-speed ride quality. Cement- and asphalt-stabilised bases and concrete pavements have been used to provide better bearing capacity for heavy traffic. The need to construct safe, durable, sustainable and low-maintenance pavement infrastructure has become very important in recent years due to environmental legislation and customer expectations. The use of low carbon materials and resilient design and construction has emerged to address environmental pressures and mitigate the risk of climate change. Pavement asset management has become a key to optimise costs, performance and risks. New technologies have recently emerged, such as telematics for data collection, smart materials and innovative constructions. Therefore, modern pavements are expected to meet users' expectation of safety, ride quality, noise and durability to minimise maintenance and avoid traffic delays, congestion and pollution.

Traditionally, pavements are designed using empirical approaches, where performance under traffic loading and environmental attack is monitored to establish the appropriate construction thicknesses. Analytical design from first principles, using an analysis of stresses and strains induced within a pavement, has been used to investigate a wide range of pavement materials, layer combinations and traffic loading. Hence, pavement engineering has become a science that combines structural engineering and geotechnics to model, analyse and design a multi-layered construction over natural ground. It requires good understanding of materials engineering, drainage, traffic loading and design reliability. Designers should consider the expected serviceability level of the pavement as factors such as deformation, cracking, longitudinal profile deterioration and loss of skid resistance occur. Highway pavements require appropriate skid resistance, minimum cracking/rutting and good ride quality for high-speed vehicle movements and should have low maintenance due to the high cost of lane closures and their impact on users and safety. Airfield pavements require limits on foreign object debris (FOD) damaging aircraft and resulting in costs and operational delays. On the other hand, limited cracking, deformation and poor skid resistance might be acceptable in industrial pavements where slow-moving vehicles are expected. Although the structural behaviour and failure mechanisms of airfield pavements are similar to those of roads, there are some major differences in loading and performance requirements. These include the wheel loading magnitude (up to 30 tonnes), wheel configuration (with up to 28 wheels in an undercarriage), tyre pressure (18 bar for civil aircraft and 30 bar for military aircraft), number of load repetitions expected during the design life and FOD damage requirements. Similarly, port container and industrial pavements are expected to carry heavy loads with large dynamic factors.

Pavement engineering is not commonly taught in undergraduate engineering courses and rarely in postgraduate courses. However, considering its significant contribution to sustainable transport, climate resilience and the economy, a good understanding of the theoretical principles and practical application of design and management is essential.

Principles of Pavement Engineering provides a much-needed document to help university students, researchers, graduate engineers, professional designers and asset managers to deliver sustainable transport infrastructure for the betterment of society. Using a simple approach, the book describes fundamental pavement engineering principles, materials characterisation, traffic loading and practical design, maintenance and asset management methodologies. It covers a wide range of pavements (including highways, streets and urban pavements, heavy-duty pavements for airfields, ports and industrial applications and low-volume roads using unbound construction), loading types and performance expectations.

The third edition of *Principles of Pavement Engineering* provides an improved structure with focus on materials and associated pavement categories covering structural design and maintenance. These include unbound, hot and cold asphalts, hydraulically bound and pavement-quality concrete. Part 1 provides a summary of the design principles, pavement types and applications, traffic loading and design reliability. The fundamental engineering properties of unbound material, including non-linear stress and moisture dependency, are not commonly presented in textbooks in the context of pavement design and maintenance.

Part 2 addresses these issues and also covers practical testing and unbound pavement structural design.

Asphalt binder types and characteristics, and the engineering properties of different asphaltic mixes including the effects of aggregate grading, binder content and sensitivity to temperature and loading time and methods of testing with practical examples, are presented in Part 3. This part also details the structural design and construction of flexible pavement, design methods, asphalt surfacing design and materials options including recycling and low carbon asphalt, pavement deterioration mechanisms and maintenance solutions. Theoretical fundamentals and practical examples and illustrations are also presented. To address sustainability, cold-mix asphalt design, testing and specification, and pavement structural design and maintenance are presented in Part 4.

Hydraulically bound material types, engineering properties and testing requirements, as well as the structural design and maintenance of composite pavements (with asphalt surfacing over hydraulically bound base) and heavy-duty modular block pavements are described in Part 5. Pavement-quality concrete types, engineering properties and testing requirements, as well as the structural design and maintenance of rigid pavements are presented in Part 6. The principles of water movement within a pavement structure and its impact on material properties and drainage provision are detailed in Part 7. Finally, pavement asset management, including whole-life cost optimisation based on network- and project-level monitoring, evaluation, surveys and testing to diagnose and propose maintenance and strengthening solutions, is presented in Part 8.

I have known and worked with the author for over 25 years and shared the same professional background in pavement engineering, research and innovation. Dr Thom has unique experience with a very strong academic background in pavement engineering, combined with research, practical design and specification. His understanding of design principles, practical constraints and risks has led to the successful introduction of many research findings into practice, which have improved value and provided better pavements.

In this book, Dr Thom manages to bridge the gap between theoretical pavement design and practical applications in a simple structured approach. The book is a good reference that paves the way for engineering students to understand design principles, for practising pavement engineers to design, specify and plan maintenance with a good understanding of materials and performance expectation, and for researchers to explore innovations such as smart materials, digitisation and pavement solutions in order to fulfil users' expectation of safe, durable, low-maintenance, sustainable and resilient transport infrastructure.

Dr Bachar Hakim
Director of Roads Asset Management, AECOM

Acknowledgements

While I take responsibility for the contents of this book, I cannot possibly take all the credit. Life is a continuous learning experience and I continue to learn much from every contact with practitioners, sometimes consultancy clients, sometimes research partners – too many to list. Much of the information in this book comes from the hard work of others, whether my ex-colleagues in industry or the large number of PhD students that I have had the privilege of working with. I have simply taken what others have done and packaged it, sometimes adding my own interpretation. For this third edition, I would specifically like to offer thanks to ex-colleagues at Aecom, Bachar Hakim (to whom additional thanks for the foreword) and Daru Widyatmoko, to current and past colleagues at the University of Nottingham, Gordon Airey, Andrew Dawson and Tony Parry, and to a generation of hard-working researchers, among whom are Kranthi Kunar, Helena Jiménez, Ahmed Nassar, Chibuzor Ojum, Ahmed Abed, Anqi Chen, Monketh Mohammed, Dmytro Mansura, Oluwaseyi Oke, Chris Chu, Imtiaz Ahmed and Van Nguyen.

About the author

Nick Thom started as a general civil engineer with Scott Wilson in 1978, working in bridges and geotechnics before being part of the supervision team for the A180 trunk road into Grimsby. He studied for his PhD (Design of Road Foundations, 1988) at the University of Nottingham and then rejoined Scott Wilson as a pavement engineer, where he was involved in numerous pavement evaluation projects as well as general design consultancy. From 1991, he lectured and supervised research part-time at the University of Nottingham in parallel with consultancy activities, becoming a full-time academic in 2014. His research interests include pavement analysis and design, the behaviour of pavement materials, for example cold-mix asphalt, surface damage, skid resistance, the use of geogrids and stress-absorbing membrane interlayers, and vehicle energy consumption.

Part 1

Principles

Nick Thom
ISBN 978-1-83549-711-1
https://doi.org/10.1108/978-1-83549-710-420243002

Chapter 1
Introduction

When all is said and done, the pavement industry deals in high-volume, low-cost materials. In the UK alone, there are approximately 14 000 km of trunk roads and motorways with a surface area of some 250 km^2, around 10% of which has to be resurfaced each year. In the USA, this can be multiplied by about six. Furthermore, these are not figures that can be dramatically reduced while motor vehicle transport remains such a key factor in the economy, whether electric or petrol powered. The choice of materials is, therefore, limited to those that can be easily and cheaply produced in large quantities – which inevitably means the raw materials of the earth, namely rock, sand and clay. Any additive used to give extra quality – such as bitumen or cement – has to be used relatively sparingly; otherwise society just could not afford it – to say nothing of the environmental cost of such additives. The job of the pavement engineer, therefore, is to maximise the potential of these cheap, readily processable materials, including their potential to be repeatedly recycled, preferably in-place. The unit cost of the bulk materials may be relatively low, but the quantities required are very high indeed, which means that a modest saving per square metre can multiply up to a very substantial saving overall. To put it another way, if the life of a road pavement can be extended by 10%, this represents a very large contribution to the local economy.

1.1. The long history of the paved highway

It is impossible to know where or when the wheel was invented. It is hard to imagine that Stone Age humans failed to notice that circular objects, such as sections of tree trunk, rolled. The great megalithic tombs of the third millennium BC, notably the pyramids of ancient Egypt, bear witness to ancient humans' ability to move massive stones, and many commentators assume that tree trunks were used as rollers – not quite a wheel but a similar principle! However, it is known for certain that the domestication of the horse in southern Russia or Ukraine in about 4000 BC was followed not long afterwards by the development of the cart. It is also known that the great cities of Egypt and Iraq had, by the late third millennium BC, reached a stage where pavements were needed. Stone slabs on a rubble base made an excellent and long-lasting pavement surface suitable for both pedestrian usage and also traffic from donkeys, camels, horses, carts and, by the late second millennium BC, chariots. Numerous examples survive from Roman times of such slabbed pavements, often showing the wear of tens of thousands of iron-rimmed wheels. Traffic levels could be such that the pavement had a finite life.

Even in such ancient times, engineers had the option to use more than simply stones if they so chose – but only if they could justify the cost! Concrete technology made significant strides during the centuries of Roman rule and was an important element in the thinking of structural engineers. Similarly, bitumen had been used for thousands of years in Iraq as asphalt mortar in building construction. Yet neither concrete nor asphalt was used by pavement engineers in ancient times, for the excellent reason that neither material came into the cheap, high-volume category. As far as the pavement engineer was concerned, economics dictated that the industry had to remain firmly in the Stone Age. Even when the industrial revolution swept through Europe in the days of Thomas

Telford and John Loudon McAdam – the fathers of modern road building in the UK – the art of pavement construction consisted purely of optimising stone placement and the size fractions used.

Times have moved on. The massive exploitation of oil has meant that bitumen, a by-product from refining heavy crude oil, is now much more widely available. Cement technology has progressed to the stage where it is sufficiently cheaply available to be considered in pavement construction. However, there is no way that pavement engineers can contemplate using some of the twenty-first century's more expensive materials – or, at least, only in very small amounts. Steel can only be afforded as reinforcement in concrete; even in such modest quantities, it represents a significant proportion of the overall cost. Plastics and glass fibre find a use in certain types of reinforcement product and polymers can be used to enhance bitumen properties. However, the driving force is always cost, which means that, whether we like it or not, Stone Age materials still predominate.

1.2. Materials for pavement construction

In introducing the various building blocks from which pavements are constructed, it will not be possible to entirely avoid the use of technical terms such as 'load', 'strength' and 'stiffness'. Definitions of such terms are given in Section 1.5.

1.2.1 Soil

Every pavement, other than those on bridges, self-evidently includes soil. The most basic design requirement of any pavement is that the underlying soil is adequately protected from applied loads. Thus, no pavement engineer can avoid the need to understand soil. The following list provides some key facts.

- Soils vary from heavy clays, through silts and sands to high-strength rocky materials.
- Soils are not usually consistent along the length of a road or across any pavement site.
- Soils are very sensitive to water content and so are critically affected by the effectiveness of drainage provision.
- Water content will vary during the life of a pavement, sometimes over quite short timescales, in response to weather patterns.
- Some soils are highly permeable; some clays are virtually impermeable.

All this leads to one thing – *uncertainty*. However clever one tries to be in understanding and characterising soils, it is quite impossible to be 100% sure of the properties at a given time or in a given location.

This uncertainty makes life considerably harder. Nevertheless, it is necessary to categorise each soil type encountered in as realistic a way as possible, and there are two fundamental areas in which soil behaviour affects pavement performance. These are

- stiffness under transient (i.e. moving wheel) load
- resistance to accumulation of deformation under repeated load, likely to be related to shear strength.

The various means of testing, measuring and estimating these properties are covered in Chapter 6 of this book, as are the possibilities of soil improvement by using additives such as cement and lime.

1.2.2 Unbound pavement materials

Unbound materials used in pavement construction include natural gravel, crushed rock and granulated industrial by-products such as slags. Soils are also unbound materials, albeit often with a very small particle size (2 μm or less for clay), but the key difference is that a soil is not, in general, 'engineered' in any way. An unbound pavement layer, on the other hand, will be selected and quite possibly deliberately blended to give a particular combination of particle sizes. It can also be mixed with a predetermined amount of water. One would therefore naturally expect that much of the uncertainty inherent in soil properties is removed in the case of unbound pavement materials. It may still be difficult to predict performance accurately, as different material sources, most commonly different rock types, might be expected to exhibit slightly different properties due to their different responses to crushing or their differing frictional properties. Nevertheless, an unbound pavement material will be a much more controlled and predictable component than the soil. Even the variation in water content will usually be more predictable, in both magnitude and effect, than in the case of soil.

However, the properties of unbound pavement materials of interest to pavement engineers are actually more or less the same as those of soil, namely

- stiffness under transient load
- resistance to accumulation of deformation under repeated load, related to shear strength.

Chapters 6 and 7 of this book will describe the behaviour of unbound materials and the design and performance of unsurfaced pavements.

1.2.3 Bitumen-bound materials

This is a material almost unique to pavement engineering, a material whose beneficial properties were discovered almost by accident, but a material that is now very much at the centre of pavement technology. There are countless stories as to when bituminous products were first used on roads, including the accidental spillage of tar outside Derby iron works in 1901. Although mastics, including natural asphalt, had been used on footways since the 1830s, they were not stable enough for roads, and it was not until around 1900–1901 that the first usages of tar-bound stone occurred at approximately similar dates in the USA and Europe.

While proportions differ around the world, typically some 90% of paved highways have a bitumen-bound surface layer; whatever the make-up beneath the surface, bitumen and bitumen-bound materials (referred to hereafter as *asphalts*) currently play a major role. And bitumen is quite different from the sorts of cement used in concrete. Whereas concrete is a rigid material that cannot deform appreciably unless it first cracks, asphalt can slowly change its shape because bitumen remains a viscous liquid at normal in-service temperatures.

An ability to change shape may seem a rather undesirable quality and it can indeed lead to deformation – hence the phenomenon known as 'rutting' or 'tracking'. However, it also overcomes some of the difficulties encountered with concrete. For a start, the expansion and contraction with day–night temperature variation is accommodated simply by a small viscous strain within asphalt, meaning that no movement joints are required and that thermally induced cracking will only occur under the most extreme temperature conditions (continental winters, deserts). Asphalts are also able to accommodate any moderate movement within the foundation, for example, minor differential settlement in an embankment, movement that might lead to the fracture of a rigid concrete slab.

Furthermore, the tendency of asphalt to change its shape can be controlled by proper mixture design such that rutting is minimised.

However, despite the flexible nature of asphalt, it can still crack. It is impossible to define a tensile strength, as this will vary with temperature and the rate of loading; the relevant parameter is the *fatigue characteristic*, which defines resistance to cracking under repeated load. The key properties required for design are therefore

- stiffness
- resistance to deformation under repeated load
- a fatigue characteristic.

Chapter 9 will explore the properties of asphalt in some detail, leading to the design of asphalt-surfaced pavements in Chapter 11. Chapters 13 and 14 will extend this to cover the increasingly important topic of so-called 'cold-mix' asphalt.

1.2.4 Hydraulically bound materials (HBMs)

Nowadays, the availability of Portland cement, and substitutes such as fly ash or ground granulated blast-furnace slag, means that it can be economical to use such a binding agent to strengthen an unbound material. These binders are known as 'hydraulic' binders, as they require the presence of water for the cementing action to take place. Cement technology is a vast subject in its own right and is evolving rapidly in response to the need to reduce the carbon dioxide emissions associated with its production. The hydration of Portland cement involves several different chemical reactions, the most important of which are the conversion of tricalcium silicate (*c.* 50% of Portland cement) and dicalcium silicate (*c.* 25%) into hydrates (forming strong solids) by reaction with water, also generating calcium hydroxide and heat. The first reaction is rapid; the second is slower. The reader should refer to specialist literature for details.

HBMs, including so-called *pavement-quality concrete* (PQC) at the upper end of the strength spectrum, introduce a quite different type of behaviour and totally different design requirements. They possess a key property that is lacking in soils and unbound materials, namely the ability to withstand *tension*, and they also avoid asphalt's annoying tendency to change its shape. Individual particles are rigidly bonded together by the binding agent, and a definite tensile force is required to break that bond. In the case of a strong concrete, all the large particles are well bonded into a continuous matrix of fine aggregate and cement paste, and the whole material is solid and rigid. It has a stiffness that is still partly governed by the contacts between the large particles, but which is also heavily dependent on the qualities of the surrounding cementitious matrix. In the case of a weaker HBM, the binding effect may be less complete and some of the particle contacts may remain unbound, giving a certain freedom of movement within the material and a reduced stiffness and strength. Nevertheless, even a weak HBM will remain as a solid, with negligible permanent deformation until the bonds are fractured (i.e. until the tensile strength is overcome). The key properties for the pavement engineer are therefore

- stiffness
- tensile strength.

Arguably, a *fatigue characteristic* should be added, as for asphalt. However, the relationship with tensile strength is so close that it is hardly a separate property. It would also have been possible to add the *curing rate* (the rate of strength gain), as this certainly affects the construction process and economics significantly, and the *thermal expansion coefficient*, as this property strongly influences the tendency of a HBM to crack under day–night temperature variation, requiring the introduction of movement joints in most concrete pavements. Chapters 16 and 19 describe all these properties and associated tests, leading to the design of pavements incorporating these materials in Chapters 17 and 20.

1.2.5 Other materials

The four material types introduced so far represent the basic building blocks available to the pavement engineer. However, it is worth referring here to a couple of materials that do not fit so easily into any of the four categories. The first is *block paving*. Blocks are often made of concrete and so could be termed HBMs. On the other hand, they can be cut from natural stone or may comprise fired clay bricks. Moreover, the discontinuous nature of block paving means that the properties of the parent material are less important than the effects of the discontinuities. The blocks themselves may have the properties of concrete, for example, a stiffness modulus of some 40 GPa, but once discontinuities are taken into account, the effective layer modulus may be as little as 500 MPa. In effect, blocks form a perfectly shaped unbound material and block paving is therefore considered in Chapter 7 as a special case of unbound pavement.

The second special case is a hybrid material, known in the UK as *grouted macadam* and in the USA as *resin-modified pavement*; it is also sometimes known as 'semi-flexible' material. This too does not fit neatly into any of the previous categories, as it combines an asphalt skeleton with a cementitious grout filling the voids in the asphalt mixture. It therefore utilises both bituminous and hydraulic binders. Having a two-stage production process, the material tends to be expensive and is used in particular heavy-duty applications such as bus lanes and industrial pavements. However, it actually resembles an asphalt much more than a concrete, and will be included as such in Chapters 9 and 11.

Block paving and grouted macadam are bulk-use materials at the expensive end of the range. There are also specialist products that are only used in small quantities to strengthen, or in some way improve, a pavement layer. Here, one could include concrete reinforcement products, either in bar form or fibres. There are also reinforcing products designed for asphalt – some are steel, others are polymeric or made of glass fibre. A similar range of products is available for the reinforcement of unbound materials. Geotextiles comprise a closely related range of products, produced in various ways and forming continuous layers separating two different pavement materials (commonly the soil and a granular layer). These too can have a reinforcing function, but their most common use is simply as a separator, ensuring that fine soil particles do not migrate up into the pavement and that stones from a granular layer do not lose themselves in the soil. It is obviously the responsibility of the pavement engineer to understand how (and whether) these products work in particular applications rather than relying solely on the, sometimes not unbiased, opinion of a supplier. They will be introduced with each relevant pavement type and also with the topic of pavement rehabilitation in Chapter 27.

1.3. Typical pavement structures

This section is principally included to introduce the relevant terminology. The first terms that should be drawn to the reader's attention are *flexible, rigid* and *composite*, all of which are used to describe certain generic classes of pavement. Basically, a rigid pavement is one with a PQC slab as the main structural layer; a flexible pavement consists entirely of unbound materials and asphalt and a composite pavement has an asphalt surface overlying a hydraulically bound (e.g. concrete) base.

1.3.1 Pavement layers

Figure 1.1 illustrates the case of a heavy-duty flexible pavement.

At first glance, pavements appear to be unnecessarily complicated. However, everything is there for a reason. The *subgrade* is inescapable of course. *Capping*, where needed, is best seen as a subgrade-improvement layer, as it is uneconomical to place high-value materials straight onto a weak soil. Capping itself is generally a cheap, locally available material, sometimes simply the subgrade soil treated with lime, cement or some other hydraulic binder.

Moving to the top, the *surface course* (not present as a separate layer in rigid pavements) is a high-quality (and therefore expensive) material, tough enough to withstand direct loading, and with surface properties designed to achieve adequate skid resistance and, where required, low noise. As it is expensive, it is not generally a thick layer, 20–50 mm being typical.

Somewhere in the middle of the structure, the *base* is the layer that gives the pavement most of its strength. It is usually relatively thick (often 200 mm or more) and therefore has to be as inexpensive as possible within the constraints of the required mechanical properties. The result is that a large aggregate size tends to be used, whether the layer is asphalt, a HBM or an unbound material, and a likely side effect of a large aggregate size in an asphalt or roller-compacted hydraulically bound base is a relatively uneven finish. This means that it is difficult to achieve a good-quality finished road surface if the surface course is applied directly; hence there is frequently a need for an intermediate regulating layer – the *binder course*. The binder course material is often very similar to the base, but with a smaller aggregate size, which means it can be laid to a typical thickness of 50–80 mm.

Finally, the *sub-base* is much more than just a fill-in layer. The performance of an asphalt or concrete base is critically dependent on the stiffness of the layer immediately beneath, because of its influence on the flexure of the base under traffic load. A firm support will limit such flexure.

Figure 1.1 Pavement layer terminology

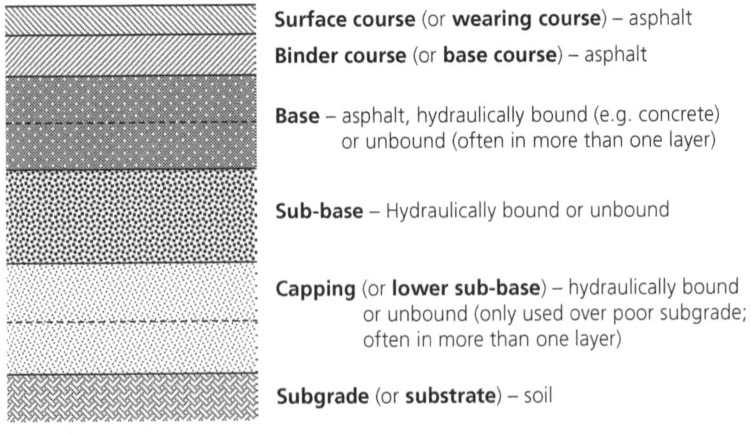

Figure 1.2 Typical highway cross-section

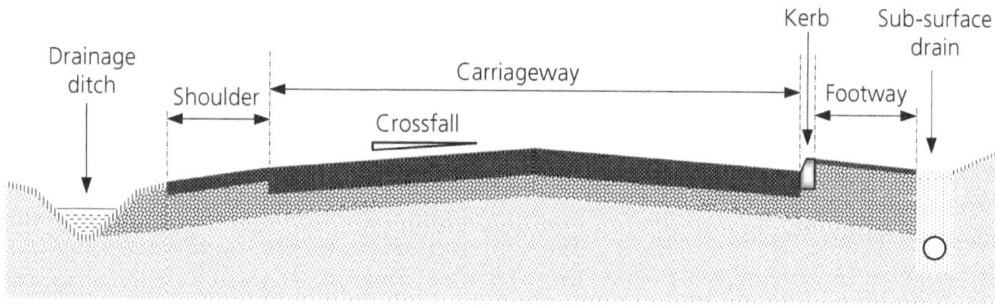

Sub-bases therefore vary from high-quality unbound materials to even stiffer hydraulically bound layers. A typical thickness is 150 mm; anything much thicker is likely to become uneconomical.

1.3.2 Pavement cross-sections

While pavements come in many different guises, there are certain common features, as illustrated in Figure 1.2. The need for effective drainage means that there will usually be a *crossfall* applying to each pavement layer. The imprecision inherent in pavement construction means that 1% is a practical minimum; anything less and there are almost bound to be locations where water will not be properly drained. A crossfall of 2.5% is typical on highways; 1.5% on airfield pavements. There will often be a section of relatively weak construction at the edge of the pavement, where trafficking is not expected to be frequent, known as a *shoulder*. This is basically an extra width of trafficable pavement, suitable for emergency use. For road pavements in urban locations, however, the edge of the pavement is likely to be defined by a *kerb*. The importance of drainage to pavement performance will be emphasised many times in this book and, as shown in Figure 1.2, *drains* are most commonly located just outside the pavement edge.

1.4. Material costs

Pavement engineering is always constrained by cost. Products with excellent performance may appear attractive, but they will only be used if the cost can be justified. Prices fluctuate of course, and will continue to do so, and price also depends heavily on geography. Taking the UK as an example, good-quality rock is plentiful in the north and west but not in the south-east, which means that it is much more economically advantageous to utilise river gravels in the south-east than elsewhere. In some parts of the world, bitumen is plentiful and relatively inexpensive; elsewhere cement is more reasonably priced. Any costs stated in this book can, therefore, only be indicative.

To start at the beginning: subgrade soil is effectively free. The natural ground beneath a pavement is there whether wanted or not. Perhaps a modest investment could improve it, for example, by installing drainage, by means of compaction if the soil is sufficiently granular or by stabilisation using a hydraulic binder, but the raw material is either free or, in the case of an embankment, is a geotechnical cost rather than a pavement requirement. However, all pavement layers above the subgrade come with a cost and it is important that pavement designers appreciate the relative expenditure associated with each type of material.

Table 1.1 shows approximate relative cost ranges, taking the price of high-quality unbound crushed rock sub-base as the benchmark. Unbound materials range in price from less than half of this for locally extracted sands and gravels, to perhaps 125% for an ultra-tightly controlled crushed rock

Table 1.1 Material costs relative to high-quality crushed rock sub-base

Material	Relative cost per unit volume: %
Soil stabilisation	60
Capping	60
Unbound sub-base	100
Unbound base	125
Weak hydraulically bound base	175
Strong hydraulically bound base	250
Asphalt	400
Unreinforced concrete	400
Continuously reinforced concrete	550

base. The absolute price of unbound materials is often largely determined by haulage (or shipping) distance; in general, it is uneconomic to transport such materials further than about 40 km by land. As shown in Table 1.1, stabilisation of a native soil using a hydraulic binder will often cost no more than about 50% of the price of a crushed rock sub-base – but of course not every soil is suited to such stabilisation.

This, of course, all means that it is economically desirable to minimise, as far as possible, the thickness of either asphalt or PQC and to maximise the use made of locally available unbound materials, even if their quality is less than one would hope for. In the case of block paving or grouted macadam, with costs up to twice that of asphalt, a very good case has to be made for their use. Similarly, while geosynthetics vary enormously in price (and effectiveness), a top-of-the-range polymer grid is likely to cost the equivalent of about 40–50 mm of asphalt, again implying that serious thought should be given before specifying such a product.

But these are the costs for new materials. If an existing layer can somehow be recycled in situ by the addition of a dose of bituminous or hydraulic binder, then significant savings, both financial and in sustainability terms, can be realised – although possibly at the cost of quality assurance.

The pressure towards the use of environmentally appropriate materials can only increase as humankind makes an ever greater mess of the planet. In pavement construction this means that 'recyclability' is an important issue. Materials that are of no further use at the end of their design life and then have to be disposed of bear a significant environmental cost. Fortunately, this is not the case for the vast majority of pavement materials, but the issue remains as to how best to maximise the value of reused materials. Bitumen, for example, is a relatively expensive component of a pavement, and the last thing one should want to do is to bury potentially valuable recycled asphalt planings in the foundation layers, where their valuable properties are not being used. This may satisfy the letter of current advice or legislation, but it certainly does not optimise the value of the resource.

1.5. Basic engineering concepts

This section is included for those who feel they need reminding of concepts studied years ago and long since forgotten under the weight of practical engineering. Various properties, quantities and concepts will be introduced, all relevant to the methods presented in the remainder of this book.

1.5.1 Basic quantities

As in any field, civil engineering is awash with technical and not so technical terms, and the exact meaning of each is sometimes far from clear. The following provides definitions of key basic quantities, most of which the reader will probably know perfectly well.

1.5.1.1 Mass

Every object has a mass, measured in kilograms (or tonnes; 1000 kg = 1 t). Mass has no particular direction.

1.5.1.2 Weight

The weight of an object refers specifically to the downward effect of a mass due to the action of the Earth's gravity. Gravity is expressed as an acceleration (9.81 m/s^2), being the rate at which an object would accelerate downwards towards the centre of the Earth in the absence of any resistance. Weight is measured in newtons (N). It is defined as the mass of an object multiplied by the acceleration due to gravity. On the surface of the Earth, therefore, 1 kg of *mass* has a *weight* of 9.81 N.

1.5.1.3 Density, unit weight and specific gravity

The density of a material is the mass per unit volume (kg/m^3). For example, water has a density of 1000 kg/m^3, while the density of most rocks is between 2500 and 3000 kg/m^3. Soils, being combinations of rock particles, water and air, tend to have densities in the range 1500–2000 kg/m^3, while crushed rock pavement layers are usually around 2200 kg/m^3. The densities of asphalt and concrete are typically around 2400 kg/m^3.

In civil engineering, where most materials are static and the effect of density is to generate weight due to gravity, it is often convenient to work in terms of unit weight rather than density. This is defined as the weight per unit volume and is expressed in N/m^3 or, more usually, kN/m^3. Pavement layers tend to have unit weights in the range 20–25 kN/m^3.

Finally, specific gravity is simply a convenient relative measure. Water is assigned a specific gravity of 1; other materials have a higher or lower specific gravity according to the ratio of their densities to that of water. Thus, bitumen, for example, being marginally denser than water, has a specific gravity of 1.02–1.03.

1.5.1.4 Force and load

The weight of an object under gravity is an example of a force. More generally, however, a force can be generated in numerous different ways (e.g. braking, acceleration, impact). Like weight, force is measured in newtons (N) or, more commonly, kN.

Load is also a force. It is usually applied in the same way as weight, but can also be used to describe any force that 'loads' an object (or a pavement for that matter). For example, load due to cornering acts horizontally and is calculated as mass × acceleration, but in this case the acceleration is not gravity but so-called centripetal acceleration v^2/R, where v is the vehicle velocity and R is the radius of the corner.

1.5.1.5 Moment and torque

Both moment and torque refer to the power of a force to cause rotation. It is force multiplied by its distance (perpendicular to the line of the force) from an axis of rotation, with units of N.m.

1.5.1.6 Stress and pressure

These two words have exactly the same meaning in engineering terms, although *stress* is most commonly used when dealing with solids, whereas *pressure* usually relates to liquids and gases. They are both defined as force divided by area, giving units of N/m^2, also known as pascals (Pa). Thus, the load from a wheel can be expressed directly in newtons, but it is also relevant to know the intensity of that load on the surface, and this is given by the contact stress (i.e. the load divided by the contact area). This will be approximately equal to the air pressure within the tyre – although the complexities of tyre–wall behaviour mean that it is a little more complicated than this in reality.

Thus, the vertical contact stress and tyre pressure of a typical heavy goods vehicle wheel are both around 600 kPa. Because of the way the pavement spreads the load, the vertical stress (due to wheel load) at the top of the subgrade may be no more than about 10 kPa.

1.5.1.7 Strain

Strain is often the consequence of stress. Strain describes the degree to which a material deforms, and is defined as the change in a dimension divided by the original magnitude of that dimension. Thus, if a 100 mm-thick layer of asphalt compresses to 99 mm, the strain is (100−99)/100, that is, 0.01. It has no units because it is a relative change. Where strains are large, they are often expressed as percentages (1% in this example). However, for the very small strains generated within a pavement under load, the unit is usually microstrain (με); 1 με is a strain of 10^{-6} (a change of one millionth in the original dimension).

1.5.2 Mechanical properties

The following paragraphs identify and explain the important engineering properties referred to throughout the remainder of the book.

1.5.2.1 Stiffness

Stiffness is a particularly loose term. However, it is in common usage among engineers and so has to be introduced. At its most basic, it refers to a ratio of load to deformation. This could be, for example, the ratio of a wheel load to a pavement deflection. It could equally refer to the load/deformation ratio of a vehicle suspension spring. However, stiffness is also used much more generally to describe the resistance of a material to stress, whether it is the bending stiffness of a pavement layer, the shear stiffness of a bitumen or the load-transfer stiffness across a concrete joint. Because it is such a loose term (and will be used as such in this book), it will not be given a specific definition.

1.5.2.2 Elastic modulus (or elastic stiffness)

This is the most common measure of stiffness used by engineers. It is the ratio of *applied stress* to *induced strain*. Thus, if a block of concrete is subjected to a stress of 1 MPa (1 megapascal or 10^6 Pa) and it develops a strain (in the direction of the stress) of 50 με, it would have an elastic modulus of $10^6/(50 \times 10^{-6})$, which is 20×10^9 Pa or 20 GPa (gigapascals).

This is an extremely useful concept, but it assumes that the behaviour of the material concerned is *linear*, meaning that if the stress is doubled then so is the strain. For concrete and other intact HBMs this is approximately true, within the normal working range; similarly for steel. However, unbound material is significantly non-linear, while the stiffness of asphalt varies with temperature and loading rate, which means these materials cannot be said to possess a single elastic modulus value. That does not stop engineers using the concept for both unbound material and asphalt – but it means they use different terms.

1.5.2.3 Stiffness modulus and resilient modulus

These terms are used to indicate that the measure used, although still stress divided by strain, does not really imply a single linear elastic property. Asphalt has a viscous (dependent on temperature and loading rate) component to its behaviour and the rather general term *stiffness modulus* is therefore used; however, at a given temperature and loading rate, the definition is exactly the same as that for elastic modulus.

The term *resilient modulus* is reserved for soils and other unbound materials. It is used to separate out the component of behaviour that is approximately elastic (although still non-linear and therefore stress-dependent) from the plastic component, where strains are non-recoverable. However, under a given set of stress conditions, the definition of resilient modulus is still exactly the same as that of elastic modulus (i.e. stress divided by strain).

1.5.2.4 Poisson's ratio, shear modulus and bulk modulus

Up until this point, it has been assumed that stress and strain act in a single direction. However, in real life the fact that the world is three-dimensional cannot be ignored, and the stresses and strains in a pavement occur in all directions.

For many purposes, the most convenient way of dealing with the interaction between effects in different directions is by means of the quantity known as Poisson's ratio. Figure 1.3 gives the definition. Poisson's ratio controls the degree to which a material compresses under load. In an incompressible material, it has a value of 0.5. Clays at high levels of saturation are nearly incompressible and so have a Poisson's ratio of 0.45 or more; unbound materials tend to have a much higher air-void content, giving a lower Poisson's ratio, typically 0.30–0.35, and the values for asphalt tend to be similar. HBMs usually have the lowest Poisson's ratio of all pavement materials, often in the range 0.15–0.20.

The shear modulus and bulk modulus are also defined in Figure 1.3. They are directly related to elastic, stiffness or resilient modulus and Poisson's ratio – through formulae given in the figure – but it is sometimes convenient to work in terms of shear or bulk properties for particular problems.

Figure 1.3 Measures of stiffness

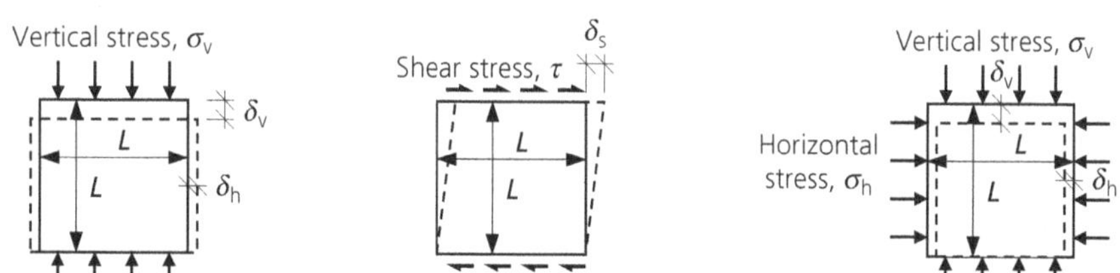

Vertical strain, $\varepsilon_v = \delta_v/L$
Elastic modulus, $E = \sigma_v/\varepsilon_v$

Horizontal strain, $\varepsilon_h = -2\delta_h/L$
Poisson's ratio, $v = -\varepsilon_h/\varepsilon_v$

Shear strain, $\varepsilon_s = \delta_v/L$
Shear modulus, $G = \tau/\varepsilon_s$

Vertical strain, $\varepsilon_v = \delta_v/L$
Horizontal strain, $\varepsilon_h = 2\delta_h/L$
Bulk strain = $\varepsilon_v + 2\varepsilon_h$
(assuming horizontal stress is same in all directions)
Bulk stress = $(\sigma_v + 2\sigma_h)/3$
Bulk modulus, K = bulk stress/bulk strain

$G = E/2(1 + v)$ $K = E/3(1 - 2v)$

Note that the underlying assumption here is *linear elasticity*. These quantities can still be used for non-linear or temperature-dependent materials, where the terms stiffness modulus and resilient modulus are used, but it has to be remembered that they will no longer be absolute and unchanging properties, although they may be valid for a certain set of conditions. Unbound materials in particular can display an enormous range of Poisson's ratio values depending on stress conditions, even values well above 0.5, meaning that the material dilates as it strains.

1.5.2.5 Modulus of subgrade reaction

This is not really a material property, but it makes sense to introduce it here because it is easy to confuse with the stiffness modulus or resilient modulus of the subgrade material. The modulus of subgrade reaction is defined as the ratio of an applied vertical stress (at the top of subgrade level) to the resulting vertical deflection. This is convenient for two reasons.

- It can be measured directly by means of a plate loading test.
- It fits nicely into a widely used computational approach for stress in concrete pavements – introduced in Part 6.

Unfortunately, although convenient, this is not a fundamentally correct way to describe a pavement foundation. Calculation Sheet 1.1 presents an approximate method for estimating the modulus of subgrade reaction from the stiffness modulus. Note the use of a load spread angle; for this, a default value of 35° is recommended as this gives surface deflections equal to those obtained theoretically for the case of a uniform elastic 'half-space' under a circular load. Taking an upper subgrade thickness of 1 m and the standard case of a 762 mm (30 in.) diameter plate test, the equation in Calculation Sheet 1.1 (with a 35° load spread angle) gives values of modulus of subgrade reaction in MN/m^3 between 54% and 84% of the stiffness modulus in MPa – typical of quoted ratios. For example, British Airports Authority (1993) quotes a direct relationship between the modulus of subgrade reaction k (in MN/m^3) and an equivalent foundation modulus E (in MPa), as $k = (E/4.97)^{1.292}$. However, the predictions are highly sensitive to the diameter of the loaded area. If a real pavement were to spread each wheel load over a 1 m diameter loaded area, these percentages would become 71% and 122%; on the other hand, a more flexible structure, spreading the wheel load over a 0.5 m diameter area, gives only 36–48%. Care is therefore needed when working with a modulus of subgrade reaction.

Calculation Sheet 1.1. Modulus of subgrade reaction

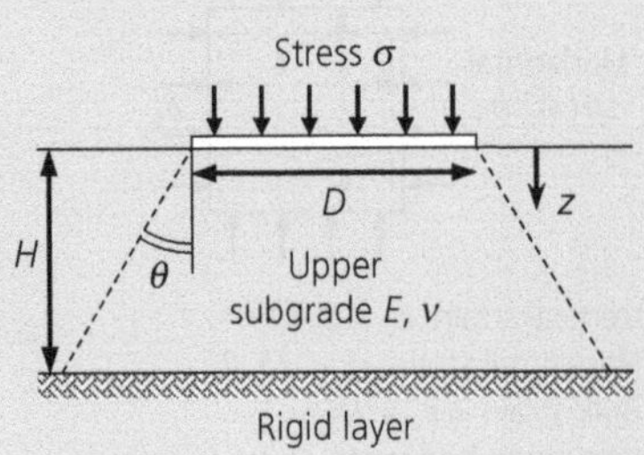

Assume:

- a circular load area
- load spreading at angle θ

- simple division of subgrade into two layers
- the upper layer is linear elastic
- the lower layer is rigid

Stress at depth z: $= \sigma(\pi D^2/4)/[\pi(D+2z\tan\theta)^2/4] = \sigma[D/(D+2z\tan\theta)]^2$

Strain at depth z: $\varepsilon_z = \sigma[D/(D+2z\tan\theta)]^2/E$

Surface deflection: $\delta = \int \varepsilon_z \, dz = \int \sigma[D/(D+2z\tan\theta)]^2/E \, dz$ (limits: $z=0$ and H)

Substitute $y = 1+2z\tan\theta/D \rightarrow$

$$\delta = (2\sigma\tan\theta/DE)\int 1/y^2 \, dy\,\delta \quad \text{(limits: } y=1 \text{ and } 1+2H\tan\theta/D)$$

$$\delta = (2\sigma\tan\theta/DE)(-1/y) \quad \text{(limits: } y=1 \text{ and } 1+2H\tan\theta/D)$$

$$\delta = (2\sigma\tan\theta/DE)[1-1/(1+2H\tan\theta/D)]$$

Modulus of subgrade reaction: $k = \sigma/\delta = (DE/2\tan\theta)/[1-1/(1+2H\tan\theta/D)]$

1.5.2.6 Viscosity and kinematic viscosity

Viscosity is a property displayed by a liquid such as water or bitumen. It is defined as the ratio of an applied stress to the resulting strain rate (i.e. strain per second), and so has units of pascal seconds (Pa.s). In effect, it parallels the concept of elastic stiffness in a solid. If related to a single direction, the property might be more properly termed *axial viscosity*; Poisson's ratio is usually 0.5, as liquids are generally incompressible, which means that shear viscosity is one-third of the axial viscosity and the *bulk viscosity* is infinite – following the inter-relationships presented in Figure 1.3. The viscosity of water at 20°C is around 10^{-3} Pa.s; that of bitumen ranges from about 10^3 Pa.s at 40°C to as much as 10^7 Pa.s at 0°C.

It is often convenient in engineering problems to work with a quantity known as *kinematic viscosity*, which is defined simply enough as the viscosity divided by the density. The resulting fundamental unit is m^2/s, but the value is more usually quoted in 'centistokes', 1 centistoke being 1 mm^2/s, giving a kinematic viscosity of water of approximately 1 centistoke.

1.5.3 Thermal properties

While the mechanical properties of pavement materials are the most influential, determining the response to traffic load, stresses and strains can also be generated by changes in temperature, which means that the thermal properties also have to be understood.

1.5.3.1 Coefficient of thermal expansion

Every material tends to change its volume slightly with temperature, which can cause problems in a continuous stretch of pavement. The extent to which this occurs is governed by the coefficient of thermal expansion, which is defined as the proportional change in a given linear dimension due to a 1°C rise in temperature. For rocks, this tends to be in the range 0.6–1.4×10^{-5}/°C, and, because rock (or rock derivatives such as slag, minestone, demolition waste) is usually the major component of pavement materials, this means that the same range is found for HBMs (and steel). Asphalts tend to have a slightly higher value, typically 1.8×10^{-5}/°C, because of the much higher thermal expansion potential of bitumen. Use of limestone aggregate minimises thermal expansion; use of gravel aggregate usually maximises it.

1.5.3.2 Thermal conductivity

If thermal expansion is a significant issue, then this means that the temperature at any depth in a pavement at a given time is also important, and this will depend on the efficiency with which the different pavement materials allow heat energy to travel. The relevant property here is thermal conductivity; it describes the flow of energy (in J/s or W) caused by a 1 K (i.e. 1°C) temperature difference across a 1 m length of material, 1 m^2 in cross-section. The units are therefore W/m.K. Typical values range from 0.25 W/m.K for dry soils to around 2 W/m.K for asphalt or concrete.

1.5.3.3 Specific heat capacity

This is the final property required in any prediction of temperature changes within a pavement. It describes how much energy is required to heat up a given mass of material by 1 K. The units are therefore J/kg.K, with typical values for pavement materials in the range 800–1000 J/kg.K, a little higher for wet soils.

1.6. Summary

The first few parts of this chapter have flitted from topic to topic, introducing some of the key issues involved in pavement engineering and trying to set the subject in its proper context. Most of these issues will be picked up in much more detail in other parts of this book. I hope you also found the definitions in Section 1.5 of use. The remainder of the book will certainly be making regular use of most of the quantities introduced.

Concerning the structure of the book, Chapters 2–5 will cover those background topics that are of general relevance to pavement engineers. Subsequent chapters will then go through each class of material and associated pavement types in logical order. Unbound material is found at some level in every pavement unless it is on a structure, and so unbound material behaviour is explored first, together with unsurfaced pavement design and maintenance (Chapters 6–8). This is followed by asphalt and flexible pavements (Chapters 9–12) and then the increasingly important topic of cold-mix asphalt pavements and in-place recycling (Chapters 13–15). Chapters 16–18 cover HBMs together with composite pavement design and Chapters 19–22 deal with PQC and rigid pavements.

This leaves Chapters 23 and 24 on water and drainage issues, and Chapters 25–28, where attention turns to the vital topic of pavement management, including condition monitoring and maintenance techniques as well as the difficult subject of managing a finite budget.

REFERENCE

British Airports Authority (1993) *Pavement Design Guide for Heavy Aircraft Loading*. BAA, Group Technical Services Aircraft Pavements, London, UK.

Principles of Pavement Engineering, Third edition

Nick Thom
ISBN 978-1-83549-711-1
https://doi.org/10.1108/978-1-83549-710-420243003

Chapter 2
Design principles

2.1. Engineering principles

Underlying each pavement design method, often out of sight of the user, are certain key principles. In reality, there are just a handful of conceptually simple tasks that a pavement has to fulfil, which can be summarised as follows.

- *Protect the subgrade.* In general, natural ground will not be strong enough to bear traffic load directly; it would deform and rut. The pavement must therefore be stiff enough and thick enough to distribute the load from vehicle wheels over a wide enough area not to cause excessive subgrade deformation.
- *Guard against deformation in the pavement layers.* At every level in the structure of the pavement, the materials used must themselves be stable enough not to suffer deformation under the stresses experienced at that level.
- *Guard against break-up of the pavement layers.* Similarly, the internal strength of the pavement layers must be sufficient not to allow excessive cracking to develop under traffic loading.
- *Protect from environmental attack.* The materials used should be such that they do not lose their properties under attack from whatever the local climate throws at them, or with the passage of time. Drainage must be adequate to prevent excessive water build-up at any level in the pavement.
- *Provide a suitable surface.* Requirements of evenness, skid resistance, noise and/or spray suppression may vary, but the design has to be suited to the provision of an appropriate pavement surface.
- *Ensure 'maintainability'.* Accepting that no pavement lasts forever, the design must ensure that it is possible to carry out necessary maintenance within acceptable limits of disruption and cost.

These six key tasks together define successful engineering design, and much of this book is devoted to the techniques by which such design can be achieved. However, design is also constrained by arguably equally important non-engineering principles. The following subsections therefore introduce the concepts of whole-life cost and sustainability – terms that may vary with time and geographic location but the principles of which remain constant.

2.2. Whole-life cost

The concept of whole-life cost is simple enough; it is the true cost of a pavement over a long period of time, including both initial and ongoing costs. However, while the concept is simple, arguments rage as to which costs should be included and how. Table 2.1 lists the key issues, taking the example of a highway project.

Table 2.1 Highway pavement costing issues

Cost/benefit	Parties involved			
	Residents (local economy)	Pavement users	Highway authority	State
Direct construction costs			✓	
Indirect construction costs			✓	✓
Indirect construction benefit	✓			✓
Direct maintenance costs			✓	
Indirect maintenance costs		✓	✓	✓
Indirect maintenance benefit	✓			✓
Indirect cost due to pavement's presence	✓			
Indirect benefit due to pavement's presence	✓	✓		✓
Accident costs		✓		✓

The problem is just where to draw the boundaries – and the answer may depend on who is carrying out the calculation. A highway authority employee who has been allocated a budget and told to stretch it as far as possible will naturally only be looking at a single item in Table 2.1, namely the direct construction cost to the authority. And this attitude on the part of an individual cannot legitimately be criticised; it is not their job to second-guess on issues that should be sorted out as policy at a much more general level, however aware that individual may be of those issues.

However, what about the highway authority as a whole? The organisation should sensibly take into account every cost item that affects it, which means both construction and maintenance costs, plus the secondary costs of road damage due to site access, aggregate transport and so on – in fact, the four ticked items in the highway authority column in Table 2.1. But there is usually no incentive for the authority to draw the net wider and examine costs and benefits to the wider community. Perhaps that is slightly unfair, and many highway authorities have an interest in issues affecting local residents – but they will not normally be concerned directly with road-user costs (i.e. travel time, fuel costs), and even accident costs only impact indirectly. If these wider issues are to be included then this has to be either done or ordered at a more general level still, which means at local or national government level. Only those removed from the day-to-day need to control highway spending have the luxury of being able to take the bigger picture into account. Of course, day-to-day decision-makers may be well aware of the issues – but they cannot be expected to take decisions that directly harm the finances of their own organisation, however great the benefit to the wider community may be.

Forgive this somewhat philosophical discussion. The aim is to get across the fact that there are wider issues that, in an ideal world, someone in local or national government should be tackling. Some of the most difficult problems are

- putting an appropriate value on travel time
- assigning a 'discount rate' to future costs (i.e. £100 in 5 years' time might only be equivalent to £80 now if that £80 were invested wisely)

- taking road damage by construction vehicles into account
- costing of accidents, particularly injury and death
- disruption to business due to restrictions on highway usage
- benefits to business due to the highway being there – or being upgraded.

One could even add in such items as the cost due to increased flood risk, as pavements generally result in additional runoff to drains, sewers and rivers; similarly the contribution of pavements and their users to climate change. The truth is that there is always a bigger picture, however wide one draws the boundaries to the problem. Nevertheless, the principle is that all of these costs and benefits have to be evaluated, and the easiest way to do this is to assign a monetary value to each. For example, the current cost of a fatality in the UK is generally put at around £2 million. There is obviously no single 'right answer' and many of the cost/benefit items will attract wildly differing estimates, particularly such items as the cost to business. However, this is no excuse for governments not to carry out such calculations – or insist that they be carried out by others. It is the only way to evaluate sensibly whether particular schemes should go ahead; it is also a logical input into the decision-making process as to just which pavement solution should be adopted in a particular case. It is not pavement engineering – but it may impact on engineering choices.

2.3. Sustainability

If whole-life costing is politically challenging, sustainability pushes the boundaries out further still. This is another subject that – though quite separate from engineering – should affect engineering choices. Like whole-life costing, it is a subject with which pavement engineers should be familiar. As in whole-life costing, the calculations are not difficult in themselves; the difficulty lies in defining the boundaries to the problem.

First of all, what is meant by 'sustainability'? It is probably foolish to try to define it in anything other than the most general terms, but basically the word represents our aspirations not to destroy the planet; that is, to make sure that our children and grandchildren actually have a reasonable future – a rather broader subject than pavement engineering! However, pavements (*a*) use vast resources in terms of tonnages of bulk products and (*b*) have a complex relationship with vehicle fuel consumption, noise, harmful emissions and so on. They therefore have the potential to add significantly to the rate at which we bring about the destruction of our environment, and so are rightly attracting the attention of government authorities trying to minimise such destruction and even reverse it by incorporating waste streams into pavement construction. The following sections outline key sustainability issues relating to pavements.

2.3.1 Material resource depletion

Traditionally, the principal component in road construction has been rock, and it is hard to argue that this resource is being significantly depleted. For instance, the entire volume of pavement in the UK (probably 1–2 billion m^3) equates to a single 1000–1200 ft (300–400 m) high mountain with 1:4 side slopes. In the cold language of overall resource, this is negligible – and the same would be true for much of the world. Locally, there may be a lack of rock (e.g. in the Netherlands), but overall there is no global rock shortage. There may be a lack of very high-quality rock, such as stone for high skid-resistant surfacing, which may necessitate some relaxation in standards compared with what is ideal. Nevertheless, rock shortage is not a serious sustainability driver.

The other obvious resource demanded by pavements is binder, whether cementitious or bituminous. Taking cement first, there are serious sustainability issues, but not under this heading – there is no lack of limestone from which to manufacture Portland cement. However, bitumen is another story. Most bitumen is derived from crude oil distillation and, while oil continues to be used as fuel and for other purposes, there will continue to be a supply of bitumen. It is, essentially, a by-product, but a by-product from a resource that is most definitely finite, namely oil. This puts it in a difficult category. As a by-product, its use fits in with sustainability goals because, if it were not used in pavements, oil companies would be desperately looking for alternative markets. However, in the long term, reliance on bitumen is unsustainable (something which is also true of many industrial by-products because industrial processes, and therefore their by-products, are continually changing). The sensible approach, therefore, is to use bitumen while it is here but to plan for a future in which it is less readily available. That future may involve the use of *bio-binders* (bitumen replacements derived from plants), but this is still firmly in the research domain.

2.3.2 Blight

The first point to make here is that the presence of a pavement implies that something else is no longer present. There is therefore a loss of agricultural productivity or of housing space or, simply, recreation space. This is not something that the pavement engineer can address of course, and it only affects the issue of new or widened pavements, but it is a subject that has to be at the heart of land-use planning. To carry out value calculations, land space has to be assigned a worth, according to its potential usefulness. Thus, prime UK agricultural land has a typical value of approaching £3 per square metre, reducing to less than £2 for rough pasture land. In contrast, urban land in the UK, where permission for housing has been granted, is valued at anything from £100 to over £500 per square metre. Of course, these are artificial figures, inflated by difficulties imposed by planning restrictions, but they provide a useful benchmark. On the other hand, if a site has been contaminated, usually due to past industrial activity, there may be a serious clean-up bill attached, effectively reducing the value to zero, even in a city environment. Compare these figures with the direct cost of a major new highway, very roughly £100 per square metre, and the real value of land can sometimes be a very significant factor indeed.

Of more direct concern to a pavement engineer is the issue of secondary blight; that is, environmental destruction caused by pavement construction. The obvious example is quarrying. There may be almost inexhaustible supplies of rock, but there are certainly not inexhaustible supplies of beautiful countryside – nor indeed of farmland. Thus, anything the pavement engineer can do to maximise the use of recycled materials, industrial by-products, construction and demolition waste and so on and therefore minimise the extraction of virgin rock (or gravel) contributes to environmental sustainability. This calls for innovation. It is necessary to look beyond tried-and-tested products and to design pavements that are able to accommodate materials that may have slightly different properties than those engineers have become used to over the years. In fact, this is one of the greatest challenges facing the pavement industry in many parts of the world.

2.3.3 Energy usage

Every component in a pavement, apart from the natural subgrade, has an energy associated with its production, transport, placement and compaction.

When the details are considered, such as the energy required to produce steel – less the saving when that steel is reused at the end of the life of a particular piece of equipment – then the calculation is clearly complex. Fortunately, researchers have painstakingly investigated these issues, narrowing down the uncertainty levels involved, so it is only necessary for those carrying out energy analyses to multiply the tonnage of each pavement material by the relevant figures for production, transportation and so on. Approximate 'embedded' energies associated with different pavement materials are given in Table 2.2, and it is worth noting that both concrete and hot-mix asphalt score poorly,

Table 2.2 Embedded energies for pavement component materials (Stripple, 2001)

	Material	Embedded energy: MJ/t
Ingredients	Sands and gravels	5–10
	Crushed rock aggregate	20–25
	Bitumen	3200–3800
	Portland cement	4500–5000
	Reinforcing steel	23 000–27 000
Mixtures	Hot-mix asphalt	600–800
	Cold-mix asphalt	150–200
	Lean concrete	450–500
	Pavement-quality concrete (PQC)	750–1000
	Reinforced concrete	1100–1500
Transportation	All materials (per journey km)	12–20

due to the high energy consumption in cement production and in heating the ingredients of asphalt. Even cold-mix bituminous products are not without their energy needs. It is also worth noting that calculations of embedded energy from different sources can differ significantly; for example, those provided in the UK National Highways carbon calculator (National Highways, 2024) are typically a little higher. Figure 2.1 illustrates the energy 'footprint' of asphalt pavement construction.

Since asphalt (especially hot-mix asphalt) and concrete are high-energy products, anything that can be done to cut back on the use of these materials and increase the engineering value of other, lower-energy, materials will improve the overall energy footprint of a pavement. Nevertheless, as no more than about 0.25% of the energy used by developed countries can be attributed to pavement

Figure 2.1 The energy footprint of asphalt

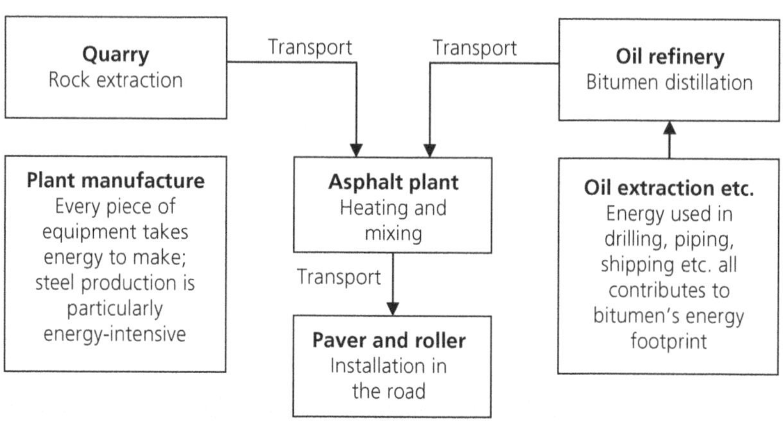

construction (or maintenance), pavement engineers would appear to have limited scope for reducing global energy consumption. Incidentally the figure of 0.25% is derived from a comparison of UK energy usage, running at about 6.5×10^{18} J/year (150 million tonnes of oil equivalent), with the energy used in pavement construction (not including major earthworks, tunnels and bridges), which is probably no more than 1.5×10^{16} J/year, extrapolated from asphalt consumption figures.

However, there is a much more significant element to the energy equation, namely the issue of vehicle energy consumption, which amounts to about 30% of all energy consumed in the UK. Many studies have shown that energy consumption differs from pavement to pavement and, summed over several decades, this is potentially a much more significant factor than embedded energy. The problem is that the studies do not all agree. There is full agreement that a rough texture leads to high rolling resistance and therefore high fuel consumption (e.g. Yap, 1987). However, there are also studies suggesting that a stiff pavement (notably a concrete surface) will reduce energy losses and lower fuel consumption (e.g. Zaniewski, 1982). Theoretical work at Nottingham University (Thom *et al.*, 2010) suggests, however, that the stiffness effect is minor compared to that of texture. Thus, a concrete pavement should be energy efficient due to its high stiffness – but not if it induces a rough ride and tyre vibration. The smoothest ride from the point of view of tyre vibration is given by asphalt surfaces with a small aggregate size and negatively textured (see Chapter 11) (e.g. porous asphalt). Logically, therefore, the most efficient pavement of all may be one with a concrete structure but a thin asphalt surface (with appropriate texture).

Research is ongoing, but claims and counterclaims sometimes speak in terms of differences in fuel consumption of the order of 10%. But how significant is this? On lane 1 of a busy highway, the traffic flow may be around 30 000 vehicles per day. If the average fuel consumption (petrol/diesel assumed) is 7 km per litre (assuming a high goods vehicle percentage), this equates to a fuel demand of around 4 litres per metre per day, approximately 140 MJ of energy per metre per day. Over an assumed lifetime for the surface course of 10 years, this gives 500 GJ/m, and a mere 1% change in fuel consumption would therefore involve 5 GJ/m, equivalent to fuel costing about £160 in real terms (i.e. ignoring government tax). A similar picture could be derived for electric vehicles.

This can be compared with the embedded energy in the pavement. For example, a typical asphalt surface course with a 6% binder content might be valued at around 800 MJ/t, equivalent to about 0.3 GJ/m of a 3.65 m wide lane 40 mm thick. On this basis, therefore, the issue of energy consumption looks to be much more significant than embedded energy. Even if the full pavement is considered, with a probable total embedded energy of something like 2 GJ per lane-metre, it is still well worth taking the effect on vehicle energy consumption into account, particularly as the life of the structure as a whole is likely to be many decades rather than the 10 years assumed for a surface course. And, in cold financial terms, the issue is, if anything, even starker. One lane-metre of major highway costs about £400; if a £160 saving to the economy is on offer over 10 years simply by designing for low vehicle energy consumption, it is an offer that is most definitely worth taking.

This issue has, frankly, been ignored for too long, and it is to be hoped that sufficient political will can be brought to bear such that the energy (and consequent financial) equation is allowed to influence spending policy.

2.3.4 Emissions

Many emissions can be tied directly to energy. The same litre of fuel, which is approximately equivalent to 35 MJ of energy, is also responsible for 2.3 kg of carbon dioxide emissions. For electric vehicles and non-transport-related energy such as that needed to run an asphalt plant, the quantity of emissions depends on the method of power generation, whether fossil fuel, hydroelectric, nuclear…, and every country will have a different and continually changing balance between different modes. In the specific case of cement production, a considerable quantity of carbon dioxide is emitted as a result of the chemical reaction taking place in addition to power demand.

The example given in Calculation Sheet 2.1 is principally based on the UK National Highways carbon tool, supplemented by Swedish data (Stripple, 2001). Hot-mix asphalt rather than warm-mix is assumed; use of warm-mix would save around 1.5 kg/m² for Pavement 1. The numbers can be compared to the typical carbon footprint for an average UK citizen, currently nearly 12 t/year.

Calculation Sheet 2.1. Carbon footprints

Designs from UK National Highways standards for 'long-life' pavements (80 million standard axles) over 5% CBR subgrade

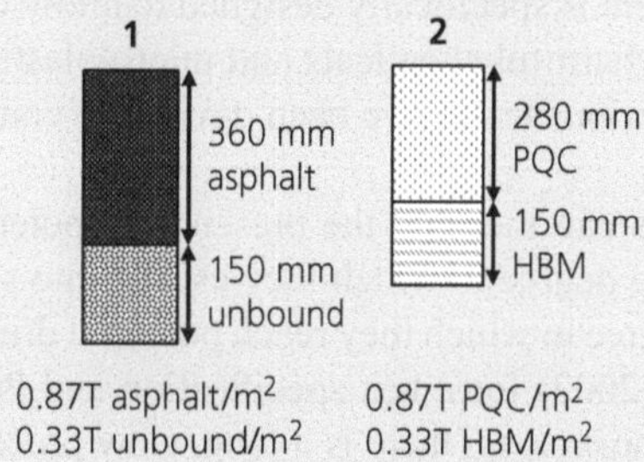

DBM50
Carbon dioxide:

Crushed rock:	1.4 kg/t × 0.955	= 1.3 kg
Bitumen:	173 kg/t × 0.045	= 7.8 kg
Production:	23 kg/t	= 23 kg
Transport:	1.1 kg/veh.km × 15 km ÷ 20 t	= 0.3 kg
		→ **33.4 kg/t**

PQC
Carbon dioxide:

Cement:	806 kg/t × 0.15	= 121 kg
Crushed rock:	1.4 kg/t × 0.5	= 0.7 kg
Sand:	0.7 kg/t × 0.35	= 0.2 kg
Production:	3 kg/t	= 3 kg
Transport:	1.1 kg/veh.km × 3 km ÷ 15 t	= 0.2 kg
Paving (+ steel):	4.2 kg/t	= 4.2 kg
Joints:	0.3 kg/t	= 0.3 kg
		→ **129.6 kg/t**

<u>HBM</u>: as PQC except 6% cement, reduced paving, no joints → **53.4 kg/t**
<u>Unbound</u>: crushed rock + transport, paving and compaction → **2.7 kg/t**

∴ **Pavement 1: 30 kg CO_2/m²**
Pavement 2: 106 kg CO_2/m²

The reason why carbon dioxide is highlighted of course is because of its role in climate change, but many other emissions are also important in the sustainability equation. Nitrous oxide, sulfur dioxide, carbon monoxide and methane are all gases that, in one way or another, can damage either the environment or human health. They will not be discussed further here, but the need to keep below dangerous limits affects the design of equipment such as asphalt batching plants and in situ recycling machines.

2.3.5 Groundwater issues

If a pavement is responsible for a loss of water resource due to harmful chemicals leaching into the groundwater, this clearly represents an unsustainable situation and must be avoided. Yet the other side of the coin is that a pavement in itself generally diverts water into a drain rather than allowing it to percolate into the soil and replenish groundwater supplies. It thus depletes resources (and contributes to local flooding in storm conditions) just by its presence. There is, therefore, justifiable pressure to avoid this in some way, and Section 24.6 is included on the subject of 'drainage pavements' – pavement engineering's contribution to sustainable drainage systems (SuDS). Nevertheless, whether a pavement is specifically designed to allow water to percolate or not, there will always be the potential for harmful chemicals and micro-plastics present on road surfaces to escape into groundwater, and various tests have been devised to combat this.

The problem is that, while tests can measure the presence of potentially harmful chemicals and solids, and can also measure the degree to which they escape into surrounding water, it is impossible to measure directly the degree to which they reach potential drinking water supplies or impact on the natural world; see BSI (2002) for a test specification and Peploe and Dawson (2006) for an overview of the subject. In most cases, there is a long, slow process of percolation, and the soil through which the water percolates tends to adsorb much (or even all) of the contaminant. Many engineers consider that rules such as are applied by environmental protection agencies have tended to be overprotective and may therefore be preventing optimised design. The point is that large stocks of waste materials have been, and continue to be, generated by industry and, as a planet, we need to treat this waste in a 'least worst' way; incorporation into pavement construction may just be that least worst solution.

2.4. Summary

The majority of this book is concerned with the engineering principles listed in Section 2.1, the sphere of expertise of the engineer. However, all engineers should have an awareness of the context in which their engineering designs are to be applied, and this chapter has highlighted both the economic context and the issues of sustainability. These are not issues with which an engineer will usually be directly concerned – any more than an economist or an environmental expert will usually concern themselves with the details of engineering. However, if optimised solutions are truly to be found then these three parties need to be able to communicate, to speak each other's technical language. It is hoped that this chapter has provided the pavement engineer with some of the appropriate communication tools.

REFERENCES

BSI (2002) BS EN 12457-3:2002: Characterisation of waste. Leaching. Compliance test for leaching of granular waste materials and sludges. Two stage batch test at a liquid to solid ratio of 2 1/kg and 8 1/kg for materials with a high solid content and with a particle size below 4 mm (without or with size reduction). BSI, London, UK.

National Highways (2024) Carbon Emissions Calculation Tool. See https://nationalhighways.co.uk/suppliers/design-standards-and-specifications/carbon-emissions-calculation-tool/ (accessed 11/03/2024).

Peploe RJ and Dawson AR (2006) *Environmental Impact of Industrial By-products in Road Construction – A Literature Review*. Land Transport New Zealand, Wellington, New Zealand, Research Report 308.

Stripple H (2001) *Life Cycle Assessment of Road: A Pilot Study for Inventory Analysis*. IVL Swedish Environmental Research Institute, Göteborg, Sweden, Report B1210E, 2nd edn.

Thom NH, Lu T and Parry T (2010) Fuel consumption due to pavement deflection under load. *Proceedings of the 2nd International Conference on Sustainable Construction Materials and Technologies, Ancona, Italy*.

Yap P (1987) *Factors Affecting Truck Fuel Economy*. Goodyear Tyre and Rubber Company, Akron, OH, USA, Report No. 862-932-513, pp. 63–79.

Zaniewski J (1982) *Vehicle Operating Costs, Fuel Consumption and Pavement Type and Condition Factors*. Federal Highway Administration, Washington, DC, USA, Report No. FHWA/PL/82/00.

Nick Thom
ISBN 978-1-83549-711-1
https://doi.org/10.1108/978-1-83549-710-420243004

Chapter 3
Pavement applications

Much of this book is concerned with the detailed engineering of a pavement structure. However, it is important to appreciate the differences between different pavement applications, leading to very different design criteria. It is certainly not necessary to apply the same standards to all pavement types. Some are for fast but relatively light traffic; others are for concentrated heavy loads. In some cases, surface properties are of key importance; in others not. This chapter is intended to give a feel for the typical pavement types expected for different applications and the factors that determine the most appropriate choice in each case.

It is not necessary to be concerned with technical details at this stage; the purpose is still to gain a broad appreciation of the subject.

3.1. High-speed highways

In general, these will be heavily used routes connecting centres of population. For the users, the purpose is to allow a fast, safe journey with a high level of reliability. The key design requirements are, therefore, as follows.

Design requirements	
	■ good ride quality ■ safe, skid-resistant surface ■ quiet surface ■ low maintenance

3.1.1 Ride quality

A good ride quality at high speed demands an accurate surface level, which in turn demands that the surface is laid using modern paving equipment. The tolerances can be achieved using either pavement-quality concrete (PQC) or asphalt, but if PQC is used then joints will usually reduce the ride quality. Non-jointed PQC is possible (continuously reinforced concrete (CRC)), but this is an expensive option and therefore suited only to the most heavily trafficked highways.

To achieve the appropriate surface tolerance in asphalt, the surface course must be placed on an underlying layer (the binder course), which itself is within relatively tight level constraints. Excessive inaccuracy in the surface of the binder course means that the surface course is being asked to accommodate an impossibly large variation. For example, if the surface course varies from 30 to 50 mm in thickness and it is placed at a temperature of 130 °C then, when the compaction roller begins work some 3 min later, the temperature of the surface of the 'mat' would already vary from about 111 °C to 120 °C (calculations are presented in Section 10.1.2), which makes an appreciable difference in mixture viscosity, and therefore in resistance to movement under the roller. In practice, asphalt pavements require at least three paver-laid courses in order to ensure the highest standards at the top of the surface course level.

3.1.2 Skid resistance

On a high-speed road, skid resistance requirements do not usually represent an onerous design condition because sight lines are good, there are no sudden stop lines or pedestrian crossings, and no sharp bends. However, high speed does present the possibility of aquaplaning if surface water is trapped between the vehicle tyres and the road surface. This means that a certain *texture depth* is required; that is, either the stones that make contact with the tyres should be set well above the level of material between the stones or the equivalent effect has to be achieved in concrete. Reasonable skid resistance can be obtained, even under water, as long as there is sufficient texture depth. This topic is covered in detail in Section 11.4.4.

If the texture depth requirement is met, the intrinsic frictional quality of the surface stones or concrete does not have to be exceptionally high in a high-speed road situation. However, it is important that the texture depth does not deteriorate excessively, and this means that the surface stones (or concrete protrusions) must have a high abrasion resistance, which usually implies the use of a hard igneous rock.

3.1.3 Surface noise

There are two quite separate aspects to surface noise. The first is nuisance to the driver and the second is nuisance to an adjacent population. Both are, of course, political rather than strictly engineering issues, but the engineer is the one who is able to help solve the political problem. However, the problem of noise is far from straightforward. Different surfaces generate noise in different frequency ranges, with concrete tending to higher frequencies, but the nuisance level depends very much on the situation of the listener. Drivers are usually annoyed by high-frequency noise from concrete, as are people close to the road. At a greater distance, lower frequency and longer wavelength sound will travel better and so will not be attenuated as efficiently as high-frequency noise. Thus, while drivers and close observers tend to prefer low-stiffness voided asphalts because they generate low-frequency noise, this is not necessarily the case for those at a greater distance. Nevertheless, certain general principles can be laid out as follows.

- Large surface protrusions generate high noise.
- High material stiffness generates high-frequency noise.

The implication is that an asphalt surface will usually be less annoying than a concrete surface, although the reader is referred to Section 21.2 for the design of relatively low-noise concrete surfaces.

3.1.4 Low maintenance

On many intensively used road networks low maintenance is a key driver in design. Maintenance is not just a direct cost for the maintaining authority, it is a severe nuisance to the road user and, more importantly, imposes a considerable indirect cost on the local economy as a consequence. Here the idea of *design reliability* is important. The point is that there is always a degree of uncertainty in the actual lifespan of a pavement, no matter how carefully it is designed and constructed. This uncertainty stems from variations in ground conditions, differences between materials from different sources (e.g. different rock types), variations in constructed thickness and uncertainty in traffic loading. This then results in a statistical distribution of possible lifespans, and one can speak in terms of a 90% or a 95% reliability, where there is only a 10% or 5% chance of the actual lifespan being less than the stated value. For major high-speed highways, the sensible approach is to design to a long life and a high reliability such that disruptive maintenance is kept to a minimum. The issue of uncertainty is pursued more fully in Chapter 5.

3.1.5 Typical constructions

As a result of these design considerations, major high-speed highways tend to have relatively thick construction, as illustrated in Figure 3.1. The designs shown are intended to be approximately equivalent, although considerable variation is possible using materials of differing qualities. The thicknesses shown are therefore purely illustrative, as is the fact that an asphalt surface course is shown overlying the PQC slab. This reflects current UK policy, which is dictated mainly by the politics of noise.

3.2. Urban roads

Urban roads present a quite different set of design challenges. Traffic is not usually travelling particularly fast; in fact there will be many locations where stationary or slow-moving traffic is expected. For the driver, there are many more potential hazards to avoid, such as pedestrians, cyclists and parked cars, and there are many reasons why it may be necessary to decelerate suddenly. For the maintaining authority there is the ever-present headache of dealing with the providers of piped services such as clean water, fibre optics, sewerage, electricity and gas, who regularly need to dig up sections of urban road as they install and then repair/upgrade their networks. The road surface also tends to be dotted with access chambers fitted with steel access covers, presenting a further cause of uneven ride quality. As traffic speeds are relatively low, the surface quality need not be of the highest standard, but the disruption caused by digging up sections of road is a significant issue.

Another specific problem is that of buses. Many cities have dedicated or partially dedicated bus lanes, which are excellent from the point of view of public transport provision, but mean that particular sections of the road carry heavy and concentrated loads and are therefore more vulnerable to deterioration, particularly rutting. The design considerations for urban roads can therefore be summarised as follows.

Design requirements	
	■ maintenance-friendly construction ■ high skid resistance in places ■ high rut resistance in places ■ low maintenance

Figure 3.1 Typical constructions – major high-speed highways

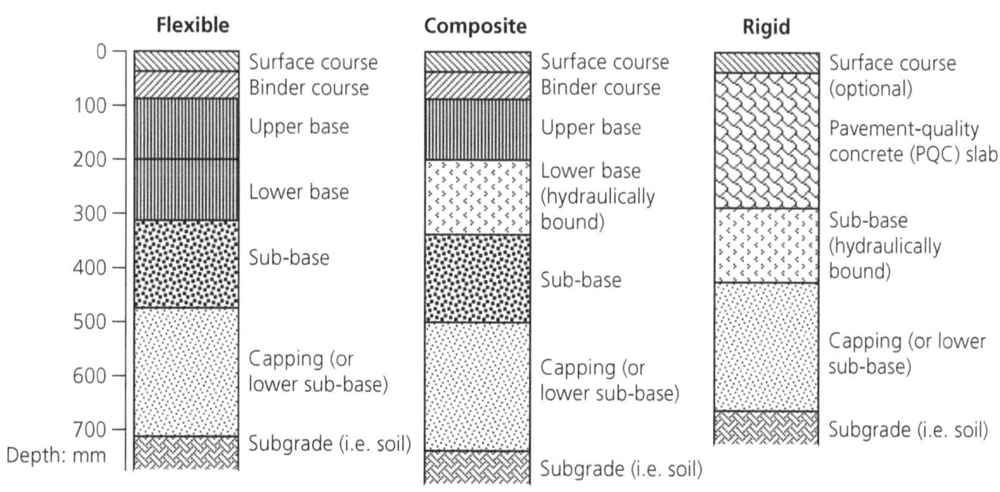

3.2.1 Maintenance-friendly construction

Here there are two main issues to be addressed.

- It should be possible to cut/break the road using standard equipment.
- Backfill to trenches should not result in a significantly poorer or less stiff construction.

The first requirement suggests that strong PQC should be avoided. Geosynthetics, particularly strong polymers, are also sometimes a significant nuisance during excavation. The act of breaking through these tends to damage the material on either side of the excavation.

The second requirement means that the pavement should be of a fairly standard type of construction, not one that relies on any specialist high-strength, high-stiffness material that cannot be reproduced when carrying out a trench repair. In practice, this means that conventional asphalt, weak hydraulically bound base and unbound material are the options most likely to be suitable. Block paving is also acceptable, but only if the maintaining authority is prepared to commit to carrying out all subsequent restoration work by carefully removing and then replacing the blocks in the area of the work.

3.2.2 High skid resistance

The principal areas where higher than average skid resistance is typically specified are at pedestrian crossings, junctions and roundabouts. This is generally achieved by means of a special treatment to the surface (high-friction stone chippings stuck down with an ultra-tough binder), and so does not affect the design of the underlying pavement, which would be the same as in other areas.

3.2.3 Rut resistance

Ruts in asphalt occur under heavy loads, particularly when those loads travel slowly. This is because bitumen is still a liquid, even at in-service temperatures, and slow loading allows the bitumen to deform and stones to slip relative to each other. The practical result of this is the rutting that is often visible in bus lanes, at traffic lights, or in bays where heavy goods vehicles load and unload. There are several potential solutions to this problem (Figure 3.2).

Figure 3.2 Typical bus lane designs

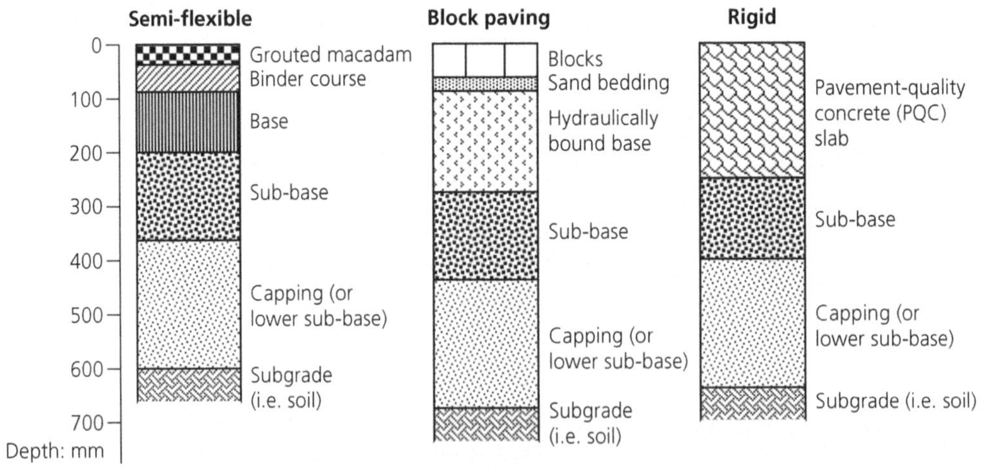

- Use PQC.
- Use block paving over a hydraulically bound base.
- Use grouted macadam.
- Take care to design asphalt mixtures to resist rutting.

PQC is rut-resistant but, as noted above, is usually unsuitable because it does not lend itself to trenching and backfilling.

Block paving has the disadvantage that any rut that develops shows up very clearly indeed. However, if the base has been properly designed and if the sand (or mortar) bedding is not too thick, rutting should never occur. Cost may, however, be an issue.

Grouted macadam has the benefit that it can be 'retrofitted' to a road with problems. As it is generally only placed as a surface course of 50 mm or less, it is only necessary to plane out the existing surface and replace it with grouted macadam. This saving in time and effort may then offset the high cost of the material itself, particularly for bus lanes. It also has advantages over concrete in that joints are not required and there is no dramatic change in construction at the interface with the adjacent asphalt. But, like concrete, it presents challenges when trying to restore the pavement after trenching.

Asphalt mixture designs are covered in Chapters 9 and 13, and it is true that, with high-quality aggregate, rut-resistant mixtures can be produced. Asphalt is therefore the usual choice, for the surface course at least.

3.2.4 Low maintenance

Accepting that an urban road is always likely to be disturbed by service providers such as water companies, there is nevertheless significant pressure not to close any part of a major urban road for longer than is absolutely necessary. Rapid maintenance such as surface dressing, slurry sealing or retexturing (see Chapter 12) is acceptable, as disruption is only brief, but major rehabilitation is highly undesirable. Major urban roads should therefore be designed to a high reliability – that is, with a suitably long design life. Of course, this point is critically dependent on the traffic level and the availability of alternative routes. Minor streets effectively come under the next heading, 'estate roads'.

3.3. Estate roads

The purpose of estate roads and minor city streets is simply to allow the passage of vehicles to and from individual residential or industrial locations. The flow is low, most loads are light, but there will also be occasional heavy goods traffic (deliveries, refuse trucks, fire engines). Speeds are (or should be) low and ride quality is relatively unimportant. As long as the road is passable and safe, its purpose is fulfilled. This leads to the following design requirements.

Design requirements	
	■ cheap construction ■ should remain passable for many decades ■ must accommodate occasional heavy traffic

3.3.1 Cheap construction

The cost of estate roads cannot be justified simply by the number of vehicles using them – they are an expensive necessity rather than a valuable economic asset. Therefore, they must be cheap.

In principle, they could be unbound pavements, as many access roads are, but for anything other than the most lightly trafficked situation this will give rise to unwanted maintenance costs, so it is usually preferable to apply a bound surface. However, this should be of minimum thickness, often consisting of two courses of asphalt, as it is hard to achieve a satisfactory finish with a single course. Concrete slabs, generally around 120–150 mm thick with light reinforcement, are also possible, but experience (and theory) suggests that deterioration may be excessive under the occasional heavy truck loads that will occur.

3.3.2 Passable for decades

It is this requirement that generally leads to the use of a bound surface. However, it should be noted that 'passable' is quite different from 'in excellent condition'. It should also be noted that there is little restriction on carrying out maintenance (except during antisocial hours) and, therefore, while it is a cost to be avoided if possible, it can be done if necessary. This means that designs do not have to be to a high reliability. Even 'failure' of parts of the road can still leave the route passable. It is therefore permissible to lower the reliability to as little as 50% (AASHTO, 2007) and accept that the lower initial cost that this allows will be partially offset by the need for future patching works. However, every effort should be made to provide surfaces that are relatively resistant to pothole development (see Section 12.1.5) since this form of deterioration has particularly high economic consequences.

3.3.3 Occasional heavy traffic

This requirement means that the use of concrete should only be allowed after careful evaluation of the wheel loads from heavy trucks. Asphalt is a forgiving material and can withstand occasional overloads; concrete cannot.

3.4. Rural roads

Rural roads present the challenge of low traffic volumes but, sometimes, relatively high speeds. Alignment constraints also often mean that there are serious safety problems. The design requirements are therefore as follows.

Design requirements	■ adequate ride quality ■ suitably high skid resistance

3.4.1 Ride quality

In developed countries, traffic speeds are likely to be of the order of 40–60 mile/h (60–100 km/h), which means that the ride quality has to be sufficient to give safe driving conditions at such speeds. This in turn dictates the use of an asphalt-surfaced pavement with multi-layer construction, as PQC would be ruled out on grounds of cost. The base could be unbound, asphalt or hydraulically bound, depending on the design traffic levels.

If the traffic volume is low enough, rural roads can be almost entirely unbound pavements, sealed with a surface dressing (a thin layer of stones spread over a bituminous spray coat; see Section 7.6). As deterioration occurs, the surface dressing treatment can then be repeated.

In the so-called 'developing world', unbound pavements are the norm in rural settings, with or without a surface dressing treatment. Such pavements are well suited to very low volumes and low to moderate speeds, but can accommodate occasional overloading with no greater damage than a slight acceleration in the rate of rut development. However, a fully unbound, unsurfaced pavement demands significant maintenance input (regular reprofiling of the surface), and so is really only appropriate for the lowest traffic categories.

3.4.2 Skid resistance

Skid resistance is only one factor in maximising the safety of rural roads, alignment being probably of much greater significance. This means that difficult decisions have to be taken as to what level of skid resistance is appropriate. It is simply not economical to apply expensive treatments to every sharp bend or junction approach, and the money could usually be much better spent improving the alignment, road markings or signage. However, at the very least, adequate texture depth should be provided to prevent aquaplaning in wet weather, and surface dressing is a convenient means of achieving this.

3.5. Pedestrian areas and cycle/scooter paths

Once again, the design challenges change. Perhaps unfortunately, the most influential design input for some highway authorities is the need to minimise claims from pedestrians who have tripped over uneven footway surfaces. Nevertheless, and however strange this may seem to those who live in cities where footways are not paved at all, it does make the point that there is little real engineering design involved. Unbound pavements are perfectly adequate structurally to carry either pedestrian or cycle loading.

The engineering, therefore, comes in developing cheaper or easier forms of construction. Many industrial by-products, for example, lend themselves to use in footways and cycleways as they allow the production of a dense, closed surface despite not having the strength to perform under vehicular traffic. Examples of industrial by-products used in pavement construction are mining waste (e.g. from coal or China clay extraction) and slags (from various metal extraction processes, notably iron), some of which possess a degree of self-cementing action. Suitable recycled materials are construction and demolition waste, notably crushed concrete, and recycled asphalt planings (RAP). Cold-mix asphalts are also ideally suited to such applications. These are described much more fully in Chapter 13, but here it should be understood that bituminous binders in cold form can be used to bind many aggregates that would not be suited to hot asphalt production. The result will be a relatively weakly bound material, ideal for footways and cycle/scooter path surfaces.

Of course, the traditional medium for paving in a city environment is block (or flag) paving. However, the justification for its use nowadays is mainly aesthetic. Blocks are not a cheap paving option, but they can look impressive and lend character to a city centre. Except in locations that will be subject to motor vehicle loading, there is no structural design involved. The blocks are bedded in sand, which is placed over an unbound base. The actual laying is a skilled operation, and poor workmanship can lead to differential settlement, ponding of rainwater and steps between individual blocks or flags – but there is little the pavement engineer can contribute.

3.6. Car, coach and lorry parks

This category of pavement presents its own special problems. The loads this time can be considerable, particularly where trailer feet are allowed to rest directly on the surface, but they are static or slow-moving. Ride quality is not usually of great importance, so a degree of deterioration is often acceptable, but most parking area owners would be unhappy if their pavement were to become too much of an eyesore. The challenge for the engineer, therefore, is to come up with solutions that are suitably inexpensive yet provide the operator with an appropriate level of service. The design considerations can be summarised as follows.

Design requirements	■ sufficient load-bearing capacity ■ maintain an acceptable surface finish

3.6.1 Load-bearing capacity

The problem with slow or static loads is that asphalt does not cope well under such circumstances. If loads are light, as in a car park, then asphalt is a perfectly acceptable solution and the design will not be dissimilar to that for a minor urban road. However, heavy static (or, even worse, ultra-slow-moving) loads impose the most damaging possible conditions on asphalt. The bitumen has plenty of time to flow, greatly reducing its effectiveness at binding the aggregate particles together and allowing permanent strain to occur, leading to settlement. Trailer feet apply a massive stress onto the surface and can cause indentations in asphalt-surfaced pavements. The solutions are similar to those proposed above for urban roads, except without the subsequent likelihood of trenching for services.

- Concrete is ideal in terms of its deformation resistance but is usually too expensive, either in actual finance or in construction time, and joints in the concrete can give rise to problems later in the life of the pavement.
- Block paving is often used, but if heavy loads are to be accommodated care has to be taken to design the full pavement structure, notably a suitably deformation-resistant base layer, otherwise the whole pavement will deform and the blocks will follow the rest of the pavement down. Of course, as soon as local deformation occurs, rainwater will pond and soak into the subgrade soil at that point – and the condition will become even worse.
- The third solution, grouted macadam, has many advantages, particularly the lack of joints. In effect, grouted macadam is a deformation-resistant asphalt. It will not be able to stop deformation occurring if there is a serious underlying weakness, but at least it will maintain a reasonably water-resistant seal.

3.6.2 Surface finish

In many applications, the cosmetic appearance of the surface is almost as important as the structural strength of the pavement, hence the popularity of block paving. If the situation demands that a high-quality visual appearance is maintained, then this logically influences the choice of construction. Block paving may be desirable, but the price tag may count against it. Concrete is a risk: while it is intact and the joints are sound it looks fine, but there are few worse-looking pavements than cracked concrete slabs, spalling at the edges. Asphalt and grouted macadam are much safer in that any blemishes that develop can be remedied without too much difficulty.

It should also be mentioned that numerous unpaved or semi-paved solutions are available, notably reinforced grass surfaces, either using hollow soil-filled concrete blocks or fibre reinforcement in the soil. These give effective and aesthetically pleasing surface finishes, although they are not necessarily cheap. Both fine polymer fibres and mesh elements have proved effective at significantly increasing the strength of sandy or silty soils. The chief difficulty (and therefore expense) lies in the mixing process.

3.7. Ports and heavy industrial pavements

The design criteria for these pavements are similar to those for lorry parks in that slow or stationary loads are expected, although loading details will usually be different. Ports and inland freight terminals generally require container stacking facilities, and container loads are transmitted through small square steel feet, generating very high pressure. A similar situation is found under racking on internal warehouse paving.

Reach-stackers (Figure 3.3) apply one of the highest axle loads of any wheeled vehicle, with individual wheel loads of 30 t or more if dynamic effects are taken into account. Fork-lift trucks range from large, high front axle load machines with conventional pneumatic tyres, to small, solid rubber-tyred vehicles which, despite the lower total load, can apply high contact pressure through the tyre. Rubber-tyred gantry cranes present another special design case; they run on certain lines only, either side of container stacks, and it is common to design the pavement immediately underneath their lines of travel to be different from that of the surrounding area.

In terms of pavement design, the key requirement is that the surface should not deform excessively, but the allowable deformation will be different for the different design cases. For solid-tyred fork-lift trucks, it is very important that the surface they operate on is even. Similarly, rubber-tyred gantries cannot accommodate any significant uneven settlement along their lines of travel. Container stack areas, on the other hand, are much more forgiving, and most pneumatic-tyred vehicles can also cope with a certain amount of unevenness. It is unlikely that aesthetics will be given any serious priority, which means that pavement cracking only becomes important if it leads to deformation. In essence, therefore, there is but one key requirement – to limit deformation.

Design requirement	■ limit uneven deformation

Figure 3.3 Wheeled vehicles for ports and other industrial areas

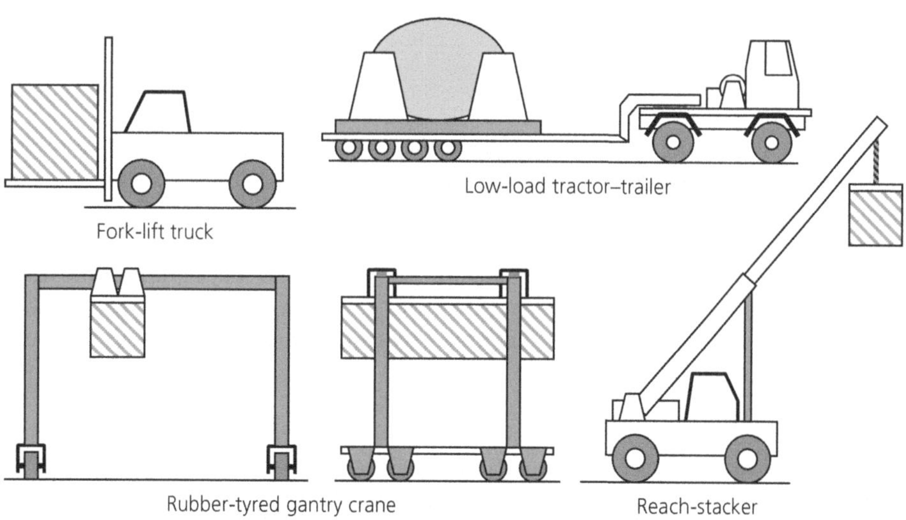

3.7.1 Uneven deformation

In practice, the full range of pavement options, including unbound pavements, is used in heavy industrial applications, reflecting the range of equipment for which designs have to be carried out. However, concrete is probably the 'default solution' because of its deformation resistance. Also, as paving is but one of a number of operations being carried out at any industrial development site, there is often less pressure on construction time, negating one of concrete's main disadvantages (i.e. the time needed to reach full strength). There is considerable variety in the types of concrete used in industrial pavements, with strengths from 10 to 40 MPa (compressive) and both wet-formed and roller-compacted placement techniques being used. Guidance is given elsewhere (Knapton, 2007).

However, use of concrete always brings with it a danger. A well-designed, well-constructed concrete pavement should have an almost indefinite and trouble-free life, but if a problem does arise, it can be serious. Concrete cracking is no problem in itself, but concrete cracking is most likely to indicate a non-uniform support, which means that settlement may occur in cracked areas and not elsewhere. The conclusion, therefore, is that it is usually worth investing in a high design reliability when using concrete.

Heavy-duty block paving is also common in industrial areas, overlying a substantial hydraulically bound base. The use of blocks overcomes the need for concrete joints, increasing its economic attraction.

Conventional asphalt is appropriate in many applications, particularly on access roads, lanes between areas of container stacking or lightly trafficked areas. It has the advantage that construction time is generally much less than for concrete, which in some contracts can be extremely beneficial. Grouted macadam has similar advantages to asphalt and, being practically non-deformable, it can, if properly supported, also be used in applications of high or static loading.

In conclusion, therefore, industrial pavements require real engineering input. It is essential that the requirements of each particular design situation are properly understood, for which both the magnitude and frequency of the loads, and also the acceptable long-term pavement condition, need to be known.

3.8. Airfield pavements

Aircraft represent yet another form of wheeled transport with its own peculiarities in terms of wheel load, arrangement and speed. The landing gear of an Airbus A380 represents just about the highest concentration of load that a pavement is ever likely to have to accommodate, around 160 t spread over six wheels – and there are four main gears in addition to the nose wheels. Selected aircraft wheel arrangements are illustrated in Figure 3.4. Such a high concentration of loads means that the stress field developed within the pavement will penetrate much deeper into the subgrade soil than is the case for other pavement types, necessitating extremely thick construction, especially over soft ground sites.

It is a common misconception that the critical loading case occurs in the touchdown zone of a runway, where the wheels impact onto the surface and are accelerated rapidly from zero to a speed of around 150–200 km/h, leaving considerable rubber deposits on the pavement surface. However, the impact is in fact a minor one, as most of the load is still being carried by the aerofoil action of the aircraft wings. The critical case is actually when the aircraft is travelling slowly with a full

Figure 3.4 Selected aircraft wheel arrangements (dimensions in metres)

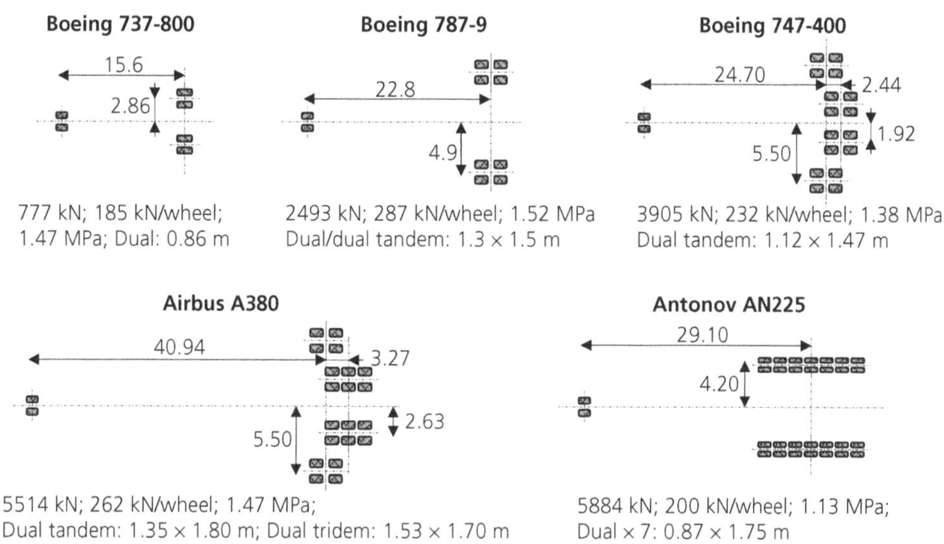

fuel load, whether on the runway, taxiway or apron, and the greatest concentration of loads commonly occurs on taxiways, where variation in the lateral position around the taxiway centreline is minimal.

The principal pavement design requirements are as follows.

3.8.1 Limiting deformation

A particular problem with deformation of airfield pavements relates to surface water drainage. Aircraft operational requirements limit the crossfall to 1.5% maximum (on major airfields) and water cannot be expected to drain effectively if the slope falls below about 1%, which means that a relatively slight degree of differential settlement can result in ponding. This is undesirable on taxiways, but it is potentially disastrous on runways if an aircraft aquaplanes. The need to avoid ponding means that the overall pavement thickness *must* be adequate to protect the subgrade and avoid deformation at that level.

Particular locations that are always likely to be subject to deformation are aircraft 'stands', as slow-moving and then stationary wheel loads are applied repeatedly to exactly the same points. The slow-moving nature of the load means that asphalt would have plenty of time for viscous flow, which can lead to indentation of the surface. It is therefore usual to use concrete on all heavy aircraft stands, although grouted macadam and block paving are alternative options.

Design requirement	■ limit deformation ■ ensure adequate skid resistance ■ avoid future closures ■ resist fuel/oil spillage ■ avoid foreign object damage

3.8.2 Skid resistance

This is another safety-critical issue. The skid resistance of runways has to be measured regularly, particularly under adverse weather conditions. Runway surfaces therefore have to be suitably textured when constructed. Concrete pavements are usually not a problem; the texture is achieved by grooving (or other similar techniques) while the concrete is still wet (see Chapter 21). Asphalt pavements, on the other hand, often have insufficient natural texture, which leads to the common practice of sawing grooves in the asphalt surface. Unfortunately, the edge of the groove then represents a weakness if there is any deficiency in the asphalt mixture, and cases of spalling at these edges are not uncommon. Because of the high stresses at the edges of the grooves, grooving is only suited to the strongest mixes.

High skid resistance is only required on runways; taxiways and aprons rarely take traffic travelling at more than about 50 km/h – other than certain usage by ground vehicles.

3.8.3 Avoiding closures

Economically, busy airports cannot afford to close; neither can they afford any long-term reduction in their capacity. This means that the critical parts of an airfield, notably the runways, have to be designed with an eye to possible future maintenance needs. Key factors are as follows.

- Major concrete repairs require time for the concrete to set. This can be as little as a few hours, but only if relatively expensive rapid-hardening cements are used.
- Asphalt overlays can be carried out at night in stages, with the ends of each night's work being ramped down to provide a sufficiently smooth ride for aircraft.
- Alternatively, work can be carried out at each end of a runway while a reduced operating length is imposed, but there will usually be a section at the centre that cannot be accessed in this way.

It is clearly best if such maintenance needs can be avoided altogether, although rapid fluctuations in air travel demand have made forecasts difficult. It is worth bearing in mind, however, that a small saving in runway materials in the short term may easily be wiped out in whole-life terms by large future maintenance costs. At major airports, design reliability should therefore be high.

3.8.4 Fuel and oil spillage

Airfield pavements are frequently subjected to spillages, most commonly at aircraft stands, where refuelling takes place. This is a serious bar to the use of asphalt in such locations, and a further reason why aircraft stands are usually constructed in concrete.

3.8.5 Foreign object damage

It is critical to aircraft safety that no foreign object is sucked into the engines. The most common such object is a bird, but pavement debris is another potential hazard. This effectively precludes any surface where stones or other surface material might work loose. Conventional asphalt and concrete (to a sufficiently high specification) are both satisfactory, but the use of rolled chippings (as, for example, in surface dressing) would not be permitted.

3.9. Guided busways and other surfaces subject to channelised traffic

All pavement types described so far have to accommodate wheels in almost any location. While distinct wheel-paths develop on most highway, industrial and airfield pavements, it is not the case that every vehicle follows these wheel-paths and so a broad area of strong pavement has to be provided.

However, guided buses follow exactly the same path every time. The same is true of automated industrial plant such as straddle carriers. And, as driverless vehicle technology advances, it is likely that wheel-path damage may also become increasingly concentrated on highways.

In cases of 100% channelised traffic such as guided busways there is no point providing a surface outside the trafficked zone, which means the 'pavement' can become two narrow beams. Realistically, this cannot be achieved in asphalt – and deformation would be excessive anyway – and so concrete is the material of choice, which can be either laid in situ or precast, almost certainly reinforced.

The more difficult case is where channelised traffic shares a pavement with vehicles that need to stray beyond the wheel-path. If driverless commercial vehicles actually become the norm in the future – and this author has his doubts – then this will bring to the fore an opportunity that already exists, namely designing or retrofitting narrow strengthening strips in the most heavily trafficked wheel-paths. This is not normally done at present, but it could be.

3.10. Summary

The aim of this chapter was to give a feel for the many and varied types of pavement out there – and the real need for engineering input. Design is always a balance between the competing needs of safety, reliability, cost, aesthetics, maintainability, ride quality and so on. Engineers must be able to evaluate where the real priorities lie.

REFERENCES

AASHTO (American Association of State Highway and Transportation Officials) (2007) *Guide for Mechanistic-Empirical Design of New and Rehabilitated Pavement Structures*. AASHTO, Washington, DC, USA.

Knapton J (2007) *The Structural Design of Heavy Duty Pavements for Ports and Other Industries*, 4th edn. Interpave, Leicester, UK.

Nick Thom
ISBN 978-1-83549-711-1
https://doi.org/10.1108/978-1-83549-710-420243005

Chapter 4
Traffic loading

As is evident from Chapter 3, the nature of the traffic using a pavement is a primary design input. Naturally, an airport runway taking large aircraft will require a different design from a highway or a port pavement because of the differences in load magnitude, number of load applications, speed of travel and also tyre pressure. Clearly, this subject has the potential to be very complicated indeed, as most pavements will have to cater for a wide variety of different vehicles travelling at different speeds. It is therefore necessary, first, to understand the importance of all the different variables and, second, to develop a means of simplifying the problem to something manageable.

4.1. Traffic variables

4.1.1 Load magnitude

Very obviously, damage to a pavement occurs much more rapidly under heavy loads than under light loads, as every part of the structure will experience higher stresses. The question is: how much more rapidly? One could look ahead to the fatigue characteristics of both hydraulically bound and bitumen-bound material introduced in Chapters 9 and 19. For bitumen-bound material (i.e. asphalt), the appropriate characteristic will be a linear relationship in logarithmic space between tensile strain and material life; this represents a power law equation. The exponent of that power law, being the slope of the characteristic in logarithmic space, is typically between −4 and −5 (or −0.2 and −0.25 depending on which way the equation is written). In the case of hydraulically bound materials (HBMs), the characteristic is a straight line in semi-logarithmic space, which is less convenient to interpret. However, in the range typically applying in the pavement, with a tensile stress between about 30% and 40% of the flexural strength, the characteristic is equivalent to a power law with an exponent of about −12.

The other principal mode of pavement deterioration, namely permanent deformation, is even harder to express mathematically. An equation for permanent deformation in unbound material will be introduced in Chapter 6, which suggests that an increase in applied stress will give an inversely proportional decrease in the logarithm of number of load applications (to reach a given level of deformation). In the normal operating range for highways, subject to between 10^7 and 10^9 heavy load applications, this is equivalent to a power law with an exponent of around −7; for airfields, with lower traffic numbers, the exponent would be nearer −4.

More complicated still is the problem of deformation in asphalt, for which no equation will be offered. The first problem is that an asphalt's resistance to rutting is derived in part from the aggregate skeleton but also from the binder. If the binder effect dominates, and if bitumen is considered as a linear viscous material, then a linear relationship (i.e. exponent = −1) is appropriate; if the aggregate dominates, then the power will rise to be closer to that for deformation of unbound material. But even these deductions only apply if the contact stress rises with increasing load, as asphalt deformation will principally be a function of contact stress. If there is no increase in contact stress (i.e. same tyre pressure; larger contact area), there will be little increase in deformation.

The reader should by now be getting an impression of a quite impractical level of complexity. It is important to be aware of these points, but it is also important to move forward. If a theoretical approach presents such difficulties, what about an empirical one? In the late 1950s, the American Association of State Highway Officials undertook an extensive series of pavement trials using controlled trafficking with known loads (HRB, 1961). Their conclusion was that, despite considerable scatter in the data, use of a fourth-power relationship (exponent = −4) between load level and number of load applications needed to cause failure was appropriate – and most of the world has opted to take this approach ever since. So ingrained is this assumption that it is even known as the *fourth-power law* and may be expressed as

$$\text{Number of equivalent design wheel loads}\left(N_{\text{eq}}\right)=\left[\text{wheel load}\left(P\right)/\text{design wheel load }\left(P_{\text{des}}\right)\right]^{4} \quad (4.1)$$

Therefore, if the wheel load (P) is twice the design wheel load (P_{des}), it will do 16 times the damage of a design wheel.

Although Equation 4.1 is expressed in terms of wheel loads, it is actually more common in road design to talk in terms of axles, where an axle represents a pair of wheels, one in each wheel-path. But of course the ratio on the right-hand side of Equation 4.1 is the same whether wheels or axles are used. For other types of pavement it is usual to select a design case – a design aircraft in the case of airfield pavement design – and to convert other load types to an equivalent number of the design case using the fourth-power law.

Some organisations have considered they can do better – for example, an exponent of 4.5 is used in the UK Transport Research Laboratory Road Note 31 (TRL, 1993). It is clearly justifiable to take a different exponent for different types of pavement or for different distress modes. A low exponent is theoretically appropriate for asphalt rutting, whereas the brittle behaviour of HBMs would theoretically demand a very high value. However, this is not a critical issue in most design approaches, as design standards will inevitably be calibrated to give an appropriate design when the specified exponent is used. It is also true that, when a typical set of axle loads found on a highway is converted to an equivalent number of design axle loads, the exponent actually seems to matter remarkably little. Calculation Sheet 4.1 spells out the required computation and illustrates the sensitivity to overloading when a high exponent is used. In general, an exponent of 4 gives an approximate minimum number of equivalent design axles for highway traffic, but exponents within the range 2–6 give variations of no more than about 15%, which makes very little difference to pavement thickness design. However, the reader is warned that the choice of exponent may be more significant for airfields or industrial pavements, or for calculations relating to occasional overloading.

In conclusion, therefore, the fourth-power law appears to be a practical and sufficiently correct means of converting a real axle load spectrum to an equivalent number of design axles for highway pavements. However, the engineer should appreciate that the real effect of load variation will be quite different depending on which mode of damage is being considered, and fatigue of HBM is the most load-sensitive mode of all.

Calculation Sheet 4.1. Use of power laws, UK motorway data

Highway traffic – 1 hour:	Load band: kN	Average: kN	Number
	0–10	5	3875
	10–20	15	1201
	20–30	25	564
	30–40	35	320
	40–50	45	254
	50–60	55	189
	60–70	65	265
	70–80	75	412
	80–90	85	365
	90–100	95	176
	100–110	105	17
	110–120	115	4
	120–130	125	1
	130–140	135	0
	140–150	145	0 or 1

Convert to equivalent 80 kN axles: $N_{eq} = N(W_{av}/80)^n$

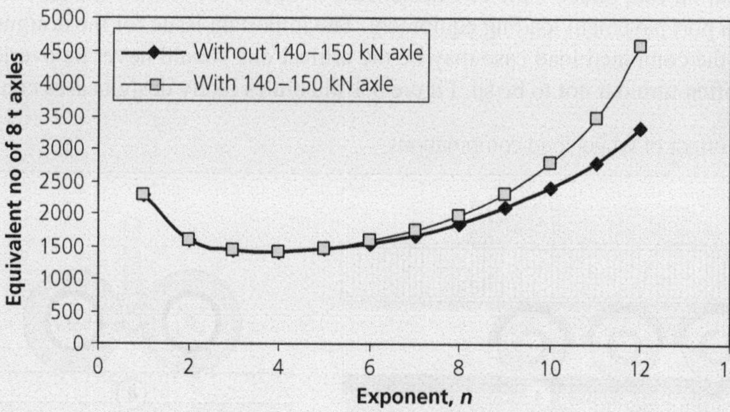

4.1.2 Contact pressure

The contact zone between a pneumatic rubber tyre and the road surface is by no means simple, because both the tyre and the surface are 'textured', reducing the contact to a series of discontinuous areas. However, this is much too complex for practical pavement engineering, even though many researchers believe that this uneven distribution of stress is responsible for crack initiation at the road surface. Practical considerations dictate that a uniform pressure distribution is assumed, and it is logical to make the assumption that the contact pressure is equal to the internal pressure of the tyre.

In fact, despite this being both logical and practical, it is far from being true. Tyre walls, reinforced with steel in the case of commercial vehicles, are stiff enough to attract load, and the result is that the pressure is commonly highest at the two edges of the contact zone and less in the middle (De Beer *et al.*, 1997; Liang *et al.*, 2021). However, although this may well influence the development of surface distress, polishing of surface aggregate and noise generation, it is not generally considered significant for the deeper pavement structure and so is generally ignored. In practice, the tyre pressure is usually assumed also to be the contact pressure, around 250 kPa for a car, 600 kPa for a large truck, 1000–1500 kPa for commercial aircraft and up to 3000 kPa for some military planes. The final simplification often made for purposes of pavement analysis is that the contact area is circular.

Although pressure is not generally explicitly taken into account when converting a spectrum of load cases into an equivalent number of design axles or design wheel loads because of the complexity this would introduce, it is nevertheless important that an appropriate contact pressure is assumed in pavement analysis.

4.1.3 Load groups

This is an important concept. Wheels are often close enough together for there to be an interaction of effects in the pavement. The case of a dual-tyre arrangement on a commercial truck is an extreme one, in which the loads are so close together that they almost act as if they were a single wheel carrying the combined load. Tandem and tridem axles on trucks tend to be spaced at least 1.3 m apart and, although adjacent wheels combine in their effect on the pavement, this combination does not usually add any significant stress at critical levels in the structure. In the case of dual, dual-tandem or dual-tridem wheel gear arrangements on aircraft, however, the increased load magnitude means that the combination can, under some circumstances, comprise the critical design case. The same is true of certain port pavement loading equipment. The important issue for the designer is that the possibility that the combined load case may be the critical one should never be overlooked, even though it may often turn out not to be so. Figure 4.1 illustrates likely design cases conceptually.

Figure 4.1 The effect of wheel load combinations

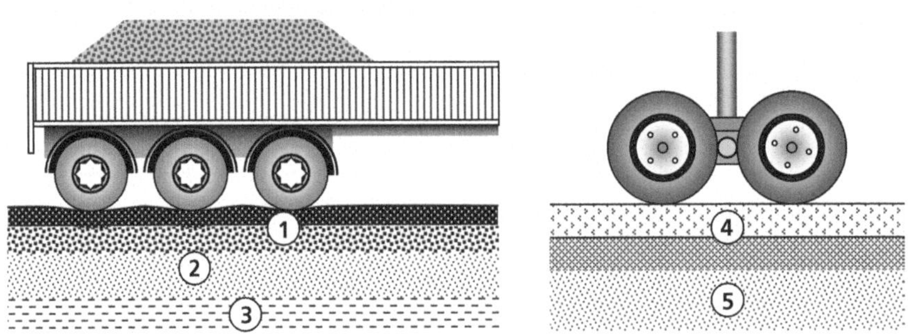

1. Asphalt layers unaffected by adjacent loads
2. Upper subgrade under asphalt also probably only influenced by individual loads
3. Lower subgrade affected by combined loads – but rarely a key design case
4. Concrete often affected by load combinations
5. Subgrade under aircraft loads usually affected by combined loads

4.1.4 Dynamic effects

No matter how efficient the suspension system fitted, all moving wheel loads fluctuate as the vehicle body oscillates vertically. Some suspension systems, notably air suspension, are more effective than others at avoiding high oscillating loads, but a certain level of dynamic load is inevitable. Furthermore, oscillation is unlikely to be purely vertical, and side-to-side oscillation results in a fluctuating imbalance between the loads on the left and right of the vehicle. When combined with cornering, particularly cornering where the radius is continually changing, the additional load on one side of the vehicle can be considerable; it is common to observe that road pavements suffer much greater damage on the outer wheel-path of a curve compared with the inner wheel-path. When considering the loading from certain industrial vehicles, such as reach-stackers, fork-lift trucks or rubber-tyred gantry cranes, the stop–start nature of the movements involved also tends to induce significant additional dynamic forces.

However, it is not the place of this book to discuss suspension systems or industrial plant operation. The pavement designer simply has to be able to estimate the degree of dynamic load to take into account. Some design approaches use a 'dynamic load factor', which is a multiplier used to convert the design static load to a design dynamic load value. But the key question is just what the effect of this dynamic load is likely to be on pavement design; Calculation Sheet 4.2 presents an analysis assuming that the load variation is a sinusoidal one, referring back to the power laws introduced in Section 4.1.1.

Calculation Sheet 4.2. Dynamic effects

Take a single dynamic cycle

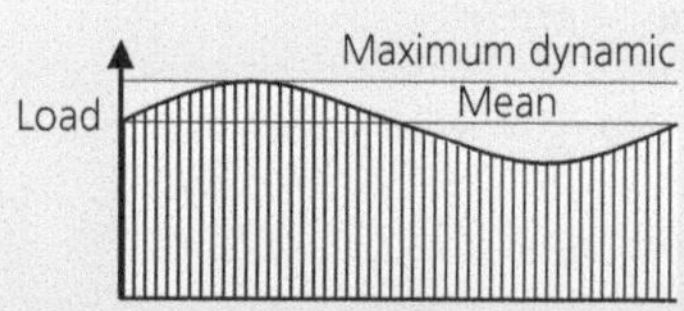

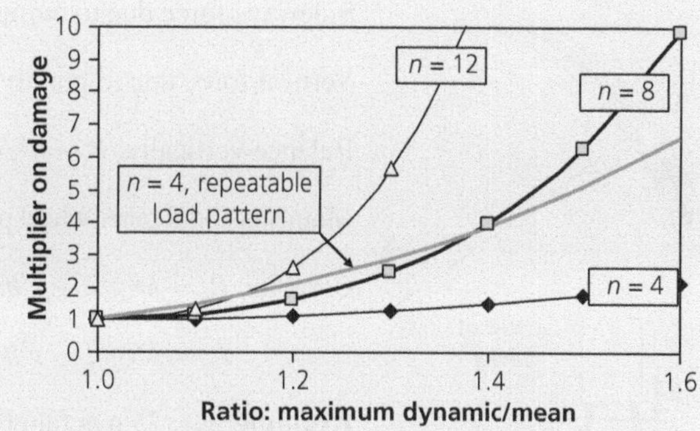

Divide into time steps

Work out relative damage in each time step

$= (W/W_{mean})^n$

Average damage over whole cycle

$N_{eq} = N(W_{av}/80)^n$

The first three curves in Calculation Sheet 4.2 make the assumption that the distribution of dynamic load is random. However, there is plenty of evidence to show that this is often not the case and that similar vehicles, having similar suspension system characteristics, will tend to apply peak loads in roughly the same locations. The fourth curve shown makes the assumption that the loading pattern is perfectly repeatable, in which case the computed damage would occur at regular intervals along the pavement, corresponding with peak dynamic load locations, and this sort of effect is sometimes visible in the form of regularly spaced areas of damage, particularly on relatively low-strength roads.

Nevertheless, on most highway or airfield pavements, vertical oscillation is small enough to be subsumed into the overall calibration of the design method used, only becoming significant if a high exponent of damage is applied (e.g. on a concrete pavement) or the dynamic load factor increases to 1.4 or more – something that will only occur where there is very poor longitudinal profile. In contrast, on industrial pavements such as at ports, dynamic load factors are commonly applied in design in recognition of the fact that peak dynamic loads often reoccur at the same points, a function of the way the plant is operated. The Interpave design manual (Knapton, 2007) makes use of dynamic factors up to 2.3 for the combined effects of cornering, braking and an uneven surface in the case of a straddle carrier.

The issue of load imbalance due to cornering is of more general significance, and this is illustrated in Calculation Sheet 4.3. In the example calculation shown, the load on the outer wheel is 68% greater than that on the inner wheel. This, if a fourth-power law is applied, implies eight times as much damage in the outer wheel-path as in the inner one. This is something that is rarely taken into account in design, on the principle that 'simplicity is elegance' – but it probably should be, especially at junctions and roundabouts, where maintenance nuisance is greatest.

Calculation Sheet 4.3. Cornering effects

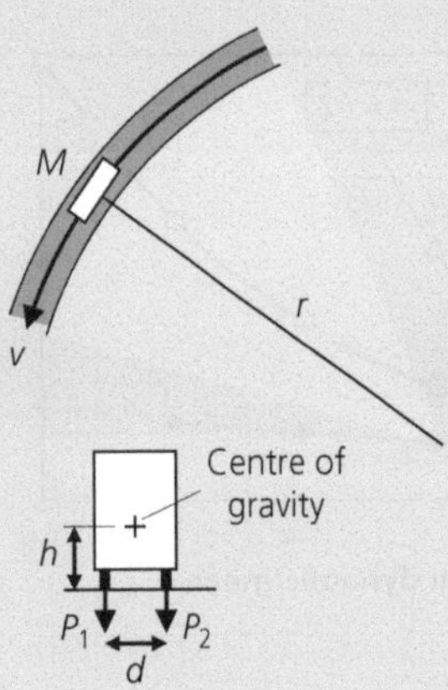

Sideways force due to cornering $= Mv^2/r$

Vertical force due to gravity $= Mg$ ($g = 9.81$ m/s^2)

Balance vertically: $P_1 + P_2 = Mg$

Moments about inner wheel path: $P_1 d = (Mv^2/r)h + Mg(d/2)$

Combine $P_1 = M(g/2 + v^2h/rd)$

$P_2 = M(g/2 - v^2h/rd)$

Example: $v = 25$ m/s (approx. 60 mph)

$r = 500$ m

$d = 2$ m

$h = 2$ m

$M = 8$ t

$\rightarrow P_1 = 49.24$ kN; $P_2 = 29.24$ kN

4.1.5 Loading speed

Vehicle speed is an important contributing factor to dynamic load; a slow rate of travel will not induce significant vertical oscillation. However, it is also an important input into design of pavements containing asphalt because the viscous nature of bitumen makes asphalt a material that is dependent on loading rate. Thus, an appreciation of the loading rate is required if appropriate design stiffnesses are to be selected. The difference between the loading rate imposed by traffic and that generated during testing must also be taken into account when interpreting such test data.

4.1.6 Lateral wander

The issue of channelisation was highlighted in Section 3.9 and the degree of channelisation affects the rate of wheel-path damage. On highways, the distribution across a wheel-path commonly has a standard deviation of around 150 mm; on airport runways this is likely to be a metre or more. It should therefore be permissible to apply a reduction factor to derive the actual number of load applications experienced by an individual point on the pavement. On roads, this reduction factor is usually neglected as too small to worry about, although some design manuals apply factors of up to 2 depending on lane width, notably the German standard (FGSV, 2012). For airfields, the effect is much more significant, on runways at least, and a factor of 2 or more is sometimes applied (e.g. Defence Estates (2006) include what is effectively a wander factor of 1.6). Designers should beware of applying this to taxiways, however, which are much more like highways in terms of the spatial repeatability of the loads.

4.2. Determining design traffic

The simple, practical way to translate real traffic loads into a convenient input parameter for design (or evaluation) is to convert everything into equivalent numbers of a single *design load*. On highways, it is common to speak of a *standard axle*, the design wheel load generally being one half of a standard axle load; on airfields, traffic is expressed in terms of a *design aircraft*. In some cases it may prove impossible to convert everything to a single design case because of the differences between different load types. Thus, port pavements are commonly checked for their resistance to container foot loading as well as vehicular traffic. Fork-lift trucks may have solid rubber tyres and thus may impose a quite different pressure on the surface compared with pneumatic-tyred vehicles; designs may have to be checked for both. However, the usual situation is for all traffic to be expressed in equivalent numbers of a design-load case.

As the damage done by a vehicle to the pavement is much more than linearly proportional to applied load, the design-load case should logically be at the heavy end of the spectrum; in the case of airfields or industrial pavements, it will commonly be the largest load case of all. Typical design cases for highway traffic characterisation are 80 or 100 kN axle loads, 80 kN being in use in the UK (with approximately the same load (18 kips) in the USA).

Having determined the appropriate design load, the next step is to determine the expected spectrum of loads during the design life of the pavement. This is sometimes carried out by direct measurement of existing traffic loads using a weigh-in-motion device, but it would be more usual to count/predict the types and numbers of vehicles and to assume typical axle loads for each. These axle loads can then be converted into equivalent design-load applications by means of the fourth-power law, or such other power law as may be preferred (see the discussion in Section 4.1).

There are also many direct methods contained in design manuals that do not require any prediction of the full axle load spectrum.

- In the UK, National Highways (2020) gives multipliers, known as *wear factors*, to convert from projected numbers of commercial vehicles or, for existing roads, from counts of different types of commercial vehicle (number of axles, rigid or articulated) to a number of equivalent standard axles. Similar techniques can be found in the design advice of most other countries, with differences reflecting the significantly different traffic characteristics (vehicle types, permitted and actual axle loads etc.). In some cases, the number of commercial vehicles can be converted directly to a 'Traffic Class', greatly simplifying design.
- Airfield design methods, such as that developed by the Federal Aviation Administration in the USA (FAA, 2021 and the computer program FAARFIELD), incorporate conversions based on the applied wheel load of each aircraft. The widely used Defence Estates guide (Defence Estates, 2006), issued in the UK, makes use of the now-superseded International Civil Aviation Organisation (ICAO) method of classifying damaging power, namely the Aircraft Classification Number (ACN) (ICAO, 2009), in which each aircraft has an ACN (actually several, depending on subgrade category and pavement type) and will only be permitted to use pavements with a Pavement Classification Number (PCN) at least as great as the aircraft ACN. Conversion now has to be made from the ICAO's new ACR/PCR (Aircraft Classification Rating/Pavement Classification Rating) system.
- The Interpave manual (Knapton, 2007) converts everything to a single equivalent wheel load.

Whichever method is used, the result will be an equivalent number of design-load cases, for example, equivalent standard axle loads (otherwise known as esals). In many cases, that will be sufficient, as empirical design methods then implicitly assume appropriate traffic speed, tyre pressure and dynamic effects. Thus, when using the 1993 design guide of the American Association of State Highway and Transportation Officials (AASHTO, 1993), for example, it is sufficient to find a number of esals; the same is true of National Highways design documentation in the UK (National Highways, 2020). It is only if an analytical design (or pavement evaluation) is to be attempted that it is necessary to consider anything further. However, as soon as computational analysis of a pavement is to be performed, then contact pressure, load enhancement due to dynamic effects and lateral wander all have to be considered. Furthermore, loading speed, while it does not affect the design load itself, should affect the stiffness moduli selected for asphalt layers.

4.3. Summary

Because of the very different nature of the loads experienced by different pavement types (highways, airfields, ports, footways, car parks etc.), it is not possible to give absolute general rules for the expression or prediction of traffic. However, the goal is to determine (*a*) a design-load case and (*b*) an equivalent number of load applications. It should always be remembered that expression of traffic in this way is approximate and that predictions are fraught with uncertainty. For these reasons, a practical approach is urged, erring on the side of caution but not being overly concerned to generate precise predictions. As will become apparent, even a significant error in predicted traffic numbers usually has only minor influence on pavement design thickness. Selection of an appropriate design-load case is, however, much more critical.

REFERENCES

AASHTO (American Association of State Highway and Transportation Officials) (1993) *Guide for Design of Pavement Structures*. AASHTO, Washington, DC, USA.

De Beer M, Fisher C and Jooste FJ (1997) Determination of pneumatic tyre/pavement interface contact stresses under moving wheel loads and some effects on pavements with thin asphalt surfacings. *Proceedings of the 8th International Conference on Structural Design of Asphalt Pavements, Seattle, WA, USA*, vol. 2, pp. 1643–1658.

Defence Estates (2006) *A Guide to Airfield Pavement Design and Evaluation*, 2nd edn. Defence Estates, London, UK, Design and Maintenance Guide 27.

FAA (Federal Aviation Administration) (2021) *Airport Pavement Design and Evaluation*. FAA, Washington, DC, USA, AC150/5370-6G.

FGSV (Forschungsgesellschaft für Straßen- und Verkehrswesen) (2012) *Richtlinien für die Standardisierung des Oberbaus von Verkehrsflächen*. FGSV, Cologne, Germany, RStO 12 (in German).

HRB (Highway Research Board) (1961) *The AASHO Road Test: History and Description of the Project*. Highway Research Board, Washington, DC, USA, Special Report No. 61A.

ICAO (International Civil Aviation Organisation) (2009) *Annex 14 – Aerodromes – Volume I Aerodrome Design and Operations*, 5th edn. ICAO, Quebec, Canada.

Knapton J (2007) *The Structural Design of Heavy Duty Pavements for Ports and Other Industries*, 4th edn. Interpave, Leicester, UK.

Liang C, Zhu D, Wang G and Shan M (2021) Experimental study on the tire-road dynamic contact pressure distribution using FTIR imaging. *International Journal of Automotive Technology* **22(5)**: 1305–1317.

National Highways (2020) CD 224: Traffic assessment. *Design Manual for Roads and Bridges. Pavement Design*. Stationery Office, London, UK.

TRL (Transport Research Laboratory) (1993) *A Guide to the Structural Design of Bitumen-surfaced Roads in Tropical and Sub-tropical Countries*, 4th edn. TRL, Crowthorne, UK, Overseas Road Note 31.

Principles of Pavement Engineering, Third edition

Nick Thom
ISBN 978-1-83549-711-1
https://doi.org/10.1108/978-1-83549-710-420243006

Chapter 5
Design reliability

A pavement is quite unlike most other engineering constructions. The level of uncertainty is always high, even for the best controlled projects, purely due to variations in material properties and layer thickness, quite apart from the fact that design traffic loading is never known with any great precision. There is, therefore, a distribution of pavement quality right from the outset, a distribution which will increase in breadth during the life of the pavement.

5.1. Selecting the appropriate reliability level

This is an economic issue. If pavement failure was associated with route closure or with multiple fatalities, then engineers would strive to get as close to 100% reliability as possible – and society would pay accordingly. However, pavements simply do not fail like that; pavement failure is a gradual process. It is unfortunate if certain areas of pavement deteriorate more quickly than expected, but it certainly is not the end of the world. It simply is not cost-effective to aim too high.

But just how high should one aim? The nominal standard for major UK highways has been to design for a 15th percentile condition, giving an 85% design reliability. This means that a pavement will have a 15% of having less life than that for which it was designed, and a 15% chance of having more. Although not explicit in current standards, this was the basis upon which the evidence of past performance was interpreted when designs were being developed (Powell *et al.*, 1984). In the USA, reliability is an explicit item in the current American Association of State Highway and Transportation Officials (AASHTO) method, as it has been for many years, with the advice that reliability should depend on the type of highway. Both the 1993 and 2007 AASHTO design guides (AASHTO, 1993, 2007) incorporate reliability as a key input to design. A 90% reliability would be the norm for major roads. A similar level of reliability should, logically, be adopted for major airfield pavements. On the other hand, pavements that can be maintained without serious knock-on effects (e.g. disruption to business), such as minor roads, industrial pavements, many airfield taxiways, car parks, footways and so on should logically be built to a lower standard of reliability, perhaps 75% or even 50%. There are no absolute rules, only principles. No one can know the full economic equation, certainly not with regard to the future, but the key point that really should be understood by pavement owners is that they are always designed in the expectation that a certain percentage of the pavement may fail early.

5.2. A statistical view of design inputs

Having selected a design reliability, say 85%, the next issue is just how this affects the individual inputs used in pavement analysis. The problem is that a combination of the 15th percentiles of asphalt stiffness, asphalt thickness and foundation stiffness (taking the example of an asphalt pavement) would be equivalent to something like a 5th percentile overall (giving a 95% design reliability), as it is unlikely that poor values of all three will occur at the same location. This approach would, therefore, give an ultra-conservative prediction of pavement life. The issue is clearly one for which practical guidance is required, rather than precise theoretical analysis.

The detailed methods of analysing all the different types of pavement – flexible, composite and rigid – will be described in subsequent chapters, as will the factors that directly influence the life (either fatigue cracking or deformation) of particular materials. Where this knowledge is required in this chapter, forward reference will be made.

5.2.1 Flexible pavements

The first task is to estimate the relative influence of each design input parameter on the different modes of pavement deterioration, and Calculation Sheet 5.1 presents an approach that is based on a 'beam on elastic foundation' analysis. This is a two-dimensional analysis and so can only be considered approximate, but it is a simple, practical way of judging how important each parameter is. And the result, assuming typical relationships between key strain values and pavement life, is that the asphalt modulus (E) is found to be relatively influential in its effect on fatigue life (with an exponent of 3) but much less so on deformation life (exponent of 1). Conversely, the subgrade stiffness (k) has much greater influence on deformation than on fatigue. The asphalt thickness (h) is highly significant for both, but particularly fatigue.

Calculation Sheet 5.1. Relative influences of asphalt pavement parameters

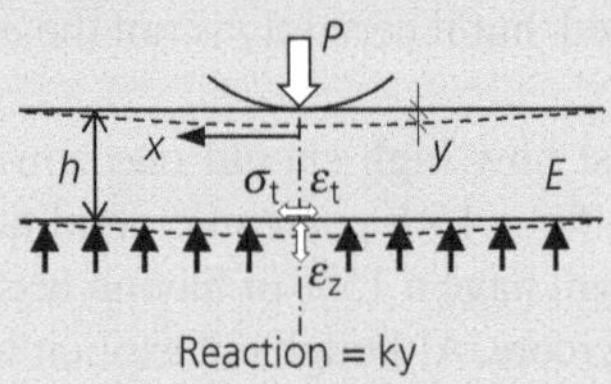

Parameters considered

Asphalt modulus, E

Asphalt thickness, h

Foundation modulus – characterised by modulus of subgrade reaction, k

Treat as a **beam on an elastic foundation**:

$$y = (P\lambda/2k)\ e^{-\lambda x}(\sin \lambda x + \cos \lambda x)$$

where $\lambda = (k/4EI)^{0.25}$

Double differentiate:

$$d^2y/dx^2 = (P\lambda^3/k)\ e^{-\lambda x}(\sin \lambda x + \cos \lambda x)$$

But

$$d^2y/dx^2 = \text{curvature} = \text{moment}/EI;\ \text{moment} = \sigma_t h^2/6$$

Therefore, at $x = 0$:

$$\sigma_t = 6EIP\lambda^3/kh^2 = 6E^{0.25}I^{0.25}P/[(2\sqrt{2})k^{0.25}h^2] = 1.14E^{0.25}P/(k^{0.25}h^{1.25})$$

and strain

$$\varepsilon_t = \sigma_t/E = 1.14P/(E^{0.75}k^{0.25}h^{1.25})$$

Assume **fatigue life** $\propto \varepsilon_t^{-4}$ (4th power law, Section 4.1.1) $\propto E^3kh^5$

Similarly, say ε_z in subgrade is approximately related to y_{max}

Therefore:

$$\varepsilon_z \propto P\lambda/2k = 0.35P/(E^{0.25}k^{0.75}I^{0.25}) = 1.19P/(E^{0.25}k^{0.75}h^{0.75})$$

Assume **deformation life** $\propto \varepsilon_z^{-4}$ (4th power law, Section 4.1.1) $\propto Ek^3h^3$

The next important matter to be established is the likely degree of variability in each of the design parameters. This is expressed in Table 5.1 as a coefficient of variance (CoV) (standard deviation divided by mean; note that $\pm$ 1 standard deviation $\approx$ 15th and 85th percentile conditions). The values in Table 5.1 are based purely on the author's subjective experience; the reader should feel free to substitute alternative values where evidence exists.

Now there is a practical problem; some of the CoVs shown in Table 5.1 are quite large and, when the effect of an exponent of 3, or even 5, is considered, as in the relationships in Calculation Sheet 5.1, there is no way that pavement life can be considered as 'normally' distributed. A sensible move, therefore, is to replace the CoVs with the multipliers (or dividers for lower limit values) shown in parentheses in Table 5.1, and to assume a log–normal distribution (this assumes that the logarithm of pavement life is normally distributed). These can now be used in a full statistical analysis of the design issue. Calculation Sheet 5.2 illustrates this approach and suggests that, based on the relationships in Calculation Sheet 5.1, in order to achieve 85% design reliability, calculations should be based on the 25th percentiles of each input parameter.

Table 5.1 Suggested coefficients of variance (CoVs) for different design parameters

Pavement type	Coefficient of variance (CoV)			
	Asphalt stiffness	Asphalt thickness	Hydraulically bound base thickness	Foundation stiffness
1. Major highways etc.	20% (×1.2)	10% (×1.1)	10% (×1.1)	40% (×1.4)
2. Minor highways etc.	25%(×1.25)	20% (×1.2)	20% (×1.2)	60% (×1.6)

Calculation Sheet 5.2. Statistical analysis of pavement life

Replace relationships from Calculation Sheet 5.1 with logarithms

Fatigue life: $N_f = aE^3kh^5 \rightarrow \ln(N_f) = \ln(a) + 3\ln(E) + \ln(k) + 5\ln(h)$

Deformation life: $N_d = bEk^3h^3 \rightarrow \ln(N_d) = \ln(b) + \ln(E) + 3\ln(k) + 3\ln(h)$

A: Apply multipliers/dividers for the two cases in Table 5.1 to each parameter, and sum. This is equivalent to using the 15th percentile of each parameter. It is **ultra-conservative**

ln(divider on N_f)

Case 1: $3\ln(1.2) + \ln(1.4) + 5\ln(1.1) = 1.36 \rightarrow N_f \div 3.90$
Case 2: $3\ln(1.25) + \ln(1.6) + 5\ln(1.2) = 2.05 \rightarrow N_f \div 7.78$

ln(divider on N_d)

Case 1: $\ln(1.2) + 3\ln(1.4) + 3\ln(1.1) = 1.48 \rightarrow N_d \div 4.38$
Case 2: $\ln(1.25) + 3\ln(1.6) + 3\ln(1.2) = 2.18 \rightarrow N_d \div 8.85$

B: Combine by squaring, summing and taking a square root. This assumes the parameters are *independent*. It will give a *realistic* equivalent 15th percentile of pavement life

ln(divider on N_f)

Case 1: $\sqrt{\{[3\ln(1.2)]^2 + [\ln(1.4)]^2 + [5\ln(1.1)]^2\}} = 0.80 \rightarrow N_f \div 2.22$
Case 2: $\sqrt{\{[3\ln(1.25)]^2 + [\ln(1.6)]^2 + [5\ln(1.2)]^2\}} = 1.22 \rightarrow N_f \div 3.40$

ln(divider on N_d)

Case 1: $\sqrt{\{[\ln(1.2)]^2 + [3\ln(1.4)]^2 + [3\ln(1.1)]^2\}} = 1.06 \rightarrow N_d \div 2.90$
Case 2: $\sqrt{\{[\ln(1.25)]^2 + [3\ln(1.6)]^2 + [3\ln(1.2)]^2\}} = 1.53 \rightarrow N_d \div 4.61$

Take ratios of dividers by methods B and A: 0.57, 0.44, 0.66, 0.52; average 0.55

∴ Use 0.55 × Table 5.1 multipliers/dividers for each design parameter to calculate equivalent 15th percentile pavement life (85% design reliability). This approximately equates to using *25th percentiles*

The same approach as shown in Calculation Sheet 5.2 can be repeated for other design reliabilities, and a more complete set of deductions would therefore be

- 50% design reliability – use 50th percentiles of individual parameters
- 75% design reliability – use 33rd percentiles of individual parameters
- 85% design reliability – use 25th percentiles of individual parameters
- 95% design reliability – use 14th percentiles of individual parameters
- 99% design reliability – use 6th percentiles of individual parameters.

In no way can these figures be considered exact (other than that for 50% reliability); however, they open the way to rational design, based on a fundamentally sound understanding of pavement performance. They allow the life of a pavement to be calculated in a reasonable way and sensible confidence limits to be placed on that calculation.

Note that the figures above assume that the same percentile is to be applied to each design parameter, but this does not have to be the case. If necessary, the relationships for the two aspects of pavement life given in Calculation Sheet 5.1 (fatigue and deformation) can be used to explore the option of using an increased percentile for one parameter and a decreased percentile for another – to achieve the same overall design reliability.

Note also that the advice in this section applies to pavements that may be considered 'fully flexible'. There may be a weak hydraulically bound base or sub-base present, but if so it should be considered as part of the foundation.

5.2.2 Composite pavements with strong hydraulically bound bases

In the case of fatigue of a strong hydraulically bound base, a stress ratio is required to predict fatigue life (see Section 19.2) – calculated at the bottom of the base layer, as opposed to the strain used to characterise asphalt fatigue. Referring to Calculation Sheet 5.1, the relationship for ε_t therefore has to be replaced by one for σ_t/σ_f, namely

$$\sigma_t/\sigma_f = \frac{1.14PE^{0.25}}{k^{0.25}h^{1.25}\sigma_f} \tag{5.1}$$

where σ_f = flexural strength

Also, the slope of the fatigue characteristic is quite different from that for asphalt – an exponent of 12 was suggested in Chapter 4 – which means that fatigue life becomes proportional to $k^3h^{15}\sigma_f^{12}/E^3$. Note that the modulus E now appears on the bottom of the equation, not the top; that is, life is inversely proportional to the cube of modulus. Conceptually, this should not be a surprise; it simply reflects the fact that a stiff material will tend to attract stress. But the consequence is that the appropriate percentiles for the modulus of a strong hydraulically bound layer, when considering fatigue, are the opposite of those used for asphalt. Thus, a 33rd percentile would be replaced by a 67th percentile, and so on. Please note that this only applies when considering fatigue of the hydraulically bound material (HBM), not when dealing with any other failure mode; neither does it apply to any parameter other than HBM modulus.

Following the calculation through for fatigue of hydraulically bound base, the recommended percentiles are

- 50% design reliability – use 50th percentiles of individual parameters
- 75% design reliability – use 35th percentiles of individual parameters
- 85% design reliability – use 28th percentiles of individual parameters
- 95% design reliability – use 18th percentiles of individual parameters
- 99% design reliability – use 10th percentiles of individual parameters.

Note that the thicknesses of asphalt and hydraulically bound base should not really be considered as independent variables; any low region on the surface of a hydraulically bound base will be corrected by additional asphalt, because the finished level of the asphalt does not depend on base levels. This means that the percentile used should be applied to the combined thickness of the two layers, individual layer thicknesses being given according to the ratio of the two design thicknesses.

Also note that the calculation is only for fatigue failure of the base. Other modes of failure should be treated as for asphalt pavements.

5.2.3 Rigid pavements

Exactly the same principles apply to concrete pavements as to strong hydraulically bound bases. The form of the relationship between fatigue life and the individual material and thickness parameters is the same, as are the recommended percentiles. Remember that the percentile for concrete modulus, as for hydraulically bound base, should be the opposite of those shown (65th instead of 35th etc.), for the reasons given above.

5.3. Design assurance

The next difficulty is just how to ensure that the required value of the appropriate percentile for each parameter is achieved. Taking, for example, the case of a 95% reliability design of a concrete pavement, as in the previous section, the design should, in theory, be based on 18th percentiles of individual parameters (or the 82nd percentile in the case of concrete modulus). Now, supposing the design were to demand that the 18th percentile flexural strength of the concrete was 4.5 MPa, this would mean that the average strength would have to be higher. A reasonable estimate for the actual CoV of concrete flexural strength might be around 15%, based on the author's experience, from which it can be calculated that the required average flexural strength would be about 5.1 MPa. This is quite straightforward and easy to incorporate in a material specification, whether of a material property or a thickness; the only real engineering judgement needed is the CoV. However, the picture can become much more confusing, as the following subsections explain.

5.3.1 Measurement variability

Measured variability is always likely to be considerably greater than real variability. In the case of laboratory testing, this is an unavoidable consequence of testing small samples of material containing relatively large aggregate particles. If reliance is placed on field tests, the level of test control tends to be poorer and the test variability is, consequently, greater. Figure 5.1 illustrates the problem. It is therefore dangerous to specify material properties other than in terms of a mean. The design may call for a certain absolute minimum value for concrete flexural strength, but one

Figure 5.1 The concept of real and measured variability

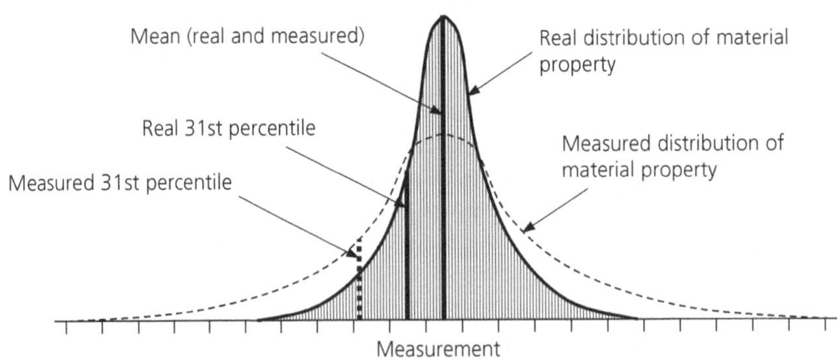

would have to allow a lower value for the measured strength, and this can only be judged if a good estimate of inherent test variability can be made. Needless to say, each test will carry its own inherent variability, and this will probably depend partly on the material being tested, particularly the aggregate gradation.

5.3.2 Other sources of error

It is not too hard to find reasons to increase the uncertainty in any prediction of pavement performance. Pavement analysis is hardly a precise science and, whichever form of distress is being predicted, there are plenty of questionable assumptions

- an assumed relationship between compressive and flexural strength of a HBM (Section 19.1.4)
- an assumed temperature and/or loading rate correction for asphalt (Section 9.3.3)
- an assumed fatigue relationship for both asphalt (Section 9.4.4) and HBM (Section 19.2)
- an assumed long-term stiffness modulus for unbound materials.

Depending on the type of pavement and the vehicle load type, there may be others, for example, concerning dynamic load effects, joint load transfer, drainage function and so on; the list goes on. Each uncertainty effectively adds an additional term to the variance of each parameter.

For example, the first mentioned above, the relationship between compressive and flexural strength, probably adds a CoV of around 30% to the strength parameter if based on compressive strength testing, dwarfing the 15% suggested for the real CoV of strength and probably also significantly exceeding that due to measurement uncertainty. A similarly large effect should probably be assigned to the fatigue characteristic of asphalt.

These are very important matters. The modest 15% CoV for strength becomes around 35% (equivalent to a factor of 1.35) once uncertainty in the compressive–flexural strength relationship is included, treating the two effects as independent (as in part B on Calculation Sheet 5.2). Thus, if 4.5 MPa is the design requirement for flexural strength and quality control is to be by means of compressive strength testing, and if the appropriate percentile is the 18th (suggested above for 95% reliability of concrete pavements), then the target mean value, converted from compressive strength tests, becomes 5.9 MPa.

5.3.3 Recommendations

As one can see, when all the various causes of uncertainty are factored in, the picture can look rather complicated. However, the key is always to assign a value for the CoV for each cause of uncertainty, and to combine them assuming that all causes of uncertainty are independent.

With regard to the specification of properties that require testing, it is always preferable to specify the mean or 50th percentile value (from a suitably large number of tests) rather than a different percentile, because this avoids most of the problems of scatter in test data.

5.4. Summary

Statistics is a subject too often ignored by pavement engineers. Yet, almost more than any other civil engineering structure, pavements demand to be treated statistically. They are inherently variable and have no clearly defined failure point, just a gradual decline in serviceability. It makes no sense at all to claim that the life of a pavement is some fixed number of years or passes of a certain design wheel load, however attractive such a simplistic view may seem. To make such a claim is to deny the pavement authority the tools it needs to predict maintenance and to manage budgets. While pavement engineers refuse to view their designs in a probabilistic way, it is no surprise to find that pavement maintenance is often a very hit-and-miss affair.

This subject will not be picked up directly in Chapters 10, 14, 17 or 20, the chapters that deal with pavement design. Those chapters will concentrate on determining pavement life as a function of inputs – but those inputs should be chosen based on a statistical view of each variable.

REFERENCES

AASHTO (American Association of State Highway and Transportation Officials) (1993) *Guide for Design of Pavement Structures*. AASHTO, Washington, DC, USA.

AASHTO (2007) *Guide for Mechanistic-Empirical Design of New and Rehabilitated Pavement Structures*. AASHTO, Washington, DC, USA.

Powell WD, Potter JF, Mayhew HC and Nunn ME (1984) *The Structural Design of Bituminous Roads*. Transport and Road Research Laboratory, Crowthorne, UK, Report LR1132.

Part 2

Unbound material and unsurfaced pavements

Principles of Pavement Engineering, Third edition

Nick Thom
ISBN 978-1-83549-711-1
https://doi.org/10.1108/978-1-83549-710-420243007

Chapter 6
Unbound material

It is easy for pavement engineers to get lost in a world of asphalt and surface issues and to forget that the most fundamental element in any pavement is the ground on which it is built – and that ground will be an *unbound* material. It is a class of material that is all around us, which we come into contact with on a daily basis, but which is very poorly understood from the point of view of its mechanical properties. Yet a reasonable level of understanding is not hard to achieve. The defining feature of an unbound material is that it has no binder. This means that it has no intrinsic tensile strength. The right water content can help – think about a lump of dryish clay – but tension is not what unbound materials cope with best. As we will see shortly, their strength is in shear. But first let us look at the way that these materials are commonly tested and specified today.

6.1. Traditional tests and specifications

6.1.1 California bearing ratio (CBR)

The California bearing ratio was developed by the California State Highways Department in the 1930s as an index test for soils and unbound pavement materials, specifically for the purpose of pavement design, as it uses relatively low-technology equipment and gives a general indication of strength, either in situ or in the laboratory. The CBR is expressed as a percentage of the value obtained from a particular Californian crushed rock. It is still in common use worldwide, and numerous specifications describe the test, with only minor variations (e.g. BSI, 2012). However, it is important to remember that it is an *empirical* measure and has no absolute and fundamental meaning. It has been described, not unreasonably, as an instrumented 'heel of the boot'. The details of the standard laboratory test are shown in Figure 6.1. The loading specification is the same in situ but there is, of course, no mould.

Figure 6.1 The California bearing ratio (CBR) test

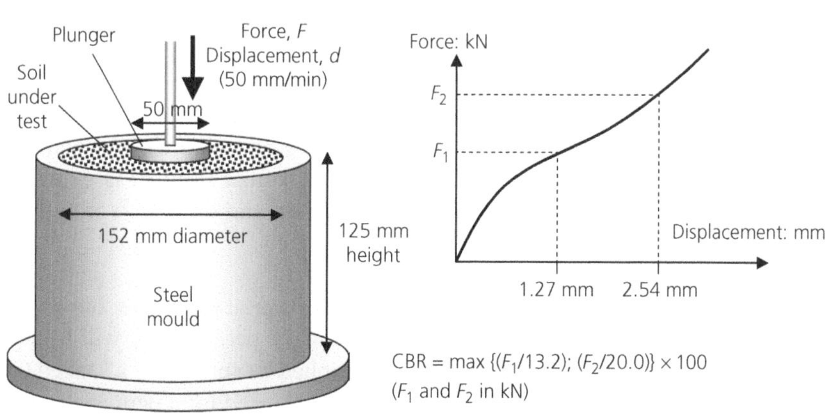

In situ test specifications commonly demand the addition of a surcharge to the surface of the soil immediately around the plunger, simulating the effect of confinement beneath a pavement, and both surcharging and the presence of a confining mould can have a marked influence on the measured CBR in the case of sandy soils with low cohesion.

Despite the non-fundamental nature of the CBR, its convenience and relationship to readily observable behaviour on site has meant that the test continues to be used and that pavements continue to be designed on the basis of a 'design CBR'. In many cases a 'soaked CBR' is specified, this being a measure of the properties of a material in a wet condition, and procedures exist in several specifications for carrying out the soaking, a 4-day soaking period being typical (e.g. TRL, 1993).

There is an obvious limitation to the laboratory CBR in that the size of the mould, 152 mm diameter, does not permit the inclusion of large particles, and specifications commonly call for the exclusion of anything over about 20 mm. This means that if the proportion >20 mm is significant the test becomes almost meaningless.

6.1.2 Plate loading tests

A logical alternative to the CBR test is to increase the size of the loaded area – impractical in the laboratory but possible on site. This provides a more uniform distribution of stress, but it also provides a deeper zone of stress, usually including effects from more than one layer. The plate loading test cannot therefore be seen as a material test like the CBR test, but rather as a composite foundation (or part-foundation) test. Furthermore, the larger the plate, the less the test can be considered a strength test and the greater the influence of material stiffness, as practical considerations dictate that the deflection induced is rarely more than a very few millimetres, representing a relatively small strain. Nevertheless, conversions between plate loading stiffness k (pressure under plate ÷ area) and CBR have been proposed. One such equation was proposed by Day (1981)

$$\mathrm{CBR}(\%) = 0.001035\left(k_{762}\right)^{1.733} \tag{6.1}$$

where k_{762} is the stiffness in kPa/mm (or MN/m^3) under a 762 mm dia. plate.

Factors have also been proposed to convert from the traditional 762 mm dia. plate to other sizes, for example a divisor of 2.23 on k for a 300 mm dia. plate. Interpretation of the test is discussed further in Section 6.3.10.

6.1.3 Particle size distribution (or grading/gradation)

The gradation of a soil cannot, of course, be specified – it is what it is and design has to accommodate it. Pavement layer gradation on the other hand is almost always specified between upper and lower limits. Figure 6.2 is an example of the format used. It is expressed as the percentage (by mass) passing through particular sieve sizes, and the size is usually shown on a logarithmic scale.

Subsequent sections will explore the effect of gradation on material properties, but it will be obvious to the reader that strength, stiffness and, most particularly, permeability will be significantly affected. Strength and stiffness might be adequately controlled by specifying a soaked CBR, but if permeability is required then gradation limits are needed. The need for a permissible range stems

Figure 6.2 Typical particle size distribution limits for a granular sub-base. The gradations are taken from the current UK National Highways specification for Type 1 sub-base, which is a high-quality crushed rock material typical of those used throughout the world (National Highways, 2021)

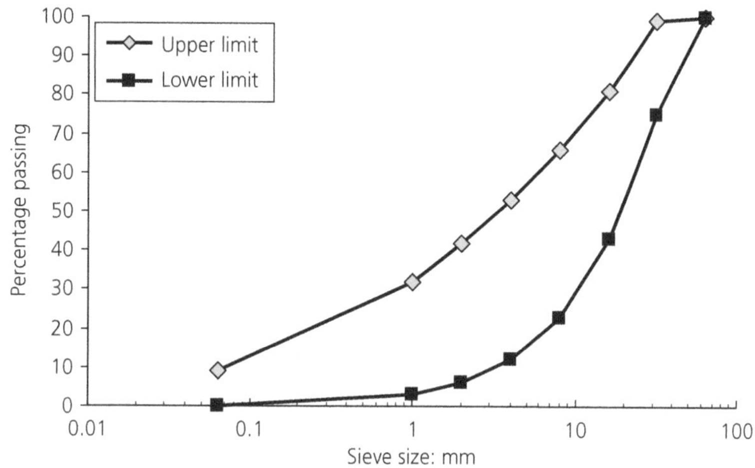

from the impossibility of hitting a specified target gradation perfectly. The tighter the limits, the greater the assurance of material performance – but the higher the cost of production.

At this stage, a concept that this book will term *critical particle size* will be introduced and, for that, reference must first be made to a so-called 'Fuller curve' (Fuller and Thompson, 1907). The idea of a Fuller curve is that, at any given particle size d, the relative proportions of different smaller-sized particles remain the same and the curve obeys the formula

$$\text{Percentage passing sieve size } d = 100\left(d/d_{\max}\right)^{n} \quad (6.2)$$

The key parameter is the exponent n; in the case of crushed rock it is generally found that the most efficient packing is given when $n \approx 0.45$. At this gradation, there are just the right number of particles of each size to fill the available void spaces in the most efficient fashion, taking into account the realities of particle shape and compaction method. A theoretical Fuller gradation would continue to an infinitely small particle size, but this is impossible to achieve of course, so deviation from the Fuller curve is inevitable. Below the point of deviation – termed here the critical particle size – there will not be enough particles to fill the voids. This means that the critical particle size is the smallest size particle within what is termed the soil (or aggregate) 'skeleton'.

Furthermore, the point of deviation from the Fuller curve really means the point where the gradation just touches a Fuller curve, as shown in Figure 6.3; above this size, particles will be larger on average than on the ideal Fuller curve. With larger but fewer particles compared with the ideal Fuller gradation, these larger particles will tend to be slightly 'lost' in a matrix dominated by particles of approximately the critical particle size.

The critical particle size concept will be referred to many times in this book. It is an approximate way of choosing a single particle size to characterise a real gradation.

Figure 6.3 The critical particle size concept

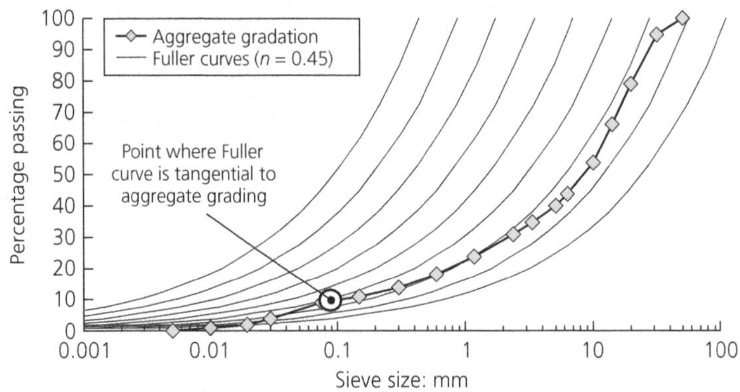

6.1.4 Water content

Water content is expressed as a percentage by mass of dry aggregate. In a wet clay soil with equal masses of water and mineral particles the water content is therefore 100%, but soils such as peat can have much higher water contents than this. Note also that while the density of water is 1000 kg/m^3, that of the minerals forming the soil or aggregate particles is likely to be in the range 2500–3000 kg/m^3, which means that equal mass of water and mineral means about 2.5–3.0 times as much water by volume. In the case of pavement layers it is usual to specify water content within certain limits since, as explained in Chapter 7, this affects material compactability.

6.1.5 Particle shape

Particle shape can be specified in various ways – using words such as 'angular' or by demanding a certain minimum percentage of crushed faces. Many specifications also place limits on the 'flakiness' and 'elongation' of particles, which are determined from the ratio of the minimum and maximum dimensions, respectively, to the mid-size dimension.

The shape of a crushed particle is a function of the equipment used to carry out the crushing and the nature of the parent rock. In general, a mixture with high shear strength is achieved with highly angular stones. Individual stones can then lock together effectively within a pavement layer and inhibit the movement of one particle relative to another. Restrictions on flakiness and elongation are included because experience has shown that, with a high proportion of such particles, handling difficulties can arise, accompanied by segregation of large particles from small ones; however, debate continues about the need, or otherwise, for such limits, as many flaky or elongated materials have been found to perform well in practice. There is no limit for flakiness or elongation in current UK standards for unbound pavement materials.

6.1.6 Particle soundness

Particles of unbound pavement layers do not have to be exceptionally strong, as the behaviour of the layer as a whole is largely a function of interparticle effects at contacts. Nevertheless, particles cannot be allowed to crumble to dust. It is therefore usual to place limits on their strength and durability. Tests exist for crushing strength, abrasion resistance, frost damage and chemical weathering, and all of these are intended to ensure durability. Unfortunately, historically, acceptable limits have often been set purely in order to exclude certain materials rather than for scientifically sound reasons, and this means that otherwise excellent secondary or recycled materials can fall foul of such limits.

The standard particle strength tests worldwide are the Los Angeles abrasion test (AASHTO, 2019; ASTM, 2016, 2020; BSI, 2020) and the micro-Deval test (AASHTO, 2022a; ASTM, 2017; BSI, 2011), both of which consist of 'shaking' a batch of material and observing its degradation. There are other tests, using magnesium or sodium sulfate, to evaluate resistance to chemical weathering (AASHTO, 2022b; BSI, 2009). This author's advice is not to over-specify. In many applications the key issue is whether degradation leads to a serious loss of permeability, affecting drainage design (see Chapters 23 and 24), and as long as the degradation does not produce clay-size fine particles there may be no significant problem.

A further issue is freeze–thaw soundness; AASHTO T103 is an example of a specification for this (AASHTO, 2022c). Again, the key is to appreciate the manner of the degradation. For example, freeze–thaw can degrade rock chalk to a clay-like material with very low permeability and very poor long-term properties, and this needs to be specified out in frost-susceptible climates. Other rock types may break down to a fine grading but, without the clay content, may retain acceptable long-term properties. Reliance is often placed on general descriptors such as 'clean, hard and durable', but any aggregate with an absorption of more than about 0.5% should be considered potentially at risk.

6.1.7 Plasticity

Plasticity is a property associated with clays, but it also occurs in any unbound material with a proportion of its particles in the clay size range – which means 2 μm or less. Measurement of plasticity provides a relatively easy method of reducing a highly complex problem – the sensitivity of soil to water – to convenient numbers. The parameters concerned are as follows.

- Plastic limit (PL): the water content at which a soil (or other unbound material) makes the transition from a dry, crumbly state to a plastic, malleable state.
- Liquid limit (LL): the water content representing the transition from a plastic, malleable state to a liquid, flowing state.
- Plasticity index (PI): PI = LL – PL.

Specifications for the LL and PL include those of AASHTO (2022d, 2022e), ASTM (2018) and BSI (2018).

The PL and LL (known as Atterberg limits) are conceptually like the freezing and boiling points of a liquid, but rather less well defined. In fact, measurement is decidedly non-scientific. The standard PL test involves rolling out a 'sausage' of material; if it crumbles, the water content is less than the PL value, if not it is higher. The traditional LL test involves tapping a small container of soil in order to close a groove cut in the surface, although nowadays it is usual to measure the penetration of a weighted needle instead. In terms of the particulate nature of a soil, the PL marks the point where there is just enough water to saturate the material with the particles in a densely packed state. The LL, on the other hand, describes the water content at which the particles are in their loosest possible arrangement while maintaining particle–particle contacts. The actual water content will depend on particle shape and gradation, although 15–25% is a common range for the PL of a clay soil and 40–60% for the LL.

A key point is that these two measures represent actual physical behavioural properties and have a clear meaning to engineers. If the water content is kept at or below the PL value, the material will have high strength, high stiffness and so on; if it exceeds the LL value, it will be soft, weak and deformable. In general, soils are 'uncomfortable' at a water content lower than the PL; the resulting non-saturated state induces suction, which means that any water that is available will be absorbed until the PL is reached. Even at the PL, the particle packing arrangement will still be uncomfortably tight and zero suction will only be achieved once sufficient water has been absorbed for the particle structure to be 'at rest'. Experimentally, it is possible to show that the 'at-rest' state is a little less than the LL. Figure 6.4 gives measured suctions for three quite different clay soils, plotted against a parameter known as the 'liquidity index', indicating the relative position on the PL to LL scale – and the data are impressively consistent. In reality, subgrade soils will vary in water content (and therefore in suction) throughout the year, and this is critically dependent on drainage design.

The concept of plasticity is widely used by geotechnical engineers in characterising soils. In particular, it is useful to distinguish between soils that are sensitive to changes in water content and those that are not. If the PI is high, then a small change in water content will have relatively little effect on material properties. On the other hand, a low PI (say 15% or less) means a small difference between PL and LL, such that a modest change in water content will drastically alter the soil's behaviour. This is as important in the context of pavement subgrades as anywhere else and should certainly be taken into consideration when pavement design decisions are being made.

Plasticity also applies to layers other than the subgrade. The tests themselves can only be carried out on the finest fractions of a mixture (material passing a 425 μm sieve in most standards), so if this fraction only represents a minor part of the whole mixture the relevance of a plasticity value becomes questionable. Once more, the concept of *critical particle size* is important. If this is larger

Figure 6.4 Relationship between suction and water content, w (data are plotted from information in Brown *et al.* (1987))

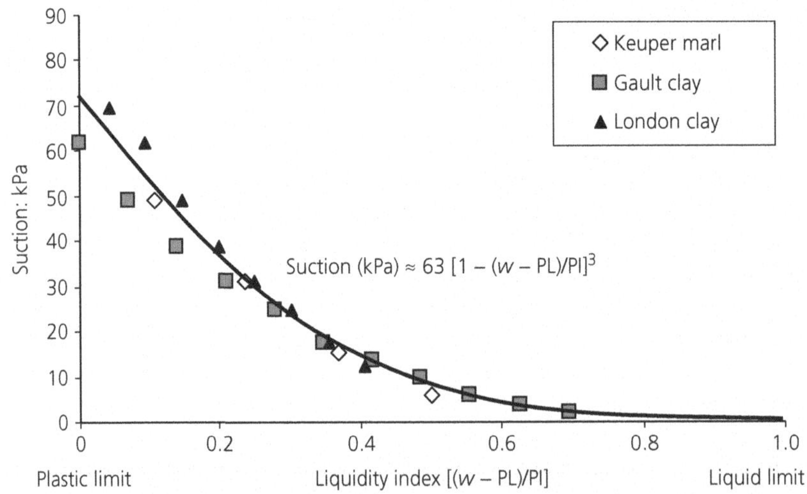

than 425 μm, then plasticity is fairly meaningless. In practice, it is common to insist that base layers are 'non-plastic', which means that it is impossible to achieve a plastic state at any water content, indicating an almost total lack of clay-size fines. For sub-bases and cappings, a small plasticity may usually be permitted without significant reduction in performance, as long as the fines fraction is not too large.

6.2. Shear strength

The CBR test is, in reality, a sort of shear strength test. The failure it induces is in the form of shear slip planes. Similarly, if a pavement subgrade fails, what this really means is that there is excess strain developing on shear planes. Therefore, we will now take a fundamental look at how unbound materials work in terms of their shear resistance.

6.2.1 Interparticle friction

Significant shear strain in an unbound material can only occur if slip takes place between particles (Figure 6.5). In order to achieve shear strain within the body of the material, at least some of the individual particles have to translate and rotate into new positions, and this can only occur through soil particles or stones sliding against one another. This, of course, is resisted by *friction* and the resistance to strain might therefore be expected to relate to the frictional qualities of the particle surface. Furthermore, the *shape* of the particles is clearly also important. A rounded particle is obviously easier to slip past than an angular one, whatever its frictional properties.

As in any frictional problem, it is the ratio of the shear force to the normal force that controls whether slip occurs. For example, particle A in Figure 6.5 could never slip along the face of particle B unless the shear force F was enough to overcome the normal force N. This frictional property can be expressed as either a coefficient of friction (μ) or as an angle of friction (θ).

$$\text{At slip: } F = \mu N = \tan\theta \times N \qquad (6.3)$$

For crushed rock particles, θ is typically around 30–35°; it is rather less for smooth river gravels.

6.2.2 Angle of internal friction

Just as the slip between individual particles is controlled by the friction at particle contacts, so shear failure of an unbound matrix can be thought of in a similar fashion. Interparticle friction, particle shape and also the particle size distribution all contribute to resisting the development of a

Figure 6.5 The mechanism of shear in an unbound material

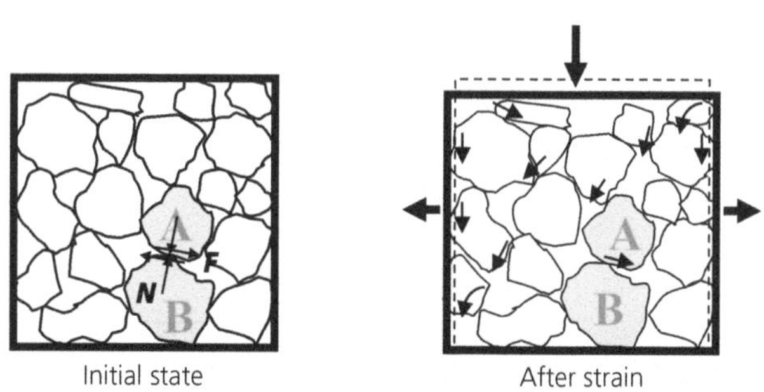

Figure 6.6 Shear failure of an unbound granular material

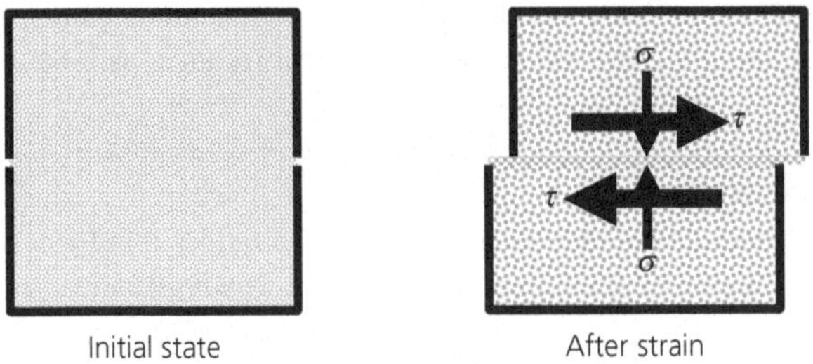

shear plane. The actual movements of particles will be a complex combination of slip, rotation and sideways displacement, but the overall effect (Figure 6.6) will be that the shear stress (τ) to cause failure on a given plane will be a function of the normal stress (σ) across that plane – just as in any frictional system – and this relationship can be described by the angle of internal friction (ϕ).

At shear failure: $\tan\phi = \tau/\sigma$ (6.4)

The situation shown in Figure 6.6 may be readily understandable, but it represents a highly constrained and artificial state for the material, a state in which the failure plane is predetermined. More generally, however, failure will take place on any plane where the τ/σ ratio exceeds its limiting value. For example, Figure 6.7 shows the stress state at a point within a material for constant σ_2 and three different levels of σ_1. This is shown in the form of 'Mohr circles' of increasing magnitude, together with a line representing the limiting τ/σ ratio. The Mohr circle construction (see Calculation Sheet 6.1) describes the stress state in every direction on a given plane, and the point on the circle that touches the limiting τ/σ ratio line represents the onset of shear failure in a particular direction. In Figure 6.7, shear failure planes would be inclined at an angle of $\alpha/2$ to the vertical – following the mathematical explanation given in Calculation Sheet 6.1.

Figure 6.7 The shear failure state on a Mohr circle diagram

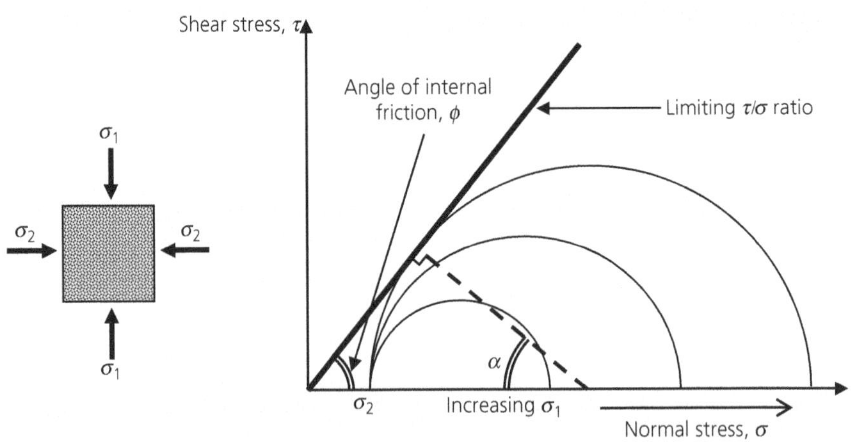

Calculation Sheet 6.1. Relative influences of pavement parameters

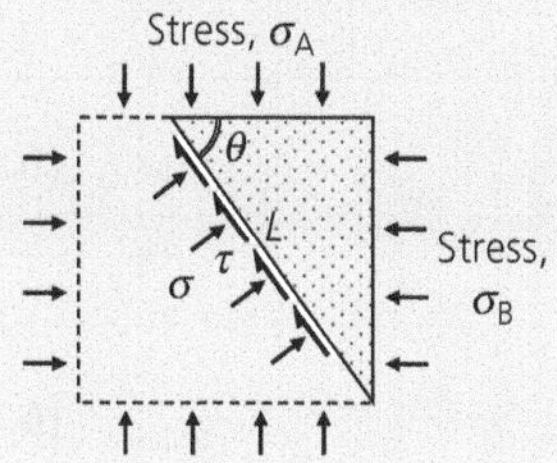

σ_A and σ_B are principal stresses (i.e. no shear in those directions)

Question: What are the values of normal and shear stress on the θ plane, at an angle of θ to the A plane?

Resolve vertically:

$$\sigma_A L\cos\theta = \sigma_\theta L\cos\theta + \tau_\theta L\sin\theta \qquad \text{(Equation 1)}$$

Resolve horizontally:

$$\sigma_B L\sin\theta = \sigma_\theta L\sin\theta + \tau_\theta L\cos\theta \qquad \text{(Equation 2)}$$

Multiply Equation 1 by sinθ and Equation 2 by cosθ and subtract Equation 2 from Equation 1:

$$\sigma_A L\cos\theta\sin\theta - \sigma_B L\sin\theta\cos\theta = \sigma_\theta L\cos\theta\sin\theta + \tau_\theta L\sin^2\theta - \sigma_\theta L\sin\theta\cos\theta + \tau_\theta L\cos^2\theta$$

$$\tau_\theta L = (\sigma_A - \sigma_B)L\sin\theta\cos\theta \rightarrow \tau_\theta = \frac{1}{2}(\sigma_A - \sigma_B)\sin 2\theta \qquad \text{(Equation 3)}$$

Divide Equation 1 by cosθ and Equation 2 by sinθ and sum Equations 1 and 2:

$$2\sigma_\theta L = \sigma_A L + \sigma_B L - \tau_\theta L(\sin\theta / \cos\theta - \cos q / \sin\theta)$$

$$\sigma_\theta = \frac{1}{2}(\sigma_A + \sigma_B) - \frac{1}{2}\tau_\theta(\sin^2\theta - \cos^2\theta) / (\sin\theta\cos\theta)$$

$$= \frac{1}{2}(\sigma_A + \sigma_B) - \frac{1}{2}\tau_\theta(-\cos 2\theta) / (\frac{1}{2}\sin 2\theta)$$

Substitute for τ_θ using Equation 3:

$$\sigma_\theta = \frac{1}{2}(\sigma_A + \sigma_B) + \frac{1}{2}(\sigma_A - \sigma_B)\cos 2\theta \qquad \text{(Equation 4)}$$

Equations 3 and 4 can be represented graphically:

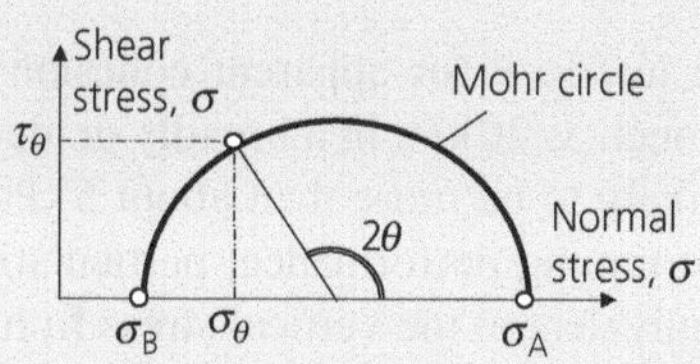

Please do not confuse the angle of internal friction (ϕ) with the angle of friction relating to interparticle slip (θ). The angle θ is a fundamental property of the surface of the aggregate particles, while ϕ is a composite, with contributions from interparticle friction, particle shape and particle size distribution. While θ has a relatively narrow range of possible values, between about 25° and 40°, ϕ can range from around 20° for some single-sized silts to as much as 60° for broadly graded, angular aggregate.

6.2.3 Stress ratio at failure

In many applications it is convenient to think in terms of a limiting ratio of two stresses at right angles to each other, σ_1 and σ_2, as in Figure 6.7. This is related to the angle of internal friction ϕ by the geometry of the Mohr circle diagram, as the sine of ϕ is the ratio of the radius of the Mohr circle to the distance of the centre of the Mohr circle from the origin, that is

$$\sin\phi = \frac{1}{2}(\sigma_1 - \sigma_2) / \frac{1}{2}(\sigma_1 + \sigma_2) \tag{6.5}$$

Rearranging

$$\sigma_1 = \sigma_2(1 + \sin\phi) / (1 - \sin\phi) \tag{6.6}$$

or

$$\sigma_1/\sigma_2 = (1 + \sin\phi) / (1 - \sin\phi) \tag{6.7}$$

Good-quality crushed stone base can have a value of over 10 for the limiting σ_1/σ_2 ratio (equating to a ϕ value of over 55°).

6.2.4 Interlock

The characterisation of shear strength by means of the angle of internal friction (ϕ) is approximately correct for most unbound materials other than those with a clay content (see the next section) – but only approximately. The way that angular particles interact creates what has been termed the 'dry stone wall' effect. The point is that dry stone walls – walls that make no use of cementitious mortar – can stand many feet in height with vertical faces, so long as individual stones are carefully selected to fit neatly into the gaps between neighbouring stones, producing a strongly interlocking structure; near the base of such a wall, the vertical stress may be 20–30 kPa, while the horizontal stress is effectively zero – giving an infinite stress ratio. Referring back to the diagram on the left of Figure 6.7, this means that it is possible to withstand a moderate level of stress σ_1 even if σ_2 is zero. The effect this has on the failure characteristic in the τ, σ space is to raise it slightly, giving an intercept on the τ axis. This intercept is akin to the effect of water in clay soil, for which the term 'cohesion' and the symbol c are used. In this case, the property is certainly not cohesion, but the term 'apparent cohesion' is still often applied. The effect on the stress state at failure is

$$\sigma_1 = [\sigma_2(1 + \sin\phi) + 2c\cos\phi] / (1 - \sin\phi) \tag{6.8}$$

The value of the interlock (or apparent cohesion) intercept is rarely more than a few kilopascals, perhaps as much as 20 kPa in a broadly graded crushed rock (Thom, 1988). In a single-sized material it is unlikely to be more than about 5 kPa, but even this modest quantity can be highly influential in determining performance, particularly in unsurfaced pavements. After all, 5 kPa is approximately equivalent to the vertical stress from 250 mm of overburden.

6.2.5 Water content, suction and cohesion

Water content affects unbound materials dramatically and this will be thoroughly explored in Chapters 23 and 24. Here it should simply be noted that a high water content is usually associated with a low unbound material shear strength, and that this may be due to two effects.

- Void spaces that become filled with water have the potential to develop positive pore water pressure under applied load. This reduces the normal forces between aggregate particles, making interparticle slip easier. Thus, the greater the proportion of voids that are filled with water, the lower the shear strength will be.
- In materials that are already saturated – generally clays – an increased water content simply means larger spaces between particles, and therefore less contact points. This too reduces shear strength.

Conversely, a low (but not zero) water content can induce suction, which increases shear strength. A greater proportion of fine material, especially clay-size fine material, induces a higher suction.

In clay soils, the effect of water is to give what is known as cohesion. True geotechnical engineers will often claim that cohesion does not exist, by which they mean that if you wait long enough it disappears, but in the world of the practical pavement engineer it certainly does exist. Cohesion is created by the need for water to flow between very small pore spaces as a soil deforms. Inevitably, as deformation takes place and particles rearrange themselves, some void spaces will want to expand while others will want to contract. If many or all of these voids are filled with water, this means that water has to flow from one void to another, and this will be resisted because of the viscosity of water and the extreme narrowness of some of the gaps available. The effect is dependent on the loading rate and would disappear if the material was given enough time for the water to flow and for pressures to dissipate. However, for most pavement engineering purposes, loading is much too quick for the small pores present in many soils, and pore pressure dissipation is therefore limited – which means that cohesion, even if it is termed 'apparent' cohesion, is a very real property.

For saturated clays under short-term (i.e. *undrained*) loading, cohesion is the dominant component of shear strength and it operates equally at all levels of confinement (i.e. all levels of normal stress). An element of soil that is saturated will be effectively incompressible. Thus, confining stress will be quite unable to cause strain in the soil skeleton, which therefore cannot increase its contribution to carrying that stress; it simply increases pressure in the water. And if there is no change in the forces between soil particles, there will be no change in the shear stress at failure. The result is a horizontal failure characteristic, as shown in Figure 6.8.

Not surprisingly, partially saturated soils have short-term (undrained) failure characteristics somewhere between that of a saturated clay and a dry granular material (see Figure 6.8). Many voids will still be full of water, which means that cohesion will therefore still exist. On the other hand, when confining stress is added, the presence of air voids means that the material is no longer incompressible and that some of the increased stress will be taken by the soil skeleton. This will give an apparent angle of friction less than the value would have been under dry conditions because a high percentage of normal stress is still being transmitted directly to the pore water.

Figure 6.8 Forms of shear strength characteristic – undrained conditions

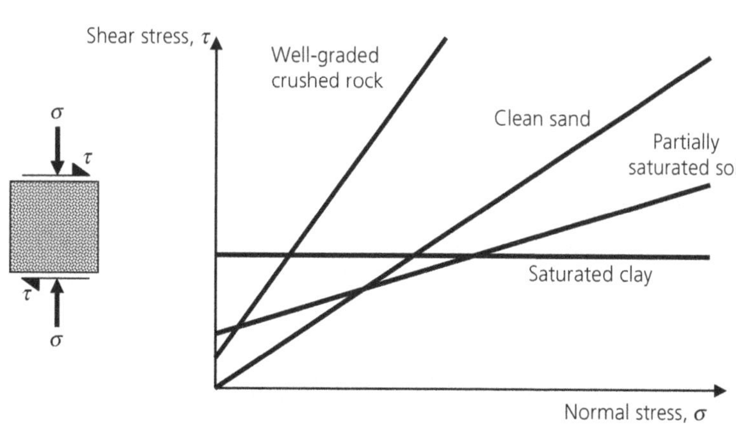

Figure 6.9 The influence of particle angularity on shear strength (Thom and Brown, 1989)

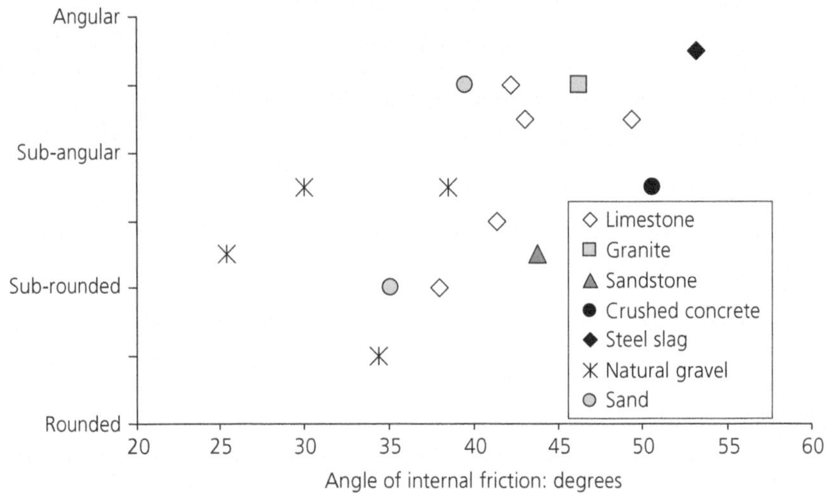

6.2.6 Particle shape and angularity

From previous sections it will already be clear that an angular material will tend to have a high angle of internal friction (and therefore a high stress ratio at failure). This is illustrated in Figure 6.9 for a range of non-cohesive pavement materials. However, the point has already been made that angularity is not a property that is readily quantified and therefore the point being made here remains qualitative. Most methodologies rely on description and visual categorisation, as in Figure 6.9.

6.2.7 Particle size

There seems little reason to expect particle size to affect the angle of internal friction of an aggregate, as long as the particle shape is unchanged. If the well-known Hertzian contact law (Hertz, 1895) between spherical particles is assumed to apply, the size effect should be zero since everything depends on ratios and these are unchanged. However, the truth is that aggregates with larger particles display a larger angle of internal friction, as illustrated in Figure 6.10 for the case of single-sized materials, and this strongly suggests that the Hertzian law, in which compression at particle contacts is proportional to force to a power of two-thirds, requires modifying for the geometry of real aggregate contacts. The effect shown in Figure 6.10 implies that the real compression at particle contacts may be approximately proportional to force to the power of a half.

There may also be an effect at large particle size because of a reduced number of particles on potential failure planes. This gives the material less choice in locating a plane of weakness on which to fail.

6.2.8 Particle size distribution

It is entirely unsurprising that a broadly graded aggregate tends to have a higher angle of internal friction than a single-sized material. Basically, the increased number of particle contacts that a broad gradation allows increases the frictional resistance and makes the development of a failure plane more difficult, increasing strength. This is illustrated in Figure 6.11 for aggregates having the same maximum particle size but different size distributions. The measure used for particle size distribution in Figure 6.11 is the *uniformity coefficient*, defined as the ratio of the sieve size through which 60% of the material passes to that through which just 10% passes.

Figure 6.10 The influence of particle size on aggregate strength (Thom, 1988)

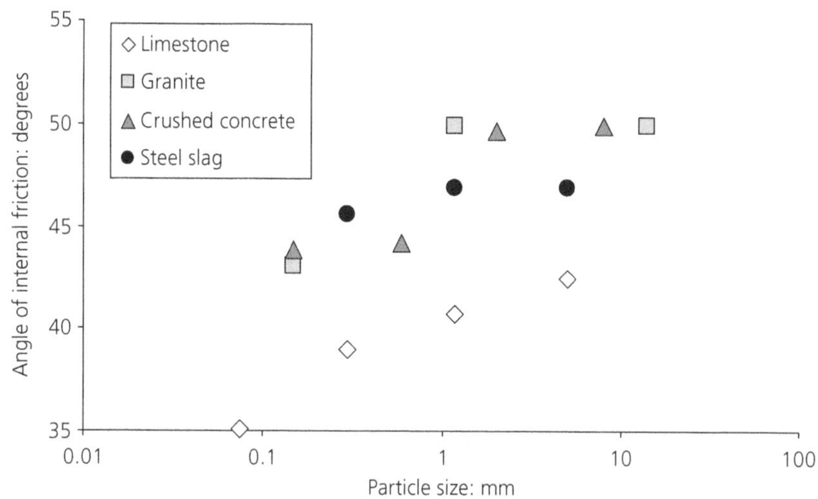

Figure 6.11 The influence of particle size distribution on aggregate strength (Thom and Brown, 1988)

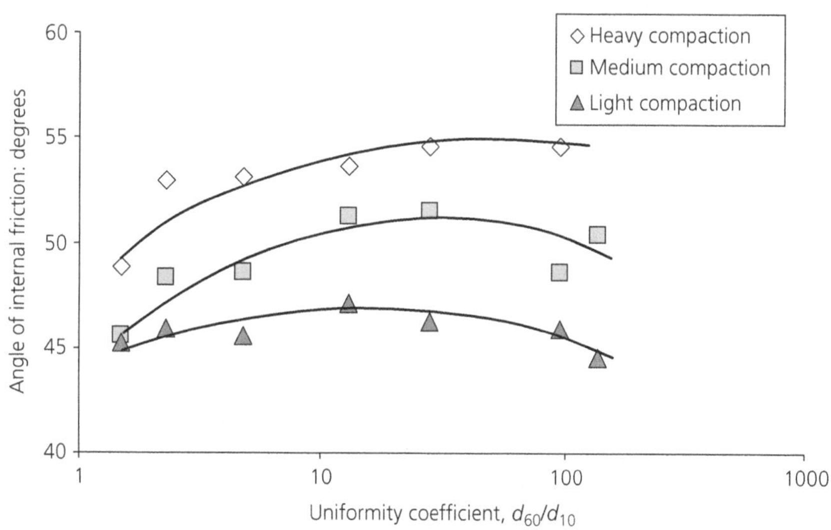

However, there is a limit, an optimum beyond which there is no gain in broadening the size distribution – a uniformity coefficient of around 40 for the 'heavy compaction' case illustrated in Figure 6.11. This is because, if too many small-sized particles are included in a mixture, they will dominate behaviour, and the larger particles will become detached from one another. This is the 'critical particle size' concept introduced in Section 6.1.3. As the slope of the grading line gets shallower, the point at which it touches a Fuller curve is shifted towards the smaller end of the scale, giving a small critical particle size and, as shown in the previous section, small particles lead to reduced shear strength.

6.2.9 Compaction

Obviously particle size distribution is an important factor controlling the efficiency of particle packing. However, packing also depends on the degree of compaction (or, for clay soils, 'consolidation'). A material following the $n=0.45$ Fuller gradation will pack well even if simply tipped out of a wagon – but it will pack substantially better if it is compacted. The beneficial effect of compaction on shear strength is illustrated in Figure 6.11.

6.2.10 Shear strength tests

The principal laboratory test is the triaxial test (Figure 6.12). Cylindrical clay specimens can be cut during borehole excavation. More usually, the specimen has to be compacted into a cylindrical mould (using a specified compaction procedure). In either case, the specimen must be surrounded by a membrane (typically latex rubber or polythene), which is sealed to top and bottom loading platens. The whole arrangement is then contained in a sealed cell, which is pressurised to a desired stress level. This stress, known as the *confining stress*, acts in all directions on the specimen. The test then consists of applying an additional stress to the top platen, known as the *deviator stress*, increasing it until the point of failure. Figure 6.12 illustrates use of the triaxial test to determine the shear strength parameters c and ϕ.

Various types of *shear box* tests (using the principle illustrated in Figure 6.6) are also available, but the stress conditions are much less uniform than those in the triaxial test, giving the result less validity.

For in situ determination of shear strength, the *shear vane* (see Figure 6.13) can be used in clays up to a certain strength. This is a quick and convenient test, ideal for testing the properties of soil exposed during a trial pit investigation. The device is thrust into an exposed clay face or into the base of the excavation, and the operator then applies a steadily increasing torque until the vane begins to rotate. This point may be taken as indicating shear failure on a circular plane.

Cone penetrometers are instruments that measure the force as a steel cone is inserted into a soil. This induces shear failure in the surrounding material and the insertion force should therefore

Figure 6.12 The triaxial test for shear strength determination

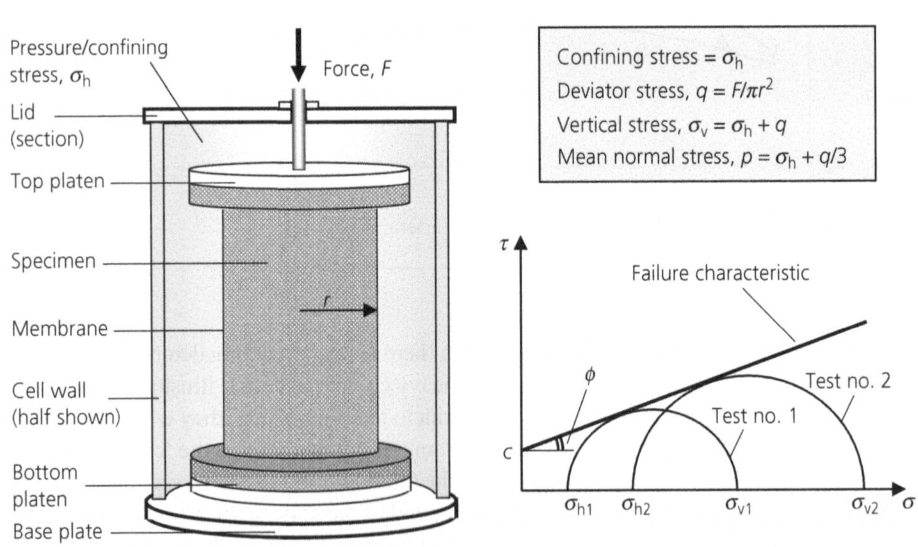

Figure 6.13 The shear vane test

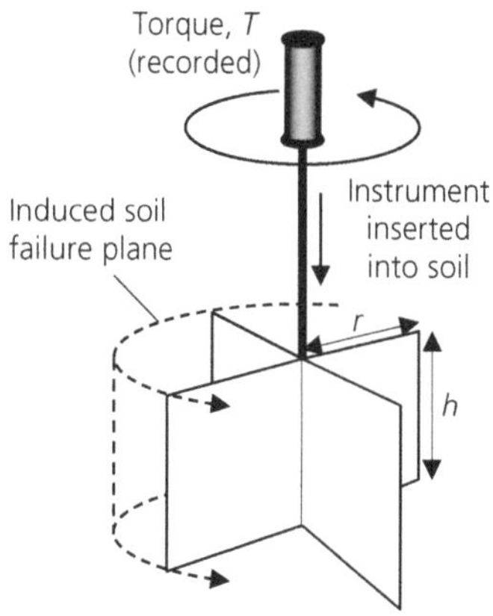

If soil shear strength is τ_f then

T = resistance from cylindrical shear plane
+ resistance from upper and lower shear planes

$$= (2\pi rh \times \tau_f \times r) + 2\int_0^r 2\pi x \, dx \, \tau_f x$$

$$= 2\pi r^2 h\tau_f + 4\pi\tau_f \times \int_0^r x^2 \, dx$$

$$= 2\pi r^2 h\tau_f + 4\pi\tau_f \times \frac{x^3}{3}\int_0^r$$

$$= \tau_f \times 2\pi r^2\left(h + \frac{2}{3}r\right)$$

Thus

$$\tau_f = T / \left[2\pi r^2\left(h + \frac{2}{3}r\right)\right]$$

relate to the material's shear strength. There are several different sizes of cone penetrometer on the market, from pocket-sized to large mechanically assisted instruments, and each has to be calibrated carefully. The number that results (either the shear strength or CBR) will only be approximate; nevertheless, the technique is extremely practical for in situ work.

The *dynamic cone penetrometer* (DCP) (Figure 6.14) is commonly used in pavement engineering. It is a conceptually simple tool, consisting of a cone on the end of a rod, which is driven into the ground under impulse forces from a dropped weight. The rate of penetration (in mm/blow) is recorded, and usually converted to an equivalent CBR. An empirically determined (i.e. based on experience) conversion equation is given in Figure 6.14. It is not really a CBR, but it is at least a measure of the strength of the material, and it means that a profile of material strength with depth can be generated, often revealing the thickness of unbound pavement layers as well as the strength of the underlying soil. The test is a relatively rapid one and the information generated is extremely useful.

Figure 6.14 The dynamic cone penetrometer. The equation was developed by the UK Transport Research Laboratory and is now incorporated in UK standards (National Highways, 2020)

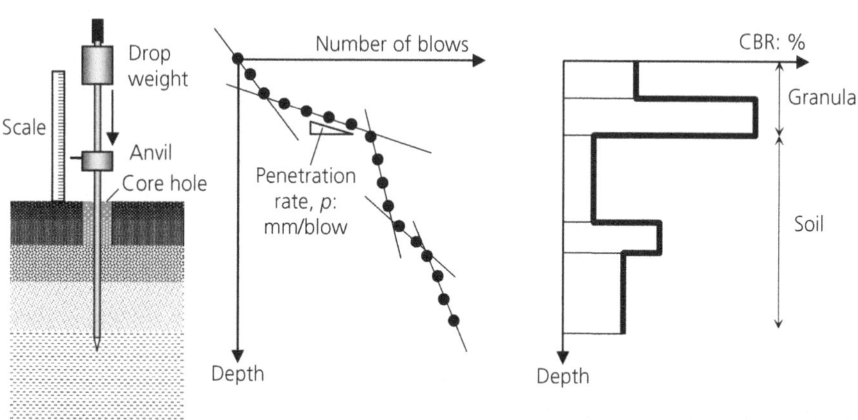

Possible equation: $\log_{10}(\text{CBR}) = 2.48 - 1.057\log_{10}(p)$

6.2.11 Typical shear strength values

This section is included in order to allow an appreciation of any data that the reader may obtain. There is no such thing as a typical material and the numbers given here should never be assumed in actual design. However, as long as this is understood, it is always useful to put data into context.

6.2.11.1 Clays

The weakest materials likely to be encountered are peats and recently deposited alluvial clays. Water contents (water mass as a percentage of solid mass) can be as high as 300% or 400% (i.e. the great majority of the volume is actually water). Cohesion values can be as little as 10 MPa, with a zero angle of internal friction (under short-term loading of course). Unfortunately, these soils tend to lie in wide, flat locations which would otherwise be ideal for the construction of cities, airports, ports and the like. Inevitably, any added vertical load will compress the soil significantly (often achieved deliberately by preloading), and cohesion would therefore be expected to increase, perhaps to 15–20 MPa.

Peats lie at one extreme of a massive range of clays of differing strengths, through soft glacial deposits, to firm and then to stiff layers. Cohesion is often in the 50–100 MPa range but, for the strongest materials, much higher values are possible.

All clays are sensitive to moisture. It is not the purpose of this book to pursue clay mineralogy any further, but the reader should remember that low-plasticity clays (see Section 6.1.7) are particularly vulnerable to wetting up during the in-service life of a pavement. A strength of well over 50 MPa under normal equilibrium moisture conditions may reduce to much lower values with just a few percentage points change in water content.

6.2.11.2 Sands and silts

In contrast to a clay, a pure sand will have a negligible value of cohesion and the strength will be provided entirely by the angle of internal friction. A value of around 30° is typical. A fine sand or silt will often have a slightly lower value, between 20° and 30°. However, in many cases, natural sands and silts include a clay content, and this will supply a cohesion of a few kilopascals, something that dramatically changes the nature of the material. Pure sands, such as wind-blown desert deposits, flow easily and are therefore intrinsically unstable; if even a small cohesion is present, stability increases greatly.

6.2.11.3 Gravels

Naturally occurring gravels are commonly used in lower pavement construction layers. In most cases, the materials are a mix of gravel, sand, silt and often clay, and the angle of internal friction is frequently in the range 40–45°. This angle is chiefly dependent on the angularity of the particles; a predominance of rounded stones results in a relatively lower angle. As with sands, a small clay content, providing a degree of cohesion, greatly increases the stability of the layer – but also its susceptibility to water content changes.

6.2.11.4 Crushed rock and other aggregates

Crushed rock is the premier pavement construction aggregate. Particles are angular (although different rock types result in significantly different typical shapes) and the faces are often rough. This, together with the fact that a broad range of particle sizes is commonly present, means that there will usually be some 'interlock' (apparent cohesion); 10 kPa is probably typical. The angle of internal friction is usually in the range 50–60°.

Although rock is the traditional source for high-quality granular layers, there are several secondary materials – materials from industrial processes and various waste products – that make perfectly acceptable aggregate when crushed. Steel slag, concrete from demolition waste, various arisings from mining operations – all such materials can logically be considered as aggregate for pavement construction. Assuming that their chemical composition does not exclude them and that the particles themselves are strong enough not to fragment excessively, the principal factors affecting shear strength will be exactly the same as for crushed rocks (i.e. particle shape, particle size distribution and surface roughness).

6.2.12 Relationship to CBR

While shear strength is a measure with fundamental meaning, shear strength tests are not always convenient to carry out. The CBR test is quick, simple and convenient, and it can be (very approximately) related to the fundamental shear strength parameters c and ϕ, as shown in Calculation Sheet 6.2.

Calculation Sheet 6.2. Relating CBR to c and ϕ

The CBR test is effectively a bearing capacity failure problem with a circular loaded area. The standard geotechnical equation for this is:

$$q_f = 1.3cN_c + 1.2p_oN_q + 0.3\gamma dN_\gamma$$

where q_f is the stress to cause failure, c is cohesion, p_o is the overburden stress, γ is the unit weight and d is the diameter of the loading platen. N_c, N_q, N_γ are bearing capacity parameters for which the following approximations are suggested:

$$N_c = 5.5 + 0.32\phi + 3\times10^{-5}\phi^4$$

$$N_q = 1.0 + 0.12\phi + 9\times10^{-3}\phi^2 + 9\times10^{-12}\phi^8$$

$$N_\gamma = 6\times10^{-4}\phi^3 + 2.33\times10^{-21}\phi^{14}$$

where ϕ is the angle of internal friction (in degrees)

To use this approach it is necessary to equate bearing capacity failure in a semi-infinite half space to a certain displacement in the confined CBR test. It is suggested here, based purely on the author's empirical experience, that this should be 1.27 mm, the corresponding 100% CBR load for which is 13.2 kN (see Figure 6.1)

Therefore: estimated CBR $\approx 100(\pi d^2/4)q_f/13.2$

Calculation Sheet 6.2 allows prediction of the CBR if strength parameters are known, but that isn't particularly useful. It would be much more helpful to predict c and ϕ from a CBR. The problem is that different combinations of c and ϕ can give the same CBR, which means that any prediction of c and ϕ requires a degree of engineering judgement depending on the type of soil or unbound material, and suggestions for this are given in Table 6.1 They are in no way fundamental; they simply represent this author's experience.

Table 6.1 Estimated c and ϕ values based on CBR

Soil type		CBR						
		2%	3%	5%	8%	12%	20%	30%
Crushed rock sub-base	c: kPa	—	—	—	—	8	8	9
	ϕ: degrees	—	—	—	—	38	42	46
Gravel	c: kPa	—	—	—	5	5	6	6
	ϕ: degrees	—	—	—	38	42	46	49
Sand	c: kPa	—	2.5	2.5	3	3	3.5	—
	ϕ: degrees	—	35	39	43	46	49	—
Silt	c: kPa	3	3.5	4	4.5	5	—	—
	ϕ: degrees	27	30	35	39	42	—	—
Clayey sand, silt or gravel	c: kPa	7	8	10	12	13	15	16
	ϕ: degrees	18	21	25	29	32	36	40
Silty or sandy clay	c: kPa	12	16	21	26	30	35	40
	ϕ: degrees	9	12	16	19	23	27	30
Clay	c: kPa	17	24	37	53	72	100	—
	ϕ: degrees	2	3	5	7	9	13	—

6.3. Stiffness

In engineering terms, stiffness means the so-called *resilient modulus* (see Section 1.5 for definition), which is the ratio of applied stress to induced strain. In a linear elastic material, having an *elastic modulus*, this is a single and fundamental property, and many engineering materials can be considered linear elastic. However, an unbound material is highly complex and certainly non-linear. This means that the resilient modulus, usually simply termed *modulus* in the remainder of this chapter, varies depending on the stress conditions applying. Numerous PhD studies have been carried out on this subject, which in itself reveals the fact that the behaviour is still not completely understood. Here, the principal factors applying will be introduced, together with a couple of approximate methods used to describe such behaviour.

6.3.1 The mechanism of unbound material strain

The stiffness of an unbound material is quite different from that of the particles from which it is made up. Rock may be considered as an approximately linear elastic material with a modulus commonly between 100 and 200 GPa; unbound materials rarely develop a modulus greater than 1 GPa, even under the most favourable stress conditions. The fact is, of course, that the stiffness of an unbound material is governed by interparticle behaviour, notably frictional slip between particles (see Section 6.2.1).

Figure 6.15 illustrates what happens throughout an unbound material when it strains (in response to applied stress). In this example, there is no change in the relative positions of particles E, F, C and D, and so no slip. At the A–C contact, there is rotation, which means the exact point of contact moves – but there is still no slip. However, as B is pushed to the left it slips against D, and it also rotates and slips against A. The material will naturally try to avoid slip wherever it can. Particles will take the path of least resistance, and this means that the number of slip points will be

Figure 6.15 Interparticle motion due to strain in an unbound material

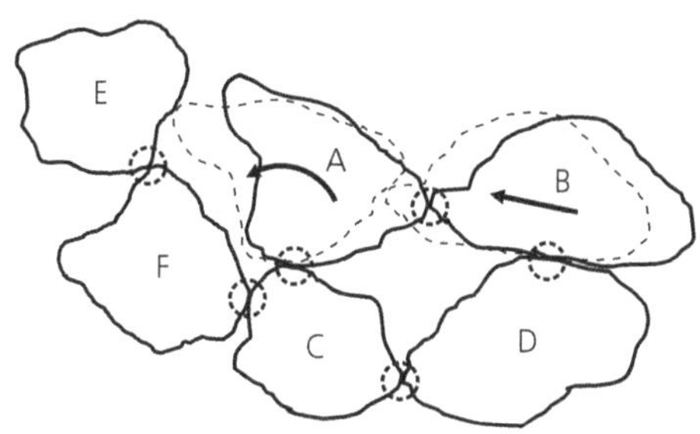

minimised as far as possible. However, it is clearly not possible to avoid them entirely. If a particle manages to rotate against one neighbour (A against C in Figure 6.15), this will usually require slip against another neighbour (A against B).

There is also one other very important factor to consider. Individual particles may be very stiff – but they are not rigid. Furthermore, the size of the contact area is extremely small, which means the contact stress can be extremely high. This causes elastic compression at the contacts, and the magnitude of this compression affects the ease with which one particle can move relative to another. For example, relatively large compression at the A–C contact in Figure 6.15 might allow particle A to rotate right past E and hit F instead. It is this elastic compression that also ensures that almost all of the movement will be reversed as soon as the load is removed; frictional slip will not reverse itself – but it can be reversed as long as there is enough stored compressive strain energy at compressed particle contacts.

6.3.2 The resulting stress–strain behaviour

Unsurprisingly, this complex interaction of compression at particle contacts, rotation and inter-particle slip cannot be expressed simply – at least not with any great accuracy. Figure 6.16 is an idealisation of the typical form of behaviour.

Figure 6.16 Idealised stress–strain behaviour of an unbound material

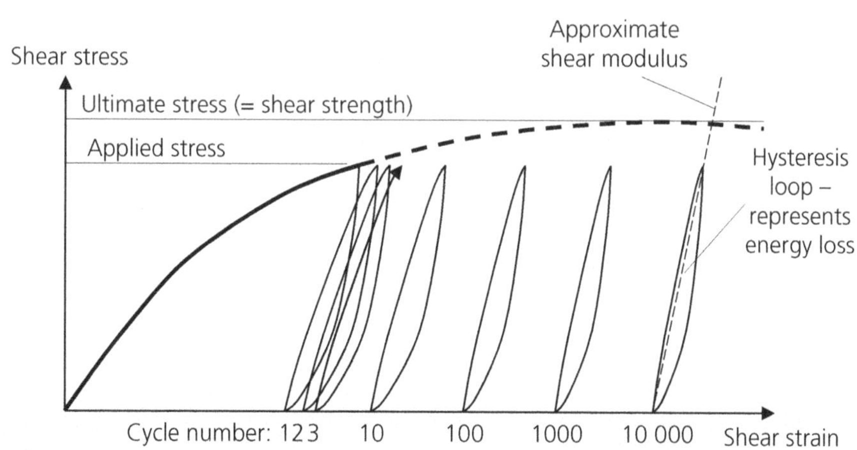

There is always a large difference in behaviour between the first load application, represented by the heavy line in Figure 6.16, and all subsequent load cycles. During that first application, the particles rearrange themselves in a major way; during subsequent cycles there are only minor, and ever-decreasing, differences from one load cycle to the next.

As pavements are structures that are subjected to large numbers of repeated loads, the first load application is of limited interest; it is the shape of the almost repeatable stress–strain characteristic (representing so-called *resilient* behaviour) that determines the stiffness of the material. In fact, the shape of the stress–strain characteristic does change slightly with increasing number of cycles, as the exact way in which the particles interact becomes ever more efficient but, to a first approximation, it may be taken to be constant.

Note the shape of the resilient stress–strain characteristic in Figure 6.16. The area enclosed represents energy loss, an unavoidable consequence of interparticle slip. However, for pavement design (and material characterisation) purposes, it is usual to ignore the detailed shape and to assume that the behaviour is equivalent to a straight line between the end points of the characteristic. The slope of this line (shear stress/shear strain) is, therefore, the *shear modulus* applying under these particular conditions.

6.3.3 The effect of differing stress conditions

Because strain within an unbound material includes frictional slip of one particle against another, it is unsurprising that the stress conditions are of critical importance. The most important influence is the magnitude of the *confining stress* (or bulk stress), as this is directly related to the magnitude of the contact forces between particles – and the resistance to frictional slip is directly proportional to the normal contact force (see Section 6.2.1). In a pavement, this means that material at depth, beneath a considerable overburden, will generally appear stiffer than material near the surface, even though it may in reality be exactly the same soil or pavement material.

The second, and less influential, factor is the effect of stress change. It appears that it is the first few slips between particles that are the hardest to generate, and that, once interparticle slips begin to occur, there is a knock-on effect, which means they become progressively easier. The result is that, all other things being equal, the strain when applying a stress change of 20 kPa is usually more than twice the strain when applying 10 kPa.

There are plenty of other complications, notably the tendency for a material to dilate as it shears, caused by particles 'riding up' over their neighbours. Table 6.2 gives a flavour of the true nature of the non-linearity of an unbound material. It presents data obtained from repeated load tests on crushed limestone sub-base aggregate using triaxial equipment, and a bewildering range of apparent moduli and Poisson's ratios under different stress conditions may be noted.

Thus, any material model that tries to capture this behaviour has to be extremely complex, much more complex than is generally required for pavement engineering purposes. Table 6.3 summarises some possible options.

For many purposes it is sufficient to assume a constant modulus. For example, in analysing a thick asphalt pavement under traffic load it is generally found that the system behaves as if the granular sub-base layer has a modulus of 100–150 MPa, depending on the material quality and moisture state. The layer is not a particularly significant one in terms of its effect on the way that stresses

Table 6.2 Data illustrating the non-linearity of a crushed limestone (Thom, 1988)

Confining stress: kPa	Deviator stress: kPa	Axial strain ($\times 10^6$)	Apparent modulus: MPa	Poisson's ratio
25	0–50	302	165	0.23
	0–150	585	256	0.47
50	0–50	115	436	0.23
	0–150	348	431	0.49
100	0–50	73	685	0.22
	0–150	209	717	0.41
250	0–50	32	1563	0.19
	0–150	111	1351	0.31

Table 6.3 Models of unbound material stress dependency

Action	Possible model[a]
Option 1: Assume fully linear	$E = \text{constant}$
Option 2: Model confining effect[b]	$E = k_1\theta^{k_2}$
Option 3: Model confining effect and stress change[c]	$\varepsilon_s = A\delta[\ln(\sigma_1/\sigma_3)]^B \delta\sigma^C$
Option 4: Model as fully as possible[d]	$G = G_1 p^{1-n}$ $K = \dfrac{(K_1/n)p^{1-n}}{1+2\beta(2-n)q^2/(np^2)}$

[a] E = (resilient) modulus; G = shear modulus; K = bulk modulus; θ = sum of any three orthogonal stresses; σ_1, σ_3 = stresses in two orthogonal directions; p = mean normal stress ($=\theta/3$); q = deviator stress ($=\sigma_1 - \sigma_3$); ε_s = shear strain at 45° to the directions of σ_1 and σ_3; δ means 'change in'; k_1, k_2, A, B, C, G_1, K_1, n and β are all constants

[b] The k–θ model (e.g. Hicks and Monismith, 1971) stemmed from suggestions put forward by various researchers in the 1960s. Additional terms have since been proposed to refine model accuracy, generally based on triaxial test evidence

[c] See Thom (1988)

[d] See Boyce (1980)

distribute in the pavement, and there is no advantage in using anything more sophisticated. The fact that the same layer may appear to have a quite different and much lower stiffness if it is trafficked (or tested) directly, without any intervening asphalt or concrete, is worth being aware of – but it does not invalidate the use of a single modulus value in full pavement analysis.

The so-called k–θ model (option 2 in Table 6.3) is quite widely used by researchers, particularly in the USA, but it immediately makes pavement analysis much more difficult. The reader should also be aware that the k–θ model is not particularly correct for locations other than directly beneath an applied load.

The third option is not even expressed as a modulus but as a direct prediction of shear strain as a function of applied stresses, and so is certainly not easy to use. However, it is much closer to a correct representation of real behaviour and is simple enough for use in certain analysis situations.

Option 4, for which several variant versions exist, is used by some European researchers. It has the merit that it gives a reasonably close match to real behaviour but it also allows input into finite-element or finite-difference quasi-elastic computations (Hornych *et al.*, 2000). Several researchers have used it to model unsurfaced or thinly surfaced pavements.

Table 6.3 does no more than give a flavour of the number and variety of stress–strain models that have been developed to represent unbound material behaviour. However, the key is to make a choice that is appropriate to the problem being tackled – and in most instances that choice will be option 1.

6.3.4 Water content

The effect of water content on stiffness is broadly the same as that on strength, which was described in Section 6.2.5. At low water contents, suction can be present, increasing the stiffness, especially in materials with a significant silt or clay size fraction. But at high water contents, suction will be replaced by positive pore pressures as voids become filled with water, decreasing normal forces between particles and so allowing easier interparticle slip. Just as this decreases the shear strength, so it will also decrease the stiffness. In the case of clay soils with high levels of saturation, an increased water content means the soil particles are further apart and this too decreases stiffness as well as strength. Reported data for a number of different soils are collected in Figure 6.17 (Chu, 2023): it should be noted that even a 1% change in water content can make an appreciable difference to the modulus applying.

Figure 6.17 Relationships between soil water content and stiffness modulus (Chu, 2023)

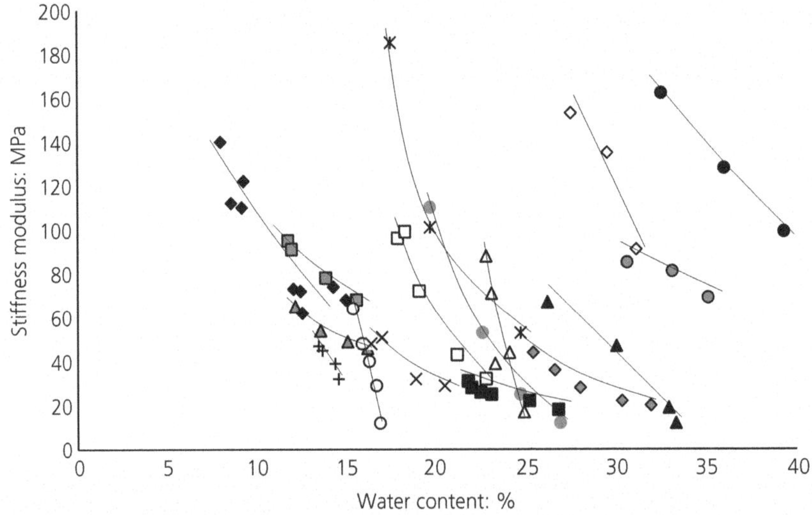

6.3.5 Particle shape and angularity

Despite the importance of particle angularity for shear strength it has very little effect on stiffness. This is because the strains that apply during resilient behaviour are extremely small, too small for awkward particle shapes to give any benefit. The shape factor that matters is particle surface roughness, which affects the friction at particle-to-particle contacts.

This is illustrated in Figure 6.18, taken from triaxial test data. Despite the scatter, there is a clear trend that materials with higher friction, notably steel slag and certain limestones in this case, display a higher modulus (expressed as an 'indicator' in Figure 6.18, as the data relate to tests under several different stress conditions). These findings have been supported by discrete-element modelling (DEM), a technique in which individual particles and the interactions between them are directly modelled. For example, DEM work at Nottingham University (unpublished) showed that a change in surface friction angle from 39° to 22° resulted in a 29% drop in overall stiffness modulus.

6.3.6 Particle size

Figure 6.19 shows sets of data on single-sized aggregates – but of different single sizes. The effect is clear, namely that the modulus decreases with decreasing particle size. The reasons are similar to those identified in relation to strength (see Section 6.2.7).

6.3.7 Parent material stiffness

No laboratory test data have been found in support of the effect of this variable, although it is essential to the explanation of granular material strain presented earlier. However, the same set of DEM results as quoted in relation to particle-to-particle friction showed that a 50% drop in particle stiffness modulus caused a 37% drop in the modulus of the system as a whole.

Figure 6.18 The relationship between stiffness and particle–particle friction – triaxial data (Thom and Brown, 1989)

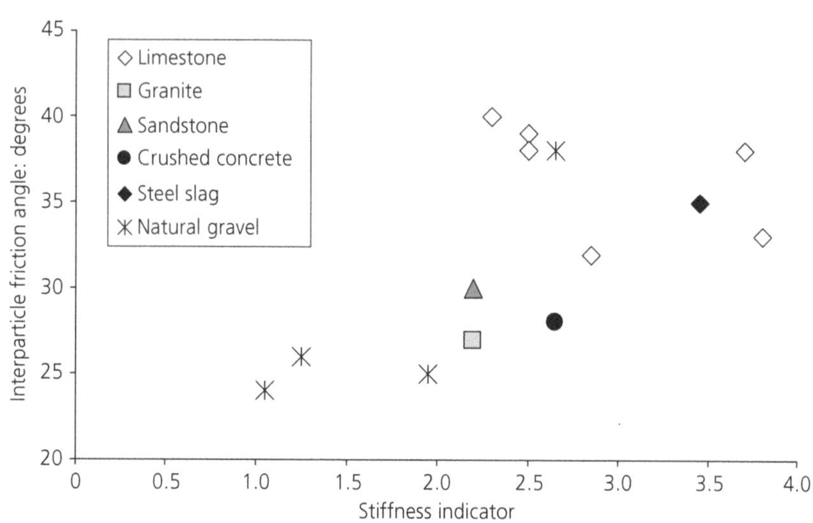

Figure 6.19 The effect of particle size on stiffness modulus – triaxial data (Thom, 1988)

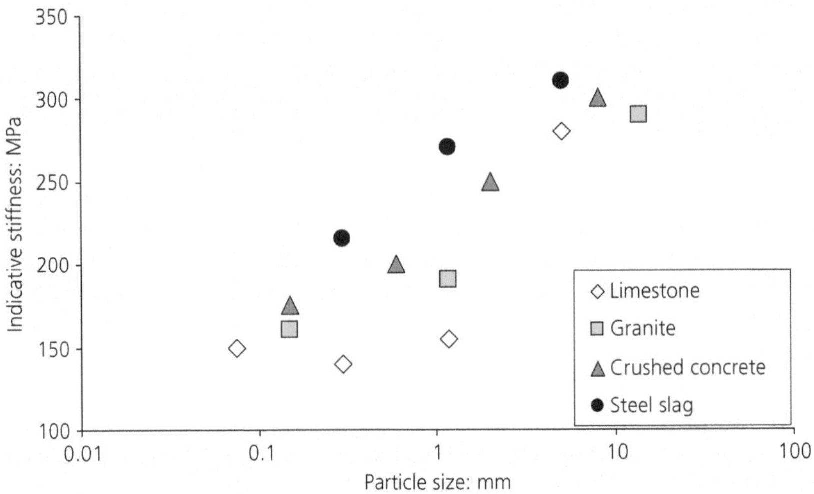

6.3.8 Particle size distribution

This is an interesting subject. It seems intuitively reasonable that a broadly graded material should have a higher stiffness than a single-sized material, particularly as it has already been noted that, up to a point, it has a higher shear strength. But, for dry materials, this is not actually the case. Figure 6.20 presents data obtained on a range of particle gradations (i.e. size distributions), from near single-sized to very broadly graded, all with a maximum particle size of 10 mm. The degree of 'broadness' of the gradation is expressed as a 'uniformity coefficient', the ratio of the sieve size

Figure 6.20 The effect of particle size distribution on stiffness – triaxial data (Thom and Brown, 1988)

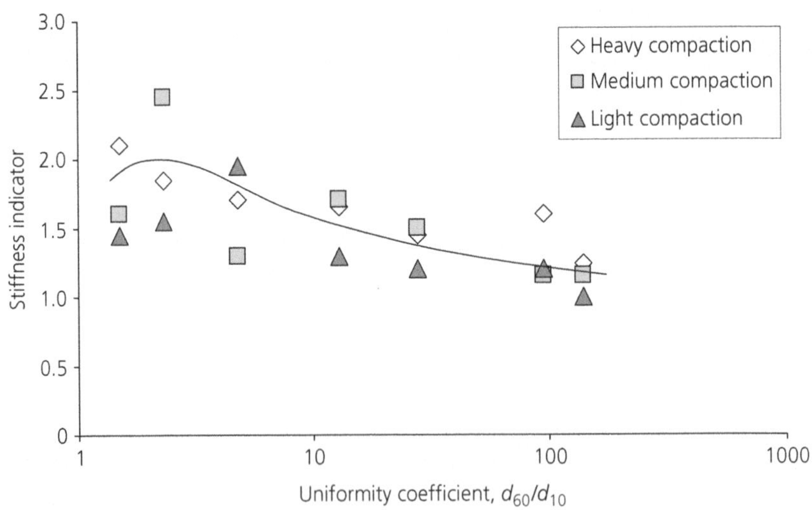

through which 60% of the material would pass to that through which just 10% would pass. Interestingly, the trend is for the more single-sized materials to be stiffer.

This unexpected outcome is, however, readily explicable. A broad particle gradation means an increased proportion of small particles and so a smaller critical particle size and, as just noted in Section 6.3.6, a smaller size gives a smaller modulus. However, this conclusion will change with the introduction of water if the water content is low enough to give rise to suction since smaller particles allow greater suction to develop (see Section 23.2). Furthermore, use of a broadly graded aggregate is generally dictated by the increased shear strength obtained.

6.3.9 Compaction

The efficiency of particle packing is affected not only by the particle size distribution but also by the amount of compaction given to a material, and most engineers would certainly expect that compaction affects stiffness. However, for a dry unbound aggregate they would be mistaken – as demonstrated in Figure 6.20. The packing efficiency has negligible effect on the stiffness of the overall aggregate matrix, presumably because the proportion of contacts that are readily able to slip when stress is applied remains almost unchanged. The potential magnitude of slip movement would logically be reduced by greater packing efficiency, but this does not seem to have any significant impact in the small-strain region relevant to stiffness. The consequence is, therefore, that the stiffness of a dry aggregate is almost unaffected by compaction. Naturally, compaction is still essential because of its effect on shear strength, but there will not necessarily be any measurable stiffness increase.

6.3.10 Stiffness tests

The triaxial test, introduced in Section 6.2.10 as a shear strength test, can also be used to measure stiffness, as long as the equipment allows repeated load to be applied. The most widely used specification for repeated load triaxial testing is given in AASHTO (2020), while BS EN 13286-7:2004 (BSI, 2004) is a European specification for a more sophisticated (and therefore less user-friendly) test. The AASHTO test measures platen movement only; the European test utilises on-sample instrumentation.

As shown in Figure 6.16, the stiffness on initial loading is not representative of that in subsequent load applications and, in fact, the stiffness tends to increase slightly as increasing numbers of loads are applied. This has led to specifications in which a large number (many thousands) of loading cycles have to be applied before the stiffness is measured, leading to a time-consuming test.

However, the main problem with the triaxial test is that it involves the use of a complex piece of equipment, and stiffness measurement demands the use of sensitive displacement measuring transducers, leading to a relatively high overall cost per test. The fact that horizontal confining stress is usually held constant also tends to result in unrealistically high stiffness moduli. Three alternatives are the K-mould (Semmelink and De Beer, 1995), Springbox (Edwards *et al.*, 2005) and PUMA (Thom *et al.*, 2012), illustrated in Figure 6.21, which might be described as 'confined compression' tests. These have been developed as tests suited to the establishment of a modulus value for pavement design purposes, and they share certain common features.

- The stress conditions are less well controlled than those in the triaxial test but they allow a form of loading that simulates much more closely a passing wheel load and they generate a stiffness modulus that is much more compatible with those measured in situ beneath a finished pavement (using the falling weight deflectometer – see Section 25.6.3).
- The tests are quicker and easier to set up and carry out than triaxial tests.

Figure 6.21 Confined compression tests

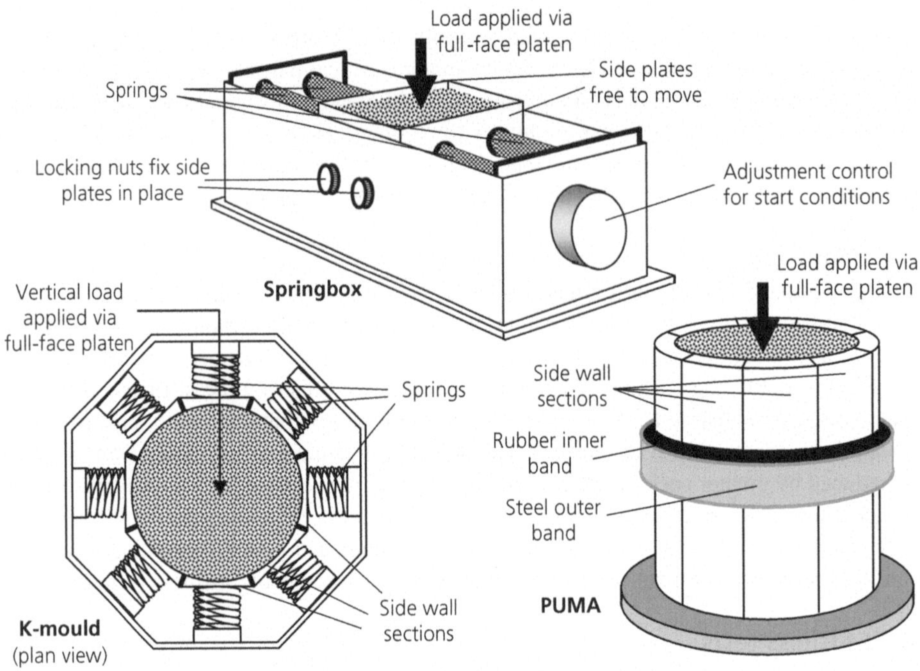

Confined compression tests therefore represent a practical compromise. The specimen sizes (150 mm diameter for the K-mould and PUMA; 170 mm cube for the standard Springbox) are intended to be suitable for most unbound base and sub-base materials. Figure 6.22 presents typical confined compression test data, in this case from the PUMA, illustrating that material stress

Figure 6.22 Typical confined compression test data (from Thom *et al.*, 2012)

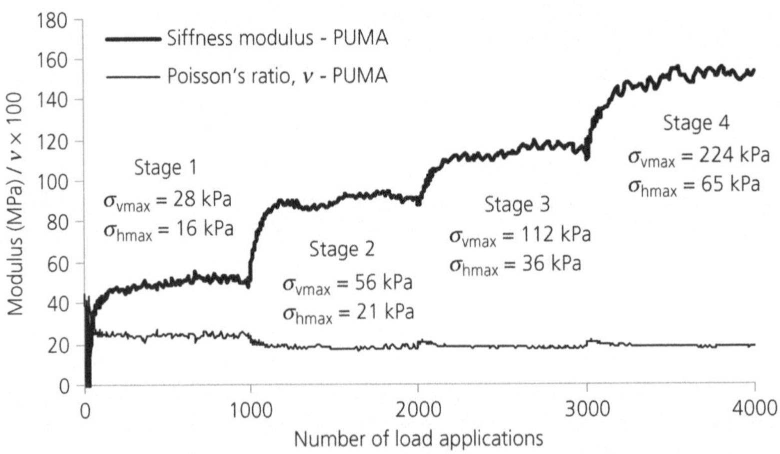

dependency can readily be explored. It is suggested that the highest stress level used in Figure 6.22 is approximately appropriate for generating a modulus for pavement design, while the lowest is more appropriate for prediction of the modulus likely to be measured in situ during construction (see below).

Calculation Sheet 6.3 predicts the difference expected between stiffness moduli deduced from triaxial testing and those deduced from confined compression tests, using a stress-dependent stiffness model for a crushed limestone sub-base aggregate.

Calculation Sheet 6.3. Differences between stiffness tests

Using the following stress–strain model:

Shear strain $\varepsilon_s = A\delta[\ln(\sigma_1/\sigma_3)]^B.\delta\sigma^C$

Volumetric strain (microstrain) $\varepsilon_v = D(\delta \ln p)^E.\delta p^F - G\{\delta[\ln(\sigma_1/\sigma_3)]^2\}^H$

where:

σ_1 and σ_3 are principal stresses (kPa)

p is the mean normal stress (kPa)

A, B, C, D, E, F, G and *H* are 75, 0.9, 0.33, 104, 0.85, 0.29, 117 and 1.0 respectively (parameters derived from tests on crushed limestone sub-base)

Triaxial: say cell pressure 20 kPa; deviator stress 0–100 kPa

$\rightarrow \varepsilon_s = 507\,\mu\varepsilon$; $\varepsilon_v = -93\,\mu\varepsilon$; $\varepsilon_{axial} = 307\,\mu\varepsilon$; $E = 326$ MPa

Confined compression: say vertical stress 0–100 kPa; horizontal stress 2–20 kPa

$\rightarrow \varepsilon_s = 521\,\mu\varepsilon$; $\varepsilon_v = 520\,\mu\varepsilon$; $\varepsilon_{axial} = 521\,\mu\varepsilon$; $E = 191$ MPa

As already alluded to, unbound material stiffness can also be measured in situ directly. The principal equipment used for this is the dynamic plate test (DPT), also known as the light weight deflectometer (LWD), several versions of which are available. Figure 6.23 illustrates the test. Interpretation generally ignores the dynamic nature of the test (i.e. the inertia effect due to acceleration and deceleration of the ground mass), and also has to assume that the material beneath the point tested is uniform, infinite and linear elastic in nature, all of which are far from the truth. Nevertheless, the test is a quick, practical method of obtaining a measure of the in situ stiffness of a pavement foundation.

Boussinesq's equations for stress, strain and displacement within a uniform half-space (Boussinesq, 1883) are enormously useful in a wide range of applications. The equations shown in Figure 6.23 are for the specific case of deflection measured immediately under the centre of the loaded area, achieved in the DPT/LWD by means of a velocity transducer sprung lightly against the

Figure 6.23 The dynamic plate test (DPT)/light weight deflectometer (LWD)

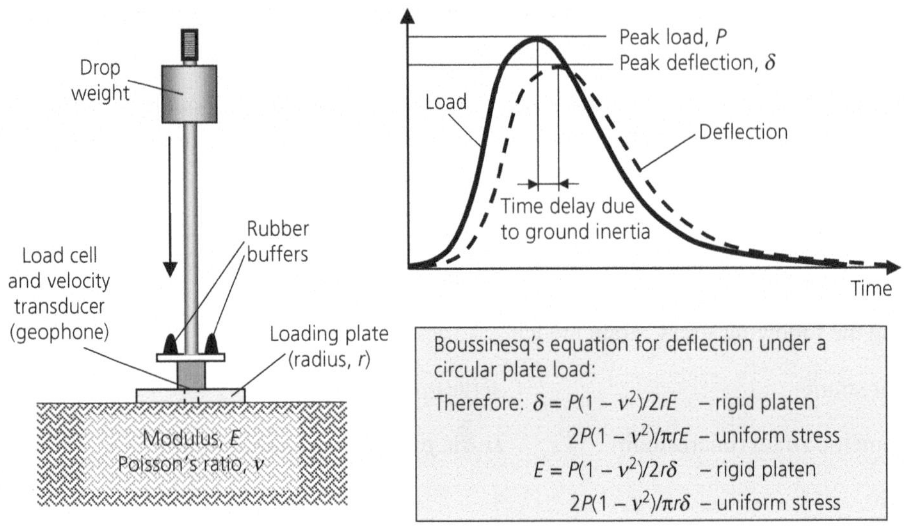

ground through a hole in the plate. Some devices measure plate deflection rather than ground deflection, and the resulting measurement is commonly significantly different because of the inevitably uneven contact between the plate and the surface, resulting in large local compressions. Plate movement is, therefore, generally greater than average ground movement, and reliance on measuring plate movement will underestimate the true ground stiffness.

When evaluating DPT/LWD data, it is important to realise the following.

- The test is of the *whole foundation*, not just the uppermost layer. The influence will decrease with depth but, in theory, all materials to a depth of about 1.5 m (the penetration of the pressure wave by the time the peak deflection is measured – typically after 10–15 ms) will influence the result to some extent.
- The calculated stiffness modulus will be lower than that pertaining once the upper pavement construction has been added, because of the lack of confinement during the test. For example, a test result of about 80 MPa may indicate an in-service modulus of at least 150 MPa. These figures match those in the UK National Highways standard CD225 (National Highways, 2020) for a high-quality unbound sub-base.
- Scatter is expected due to the impossibility of ensuring a perfect and repeatable contact between the plate and the ground. Decisions should therefore only be made on the basis of statistical treatment of the data. For example, the requirement in CD225 (National Highways, 2020) relates to the running mean of five consecutive tests, as well as the absolute minimum value.
- A decision has to be taken as to whether to treat the plate as rigid (e.g. solid steel) or flexible (e.g. if a rubber underlay is used) and, using the equations in Figure 6.23, the difference can be fairly significant.

A lower technology alternative to the DPT/LWD is the *plate loading test*, introduced in Section 6.1.2 as an alternative to the CBR test. However, it can also be used to derive an estimate of in situ modulus. Figure 6.24 illustrates a typical test result including two load applications. The slope of

Figure 6.24 Typical plate load test data

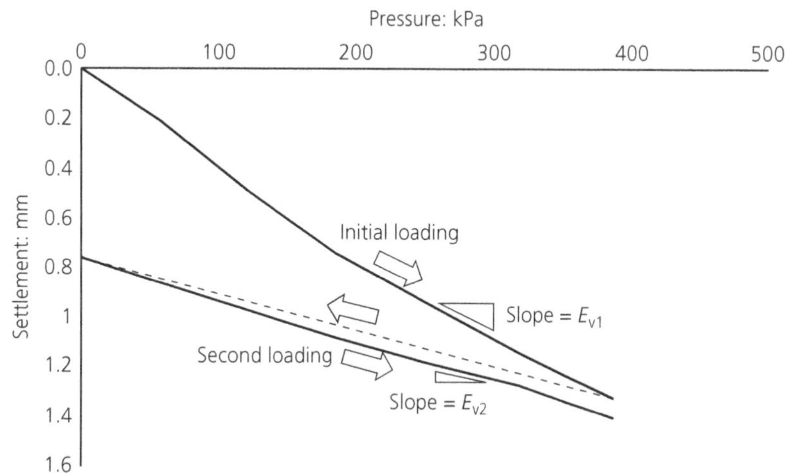

the initial loading line, E_{v1}, can be related to strength and this is what is sometimes converted to CBR (as in Section 6.1.2); the slope of the unload–reload line, E_{v2}, is commonly taken as a measure of ground stiffness. E_{v2} is generally termed the modulus of subgrade reaction and is given the symbol k (see Section 1.5). It is defined as the pressure required to induce unit deflection, and so has units of MN/m^3 although the plate size will affect the value obtained, as discussed in Section 6.1.2. Alternatively, it can be interpreted using Boussinesq in just the same way as a DPT/LWD.

6.3.11 Typical stiffness modulus values

As in the case of shear strength, this section is included purely in order to set into context any data obtained. In reality, materials vary greatly and, of course, are enormously affected by their moisture state. Nevertheless, the figures in Table 6.4 are believed to represent typical values.

Table 6.4 Typical stiffness modulus data

Material	Stiffness modulus: MPa		
	Triaxial (confining stress 20 kPa; deviator stress 0–100 kPa)	DPT/LWD (100 kPa contact pressure)	In the pavement (K-mould, Springbox and PUMA generally give similar results)
Very soft clay soil	10	5	15
Firm clay soil	50	30	80
Sandy soil	75	30	50
Gravel capping	125	50	80
Sub-base	250	75	150
Granular base	500	100	250

The principal reason for the discrepancies in the data in the table is the differing degree of confinement. However, confinement is much less influential for clay materials, whereas the magnitude of applied stress is much more so. The advice here is that both the triaxial test and the DPT/LWD should be interpreted with caution.

Mention should also be made of the widespread practice of converting CBR to stiffness. This is a practice that should fill the reader with horror, bearing in mind that the CBR is essentially a strength test and that stiffness and strength depend on such entirely different material features. Nevertheless, one should be aware of the following equations.

$$E(\text{in MPa}) = 10 \times \text{CBR} \tag{6.9}$$

$$E = 17.6 \times \text{CBR}^{0.64} \tag{6.10}$$

Equation 6.9 is internationally applied (e.g. AASHTO, 1993, 2007). Equation 6.10 is used particularly in the UK, having been proposed by the Transport Research Laboratory (Powell *et al.*, 1984). Thus, for example, if the 'very soft soil' in Table 6.4 had a CBR of 1.5%, Equations 6.9 and 6.10 would give moduli of 15 MPa and 22 MPa, respectively, while a 'firm clay soil' with a CBR of 5% would have predicted moduli of 50 MPa and 48 MPa. Clearly the order of magnitude is right – and so it should be, as the equations are based on experience. Nevertheless, this approach cannot be recommended unless all else fails – the level of uncertainty is much too great. Brown *et al.* (1987) report data showing a factor of over 3 between the highest and lowest moduli (24–80 MPa) for different soils of the same CBR (5%), tested under the same stress conditions.

Finally, in any linear elastic analysis a value of Poisson's ratio will also have to be selected, and it is common to adopt 0.40–0.45 for clay soils and 0.30–0.35 for sands and unbound pavement materials. For example, the default assumption for Poisson's ratio used in developing UK National Highways pavement designs is 0.35 for all unbound materials.

6.4. Deformation under repeated load

It is an unfortunate fact that all unbound materials deform progressively as more and more loads are applied, which means that there is a significant danger of a rut forming under repeated wheel loading. This form of behaviour was illustrated in Figure 6.16.

6.4.1 The mechanism of plastic deformation

The point has been made that normal straining of a granular matrix involves frictional slip between particles as well as compression at particle contacts. The energy stored as contacts compress is released once the load has moved on, and the material returns approximately to its original arrangement. It is also noted that the efficiency with which the material strains increases from one load application to the next (reducing the energy dissipated due to friction), and this implies a difference in the arrangement of particles after unloading compared with that applying before the load arrived. This, in turn, implies *non-recovered deformation* and, if possible, this important phenomenon is even more complex than that of stiffness. This type of deformation is often termed 'irrecoverable' or 'permanent', although these are not accurate descriptors, as it is always possible to reverse such strains if appropriate stress is applied. However, these strains are most certainly *plastic* rather than *resilient*.

6.4.2 Modelling plastic deformation

This is a subject on which numerous doctoral theses have been produced. Xiao *et al.* (2015) give a useful overview of modelling options. Here, the different types of behaviour that can occur will

Figure 6.25 Different forms of plastic strain accumulation behaviour

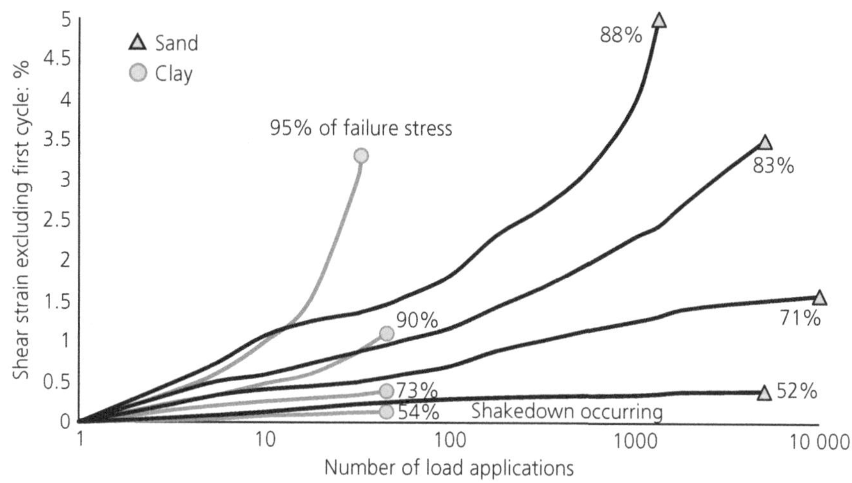

be illustrated and a possible model is presented. As illustrated in Figure 6.16, the tendency is for plastic strain to occur at an ever-decreasing rate as more cycles of load are applied. The effect is approximately logarithmic; that is, similar permanent strain occurs between cycles 10 and 100 as between cycles 100 and 1000 or 1000 and 10 000. However, many researchers have noted that, under appropriately low stress conditions, plastic strain can almost cease altogether. This is the phenomenon known as 'shakedown' and the stress below which this occurs is known as the 'shakedown limit'. On the other hand, it has also been observed that, under stress conditions closer to the failure stress, plastic strain accelerates towards failure. Figure 6.25 presents triaxial test data illustrating these different forms of behaviour, where the strain in the first cycle has been ignored because (*a*) it is somewhat variable and (*b*) in a real pavement the first cycle is effectively represented by layer compaction. The data in Figure 6.25 suggest, for these two materials at least, that if the applied stress is less than 50% of the failure stress then the materials will 'shake down' and strain will stop accumulating; at over about 75%, strain accelerates to failure.

The two examples in Figure 6.25 are from quite different soils, and the closeness in the data suggests that, to a first approximation, it may be possible to formulate a general deformation law in terms of the ratio of applied stress to failure stress. The following model (Thom, 1988) was originally derived from tests on unbound pavement materials and it is presented here because the author has found it extremely useful when developing predictive techniques for rutting in unsurfaced roads (see Chapter 7). The relevant equations are as follows.

The total plastic shear strain is

$$\varepsilon_{sp} = -A\ln[1-\sigma/\sigma_f - (\sigma/B)\ln(N)] \tag{6.11}$$

and the plastic shear strain *after* the first cycle is

$$\varepsilon_{sp} = -A\{\ln[1-\sigma/\sigma_f - (\sigma/B)\ln(N)] - \ln[1-\sigma/\sigma_f]\} \tag{6.12}$$

where σ is the applied stress, σ_f is the stress at failure, A and B are constants and N is the number of load applications.

Equations 6.11 and 6.12 have the expected ln(N) term as well as the stress ratio σ/σ_f, but also contain a further σ term, which means that it isn't just the stress ratio that matters but also the absolute magnitude of stress. In fact, it is clear that the strain in the clay is consistently a little less than that of the sand for a given stress ratio and this is at least partly because the clay was weaker (lower σ_f) and so a lower absolute stress σ achieved the same stress ratio.

Figure 6.26 shows the goodness of fit between predictions using the above equations and the test data with $A = 1.7\%$, B = 1500 kPa and $\sigma_f = 20$ kPa for the clay and 30 kPa for the sand, remembering that the data are for strain following the first cycle (i.e. ε_{sp} at N less ε_{sp} at $N = 1$). While the match is obviously not perfect, this is about as good as could possibly be expected given the uncertainties involved (e.g. in the actual failure stress for an individual test specimen). In the case of a crushed rock pavement layer then, following Thom (1988), the A value should probably be rather lower and the B value somewhat higher, but for conservative prediction it is suggested that the same values should be used as for soil. Use of this model in practice is described in Chapter 7.

6.4.3 The effect of material variables

In the model presented in the previous section, resistance to plastic strain under repeated load is a function of the shear stress ratio and so is directly affected by the shear strength (σ_f). This means that a large particle size, angular particles and a broad particle size distribution all contribute – which of course means that a crushed rock will be predicted to have much smaller deformation than a soil under the same stress conditions, despite using the same predictive equation with the same values for the constants. The water content will also influence behaviour, as it significantly influences shear strength.

6.4.4 Tests for plastic strain under repeated load

The appropriate tests here are the same as those already described for stiffness measurement, namely the triaxial, Springbox, K-mould and PUMA tests. However, only the triaxial test and

Figure 6.26 Deformation prediction using the proposed model

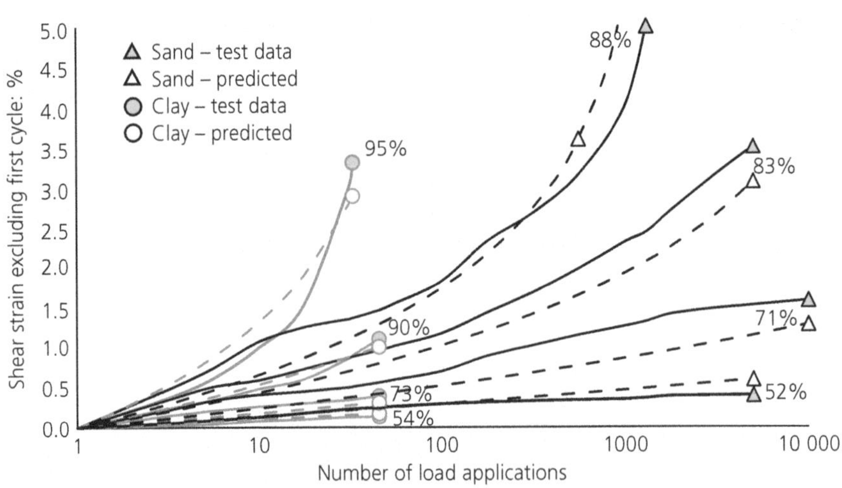

other complex research equipment such as the hollow cylinder apparatus have sufficiently well-controlled stress conditions to enable derivation of model parameters. The hollow cylinder apparatus is a useful research tool that allows the combinations of stress under a moving wheel to be reproduced, but it is much too time-consuming for general use. The K-mould, Springbox and PUMA tests can give a relative idea of performance under reasonably realistic stress conditions, but this then needs to be calibrated against performance in the field in order to be interpreted.

6.5. Summary

The foregoing sections have given the (correct) impression that unbound materials are extremely complex, both in terms of their mechanical properties and also with regard to the effect of water. However, for practical engineering purposes, it is important to remember a few key points.

- All unbound materials are stress dependent. This means that the strength or stiffness measured under one set of conditions will differ from those measured under a different set of conditions. Care is therefore needed when assessing properties for pavement design.
- Aggregates with large particle size tend to be stronger, stiffer and more resistant to deformation. Nevertheless, it is beneficial to increase the fines content (reducing the critical particle size) up to a point, because of the increased potential for suction to develop.
- Water is almost always the principal cause of problems in unbound materials, and pavement design therefore has to recognise this and take measures (i.e. drainage (see Chapters 23 and 24)) that limit water content to within a suitable range.

REFERENCES

AASHTO (American Association of State Highway and Transportation Officials) (1993) *Guide for Design of Pavement Structures*. AASHTO, Washington, DC, USA.

AASHTO (2007) *Guide for Mechanistic-Empirical Design of New and Rehabilitated Pavement Structures*. AASHTO, Washington, DC, USA.

AASHTO (2019) T96-02: Resistance to degradation of small-size coarse aggregate by abrasion and impact in the Los Angeles machine. AASHTO, Washington, DC, USA.

AASHTO (2020) T307-99: Standard method of test for determining the resilient modulus of soils and aggregate materials. AASHTO, Washington, DC, USA.

AASHTO (2022a) T327-22: Standard method of test for resistance of coarse aggregate to degradation by abrasion in the micro-Deval apparatus. AASHTO, Washington, DC, USA.

AASHTO (2022b) T104: Standard method of test for soundness of aggregate by use of sodium sulfate or magnesium sulfate. AASHTO, Washington, DC, USA.

AASHTO (2022c) T103: Standard method of test for soundness of aggregates by freezing and thawing. AASHTO, Washington, DC, USA.

AASHTO (2022d) T90: Standard method of test for determining the plastic limit and plasticity index of soils. AASHTO, Washington, DC, USA.

AASHTO (2022e) T89: Standard method of test for determining the liquid limit of soils. AASHTO, Washington, DC, USA.

ASTM (2016) C535-16: Standard test method for resistance to degradation of large-size coarse aggregate by abrasion and impact in the Los Angeles machine. ASTM International, West Conshohocken, PA, USA.

ASTM (2017) D6928-17: Standard test methods for resistance of coarse aggregate to degradation by abrasion in the micro-Deval apparatus. ASTM International, West Conshohocken, PA, USA.

ASTM (2018) D4318-18: Standard test methods for liquid limit, plastic limit and plasticity index of soils. ASTM International, West Conshohocken, PA, USA.

ASTM (2020) C131/C131M-20: Standard test method for resistance to degradation of small-size coarse aggregate by abrasion and impact in the Los Angeles machine. ASTM International, West Conshohocken, PA, USA.

Boussinesq J (1883) *Application des Potentials à l'Étude de l'Équilibre et du Mouvement des Solides Élastiques*. Gautier-Villars, Paris, France (in French).

Boyce JR (1980) A non-linear model for the elastic behaviour of granular materials under repeated loading. *Proceedings of an International Symposium on Soils under Cyclic and Transient Loading*. Balkema, Rotterdam, the Netherlands, pp. 149–167.

Brown SF, Loach SC and O'Reilly MP (1987) *Repeated Loading of Fine Grained Soils*. Transport and Road Research Laboratory, Crowthorne, UK, Report CR72.

BSI (2004) BS EN 13286-7:2004: Unbound and hydraulically bound mixtures. Cyclic load triaxial test for unbound mixtures. BSI, London, UK.

BSI (2009) BS EN 1367-2:2009: Tests for thermal and weathering properties of aggregates. Magnesium sulfate test. BSI, London, UK.

BSI (2011) BS EN 1097-1:2011: Tests for mechanical and physical properties of aggregates. Determination of the resistance to wear (micro-Deval). BSI, London, UK.

BSI (2012) BS EN 13286-47:2012: Unbound and hydraulically bound mixtures. Test method for the determination of California bearing ratio, immediate bearing index and linear swelling. BSI, London, UK.

BSI (2018) BS-EN 17892-12: Geotechnical investigation and testing. Laboratory testing of soil. Determination of liquid and plastic limits, BSI, London, UK.

BSI (2020) BS EN 1097-2:2020: Tests for mechanical and physical properties of aggregates. Methods for the determination of resistance to fragmentation. BSI, London, UK.

Chu X (2023) *Road and Railway Foundation Response to Moisture Content and Drainage*. PhD thesis, University of Nottingham, Nottingham, UK.

Day JBA (1981) Proof testing of unbound layers. *Proceedings of the 1st Symposium on Unbound Aggregates in Roads, Nottingham, UK*, pp. 15–20.

Edwards JP, Thom NH, Fleming PR and Williams J (2005) Accelerated laboratory based mechanistic testing of unbound materials within the newly developed NAT Springbox. *Transportation Research Record* **1913**: 32–40.

Fuller WB and Thompson SE (1907) The laws of proportioning concrete. *Transactions of the American Society of Civil Engineers* **59(2)**: 67–143.

Hertz H (1895) Über die Berührung fester elastischer Körper. In *Gesammelte Werke* (Leonard P and Barth JA (eds)). JA Bart, Leipzig, Germany, vol. 1, pp. 155–173 (in German).

Hicks RG and Monismith CL (1971) Factors influencing the resilient response of granular materials. *Highway Research Record* **345**: 15–31.

Hornych P, Karzai A and Quibel A (2000) Modelling of a full scale experiment of two flexible pavement structures with unbound granular bases. *Proceedings of the 4th International Conference on Bearing Capacity of Roads and Airfields*. Balkema, Rotterdam, the Netherlands, Vol. 3, pp. 1277–1287.

National Highways (2020) CD 225: Design for new pavement foundations. *Design Manual for Roads and Bridges. Pavement Design*. Stationery Office, London, UK.

National Highways (2021) *Manual of Contract Documents for Highway works. Volume 1: Specification for Highway Works. Series 800: Road Pavements – Unbound, Cement and Other Hydraulically Bound Mixtures*. Stationery Office, London, UK.

Powell WD, Potter JF, Mayhew HC and Nunn ME (1984) *The Structural Design of Bituminous Roads*. Transport and Road Research Laboratory, Crowthorne, UK, Report LR1132.

Semmelink CJ and De Beer M (1995) Rapid determination of elastic and shear properties of road-building materials with the K-mould. In *Unbound Aggregates in Roads. Proceedings of the 4th International Symposium on Unbound Aggregates in Roads, University of Nottingham, Nottingham, UK*, pp. 151–161.

Thom NH (1988) *Design of Road Foundations*. PhD thesis, University of Nottingham, Nottingham, UK.

Thom NH and Brown SF (1988) The effect of grading and density on the mechanical properties of crushed dolomitic limestone. *Proceedings of the 14th Australian Road Research Board Conference, Canberra, Australia*.

Thom NH and Brown SF (1989) The mechanical properties of unbound aggregates from various sources. In *Unbound Aggregates in Roads* (Jones RH and Dawson AR (eds)). Butterworths, Sevenoaks, UK, pp. 130–147.

Thom NH, Cooper C, Grafton P *et al.* (2012) A new test for base material characterisation. *Proceedings of International Symposium on Heavy Duty Asphalt Pavements and Bridge Deck Pavements*. International Society for Asphalt Pavements, Nanjing, China, Paper 12-097.

TRL (Transport Research Laboratory) (1993) *A Guide to the Structural Design of Bitumen-surfaced Roads in Tropical and Sub-tropical Countries*, 4th edn. TRL, Crowthorne, UK, Overseas Road Note 31.

Xiao Y, Tutumluer E and Mishra D (2015) Performance evaluations of unbound aggregate permanent deformation models for various aggregate physical properties. *Transportation Research Record* **2525**: 20–30.

Nick Thom
ISBN 978-1-83549-711-1
https://doi.org/10.1108/978-1-83549-710-420243008

Chapter 7
Unbound pavement – structural design

An unbound pavement has essentially a single possible failure mode – deformation. Deformation can originate in the subgrade or in the pavement layers themselves. Thus, the design task is to

- evaluate the amount of expected traffic loading
- evaluate the relevant properties of the subgrade
- select a combination of layers that adequately protect the subgrade without themselves deforming excessively.

However, the first step in any design process is to appreciate the practical constraints applying.

7.1. Construction practicalities

This book is not a construction manual. Nevertheless, it is important to understand some of the practicalities of material production and pavement construction if sensible pavement design decisions are to be taken, and this section gives a brief overview in relation to unbound materials.

7.1.1 Subgrade soils

The subgrade is the most vulnerable element of any pavement, and it needs protecting at all costs. Poor construction practice can easily turn a basically sound material into a muddy soup. It is therefore good practice on any site to avoid disturbance to the pavement subgrade, leaving a thin layer (e.g. 100 mm) of overlying material until the very last moment, thus protecting the subgrade soil from the effects of both weather and construction traffic. Only when the contractor is ready to place the first of the pavement layers, usually an unbound granular layer, should the actual top of the subgrade be exposed.

7.1.2 Unbound pavement layers

Some sands and gravels can be used 'as dug' from the ground, without the need to alter the proportions in any way; however, these materials are unlikely to be suitable for high-quality sub-base or base applications. Most unbound layers will be formed of crushed rock or suitable recycled material and will often have been blended from different size streams. The factors affecting the strength of an unbound material have been discussed in Chapter 6, where it was shown that particle properties and size distribution play major roles in determining the strength and stiffness of the material as a whole. However, the maximum particle size is also constrained by the thickness of the layer being constructed, as roller compaction is only effective up to a depth of around 250 mm and materials become very difficult to work if stones larger than about one third of the layer thickness are present. Actual material specifications can be found elsewhere (AASHTO, 2021; ASTM, 2020; National Highways, 2021).

7.1.3 Water content

As noted already, unbound materials are sensitive to water. If the amount of water is relatively low, it will tend to accumulate at contact points between particles because of surface tension effects at the air–water interface. This results in a difference in pressure between the atmospheric pressure of the air and the pressure within the water, the water pressure being slightly lower, and this is the effect known as *negative pore pressure* or *suction*. This will induce a small additional compressive force between particles (increasing what is known as *effective stress*), making it harder for one particle to slip over another and, once in the pavement, this is a highly desirable state. Unfortunately, it is a highly undesirable state during compaction, as it makes it more difficult to rearrange the particles into a dense packing.

If the water content is increased, suction is reduced, as some of the voids become totally filled with water. These water-filled voids then have the potential to develop *positive pore pressure* because, when load is applied, the water will resist any reduction in the volume of the void. Positive pore pressure reduces the compressive forces at particle contacts (reducing effective stress), and therefore makes it easier for one particle to slip over another. In an in-service pavement this is a most undesirable condition, as easier interparticle movement means lower strength and stiffness. However, during compaction it is most desirable, as particle rearrangement becomes much easier. Figure 7.1 shows a schematic diagram of the influence of water content on the density achieved during compaction; the *optimum water content* is the condition in which compaction is easiest and the highest particle packing density (the density discounting water, known as *dry density*) is reached. At even higher water contents there is simply no room for all the water in the mixture without a reduction in density. As illustrated in Chapter 6, maximising density is critical in maximising material strength and reducing deformation.

The most common laboratory compaction tool worldwide is the Proctor hammer (BSI, 2010) – a 2.5 kg weight, 50 mm dia. steel cylinder that is allowed to drop repeatedly through a height of

Figure 7.1 The effect of water content on achieved density

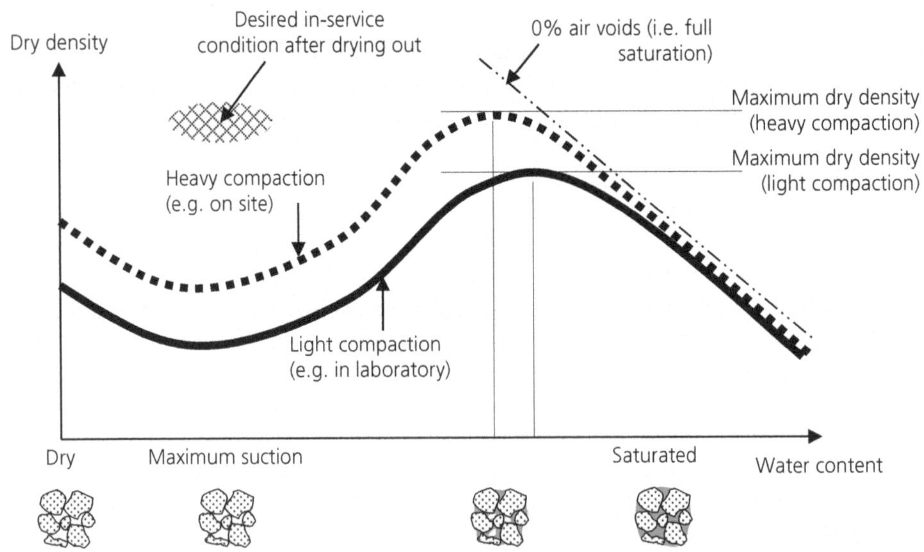

305 mm. A standard number of drops (e.g. 25) is specified for each compacted layer. Alternatives are the heavy Proctor hammer (4.5 kg weight, 457 mm drop height (BSI, 2010)) or vibrating hammer (BSI, 2021). The aims are always (*a*) to simulate site conditions, (*b*) to obtain repeatable measurements and (*c*) to achieve these quickly and cheaply. It seems to be generally acknowledged that both the heavy Proctor and the vibrating hammer are capable of approaching in situ compaction levels. The outputs from a laboratory compaction trial are the optimum water content (OWC) and maximum dry density (MDD). Density on site, which can be measured either by a sand replacement technique or a density meter (BSI, 1990; ASTM, 2016, 2023), is commonly specified in terms of percentage relative to MDD.

A proper appreciation of these water-related effects is crucial if a high-quality unbound layer is to be achieved, and this means that the following procedure is needed.

- Supply material at OWC (or slightly above to allow for evaporation during handling).
- Compact at OWC.
- Allow to dry out as drainage and evaporation take place, maximising suction.

7.1.4 Placement and compaction

The method of placement depends on the degree of level control required. Unbound material can be tipped and spread with a blade; the finish can then be improved by means of a grader – a machine with a blade set at an angle to the direction of travel. However, for unbound bases where level control is essential to avoid problems in achieving tolerances in overlying bituminous layers, an asphalt paver can also be used (Figure 7.2).

Compaction is achieved by repeated stressing of the layer, usually by means of a roller. Vibratory compactors allow very rapid load repetition, leading to rapid densification. However, the stress level applied is important, and this places a limit on the thickness of layer that can be effectively compacted by any given equipment. In general, 250 mm is a realistic limit; anything greater than

Figure 7.2 Equipment for placement of unbound layers

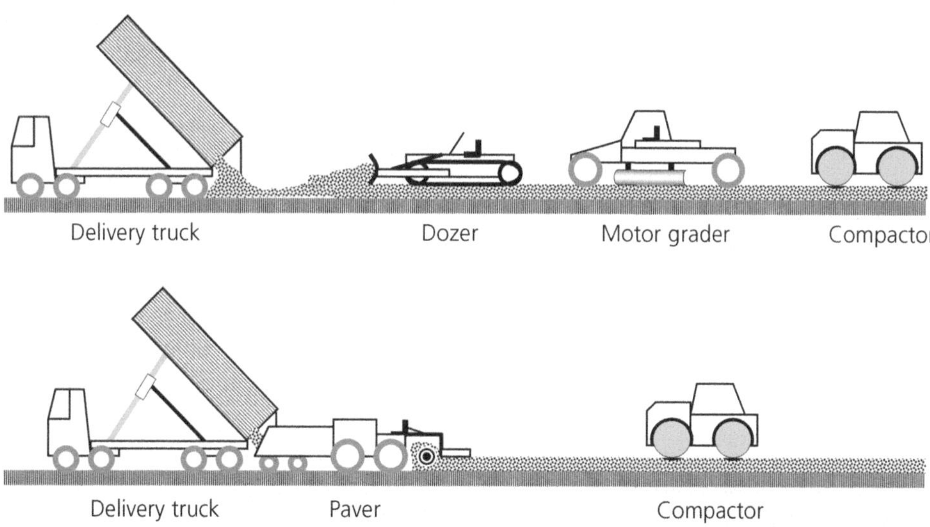

this and the bottom of the layer will tend to be under-compacted. For example, the usual UK limit (National Highways, 2021) for compaction of unbound materials is 225 mm. The minimum realistic thickness is about three times the maximum particle size, although for all materials there will be a practical lower limit, often around 100 mm; a function of economics and also of the tolerances allowed.

'Tolerance' describes the closeness that the contractor has to achieve to the specified level since perfection is impossible. In principle, the specified tolerance is there for three reasons.

- The finished pavement surface has to be acceptably even.
- Variable pavement deterioration (e.g. rutting) is undesirable.
- Overlying materials should not be placed too thin or the design assumptions will not be realised.

Naturally, the actual tolerances permitted will be a function of the pavement type, design traffic speed and so on. They should be relatively generous at the bottom of the pavement, becoming tighter towards the top. Typically $+20$ to -50 mm might be allowed at top of subgrade level, reducing to ± 10 mm or less at the surface.

7.2. Current design practice

Several design standards include unbound pavement designs for low traffic levels, of which the following are examples.

The Austroads guide (Austroads, 2021) specifically gives design thicknesses for unbound bases overlying different strength (expressed as the CBR) substrates for traffic levels up to 100 million standard axles (msa), assuming a sealed surface (see Section 7.6).

The South African pavement engineering manual (SANRAL, 2014) gives a highly scientific set of guidance, including consideration of the different properties of a range of material options, all in the context of a so-called 'mechanistic-empirical' design approach. This allows plenty of flexibility but does demand a certain amount of understanding from the user. Alternatively, the South African standard TRH4 (DoT, 1996) presents an easy-to-use menu of design choices for sealed roads, but only up to a traffic of 3 msa. A useful feature is that designs are included for both wet and dry climates.

For advice on the design of unsurfaced roads (i.e. without any seal), the US Transportation Research Board provides a useful section on design thicknesses for different material types (TRB, 1981).

7.3. Design against deformation

Permanent, plastic deformation under repeated load is a fact of life for all unbound materials. The trick is to make sure that the strains involved (and therefore ruts in the pavement surface) are small enough not to matter. Chapter 6 introduced typical forms of behaviour and some of the tools available to simulate and predict different aspects of performance. The main task of this section is to make use of these tools in design.

7.3.1 The subgrade strain criterion

Most current analytical pavement design methods, surfaced and unsurfaced, use the so-called 'subgrade strain criterion', where subgrade strain means the vertical elastic (more properly resilient) strain calculated at the top of the subgrade under the design wheel load. This represents a massive simplification – and relies on a massive assumption. The assumption is that the strength of a soil is directly related to its stiffness. That this is not generally true of unbound materials has already been shown in Chapter 6, because strength depends principally on parameters such as particle shape and voids content, whereas stiffness depends more on the frictional properties at particle contact points. Nevertheless, over a limited range of soils it may be close enough to being true to be usable. The argument goes

- life (to a limiting rut depth) = *f*(stress/strength ratio) – as in Figure 6.26
- if: strength = *f*(modulus)...
- ...then: life = *f*(stress/modulus ratio) = *f*(elastic strain).

The great advantage of this assumption is that it is only necessary to calculate the elastic strain value where it is greatest, which is at the top of the subgrade.

Various relationships between subgrade elastic strain and 'life' have been proposed in numerous different design manuals. In most cases, they are based on real evidence of performance, and so, within the range of subgrades and pavements encountered in gathering that evidence, they should prove to be practical.

For example, the equation proposed by the UK Transport Research Laboratory (Powell *et al.*, 1984)

$$N_f \text{ (in millions)} = 3.09\times10^{10}\varepsilon_z^{-3.95} \quad (\varepsilon_z \text{ in microstrain}) \tag{7.1}$$

is likely to be approximately suitable for roads on UK clay soils.

Similarly, the equation adopted by the National Association of Australian State Road Authorities (NAASRA, 1987)

$$N_f \text{ (in millions)} = 8.511\times10^{-3}\varepsilon_z^{-7.14} \quad (\varepsilon_z \text{ in millistrain}) \tag{7.2}$$

is likely to be suited to typical Australian soils and Australian pavement structures.

And the equation provided by the British Airports Authority (BAA, 1993, following Brown *et al.*, 1985)

$$N_f \text{ (in millions)} = 7.6\times10^{8}\varepsilon_z^{-3.7} \quad \text{(in microstrain)} \tag{7.3}$$

has been calibrated to give a sensible prediction for UK airfields.

It should also be appreciated that the calibrations for all these have been for asphalt-surfaced pavements and that the 'life' predicted relates to subgrade-controlled deformation. In the case of an unsurfaced pavement, the prediction will only be meaningful if internal deformation within the pavement layers themselves is minimal. On the other hand, on an unsurfaced road a larger rut is probably acceptable than the 20 mm typically taken as a limit on asphalt-surfaced roads, and so the life before subgrade-controlled rutting failure should be substantially larger than these equations would predict. If, for example, the subgrade strain under the design load was calculated as

800 microstrain, the predictions from Equations 7.1, 7.2 and 7.3 would be for lives of 105 000, 42 000 and 14 000 load applications, respectively; but the acceptable life for an unsurfaced road might be more than the 105 000 given by Equation 7.1.

In this book, the decision was taken not to specifically recommend any of the many available equations relating subgrade strain and life; each is geared to the particular conditions applying over a range of pavements carrying specific types of traffic in a specific climate over a specific range of soils. It is a reasonable course of action to use them in design, but it is not a reasonable course of action to believe them unquestioningly. Thought should always be given to the actual type of soil that applies. It should be appreciated, for example, that sandy soils will tend to have a quite different relationship between strength and stiffness modulus from those of clay soils, the effect of which is that they can carry many times the number of load applications – for a given level of computed elastic strain. For example, this author has seen evidence for using a factor of 10 increase in life (at a given calculated strain level) for roads on sandy Middle Eastern soils compared with pavements on UK clay soils.

But the question now arises: how to calculate the subgrade strain?

7.3.2 Multi-layer linear elastic analysis

This method of analysis forms the backbone to analytical pavement design and underlies many of the design methods in use around the world for all types of pavement. However, this book will not delve into the actual techniques by which the analysis is performed – it is sufficient to be aware that there are a number of computer programs on the market, all of which are capable of solving the problem. Widely used pavement analysis systems that incorporate multi-layer linear elastic analysis include BISAR, CIRCLY, JULEA, ELSYM, CHEVPC, WESLEA and KENLAYER. Some also include other functions or options. For example, KENLAYER allows a stress-dependent modulus based on the k–θ model (see Section 6.3.3) to be used, suited to unbound materials. The solutions are not exact, but they are close enough to be accepted as such for all practical purposes.

Multi-layer linear elastic analysis assumes

- layers are composed of homogeneous linear elastic material
- the horizontal extent is infinite
- the layer thickness is uniform
- load of uniform pressure is applied over a circular contact area.

In most analyses, perfect bond is assumed between layers, although some programs allow this to be varied. Similarly, most analyses consider only vertical load, but some programs give the option of horizontal load also. As the system is purely linear elastic, it is always possible to simply add up the effects of different loads, and this allows computation of behaviour under a wheel group, or to approximate a real non-circular load contact zone using a series of circular loads. In fact, as long as the idealised nature of the system is appreciated, it is possible to model a pavement structure reasonably effectively. The input data required are

- the magnitude and contact area of all applied loads
- the elastic modulus and Poisson's ratio of all layers
- the thickness of all layers
- the bond condition (slip stiffness) between layers – usually considered infinite.

Therefore, returning to the problem of predicting deformation behaviour in a subgrade, it is necessary to estimate the elastic modulus and Poisson's ratio for all pavement layers, including the subgrade itself, and then to undertake the analysis under the action of the design load (standard axle, design aircraft wheel etc.). The program will calculate any desired stress, strain or deformation within the entire multi-layer structure and, for purposes of this design element, the outputs required are the stresses and strains at the top of the subgrade.

7.3.3 Calculating closeness to failure

The subgrade strain criterion, although approximate in nature, is convenient when subgrade strength details are not known. However, in high-value pavements it may be worth taking a more rigorous approach. The traffic-imposed stresses at the top of the subgrade calculated using multi-layer linear elastic analysis, together with any additional stress due to overburden, can be expressed by means of a Mohr circle, and then compared with a second Mohr circle representing the stress condition at failure. Calculation Sheet 7.1 outlines the procedure.

Calculation Sheet 7.1. Defining stress ratio at top of subgrade

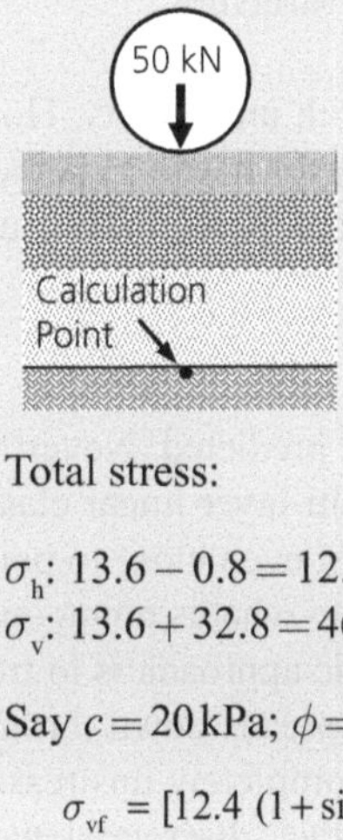

Overburden stress

$= (0.1 \times 2100 + 0.2 \times 2200 + 0.35 \times 2100) \times 9.81 = 13587\,\text{Pa} = 13.6\,\text{kPa}$

Traffic-imposed stress (from computer analysis)

$\sigma_h = -0.8\,\text{kPa}$ (tensile)
$\sigma_v = +32.8\,\text{kPa}$

Total stress:

σ_h: $13.6 - 0.8 = 12.4\,\text{kPa}$
σ_v: $13.6 + 32.8 = 46.4\,\text{kPa}$

Say $c = 20\,\text{kPa}$; $\phi = 10°$

$$\sigma_{vf} = [12.4\,(1+\sin 10°) + 2\times 20\cos 10°]/(1-\sin 10°)$$
$$= 65.3\text{ kPa}$$

Stress ratio:
$\sigma/\sigma_f = 46.4/65.3 = 0.71$
(i.e. 71%)

Using Equation 6.12 in Section 6.4.2:

Plastic shear strain after the first cycle:

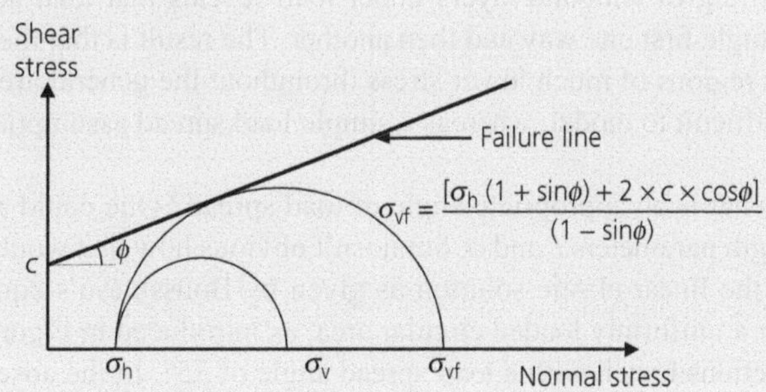

$$\varepsilon_{sp} = -A\{\ln[1 - \sigma/\sigma_f - (\sigma/B)\ln(N)] - \ln[1 - \sigma/\sigma_f]\}$$

where σ is the applied stress (46.4 kPa), σ_f is the stress at failure (65.3 kPa), A and B are constants (Section 6.4.2 proposed $A = 1.7\%$ and $B = 1500$ kPa) and $N =$ number of load applications.

Therefore:

@ 1000 load cycles, strain = 2.3%

@ 10000 load cycles, strain = 7.1%

The approach in Calculation Sheet 7.1 may give a reasonable prediction – but it definitely does not represent a safe design. The consequences of subgrade failure are so serious that there really ought to be a significant factor of safety built in, covering the fact that not all soils will behave as the two shown in Section 6.4.2, estimation of soil strength parameters is not an exact science and none of them are linear elastic. For example, if 7.1% is considered as failure and if 10 000 load applications are required, for which Calculation Sheet 7.1 gives a σ/σ_f ratio of 71%, this should probably be reduced to no more than 60% for confident design, requiring a rather thicker structure.

It may appear that this method of prediction is both complex and fraught with uncertainty. However, multi-layer linear elastic analysis is nowadays a readily available and user-friendly method. The chief difficulty undoubtedly lies in estimating the relevant soil parameters, particularly c and ϕ, making allowance for the long-term condition.

7.3.4 Load spread angle calculations

Pavement deformation is a complex phenomenon, which means it is rarely predicted. Nevertheless, this section will push the boundaries of prediction as far as it can. Multi-layer linear elastic calculation is an excellent tool; the problem is that unbound materials are not even close to being linear elastic. Furthermore, the subgrade strain criterion doesn't really relate to what is going on in all the different pavement layers. An alternative and arguably no less realistic approach is to treat unbound materials as what they really are – piles of stones. The load then spreads down through an ever-widening pyramid of particles, leaving those outside the pyramid completely unstressed. It's not 100% true, but it's no worse than the linear elastic assumption. In fact, discrete element modelling of unbound layers under load reveals that load actually spreads along discrete paths that angle first one way and then another. The result is that there are zones of high stress in among other regions of much lower stress throughout the general area under the load. In short, it's much too difficult to model, whereas a simple load spread assumption is at least possible to work with.

But what is an appropriate angle of load spread? One could argue that it should be related to the strength parameters c and ϕ, but it isn't obvious how that would work. An alternative is to compare with the linear elastic solution as given by Boussinesq's equation for deflection of a half-space under a uniformly loaded circular area, as introduced in Figure 6.23, and the result is that the two deflections match with a load spread angle of 35°. In the absence of any definitive data, it is suggested that a load spread angle of 35° is a sensible assumption.

This simple approach allows the approximate vertical stress beneath the loaded area to be calculated at each depth. This can then be compared with the ultimate stress required to overcome the overburden stress just outside the loaded zone. This stress ratio (applied stress/ultimate stress) can then be inserted into a permanent deformation prediction equation, such as that shown in Calculation Sheet 7.1.

For practical purposes, the pavement should be divided up into thin layers, each with its own vertical stress, its own stress ratio and so its own plastic strain as a function of the number of load applications (N). Calculation Sheet 7.2 outlines the method while Figure 7.3 presents example predictions using this approach.

Calculation Sheet 7.2. Rut prediction: load spread angle method

σ_o is the overburden stress due to self-weight

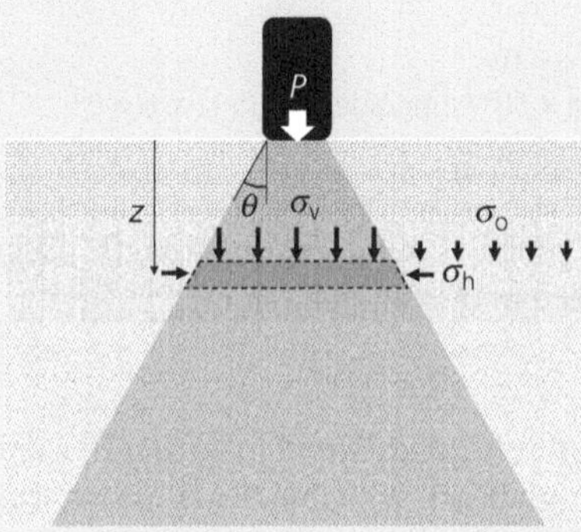

σ_{hmax} is calculated from the strength parameters of the relevant material, c and ϕ:

$$\sigma_{hmax} = [\sigma_o(1+\sin\phi)+2c\cos\phi]/(1-\sin\phi)$$

σ_{vmax} is calculated in a similar way:

$$\sigma_{vmax} = [\sigma_{hmax}(1+\sin\phi)+2c\cos\phi]/(1-\sin\phi)$$

σ_v is calculated from wheel load P, the load spread angle θ (in three dimensions) and the depth z

Apply the permanent strain equation:

$$= A\ln[1-(\sigma_v/\sigma_{vmax})-(\sigma_v/B)\ln(N)$$

where N is the number of load applications

Sum the displacement over the depth of the pavement

Recalculate σ_v as rut develops

Of course, while this approach does indeed allow the development of rutting to be predicted, the strength parameters used are critical inputs, and these will commonly vary through the seasons as water contents change, so it is important that sensibly conservative estimates are made.

7.4. Geosynthetic reinforcement

Steel reinforcement of concrete is a familiar concept. However, the range of possible reinforcing products for use in pavements is much greater than this, and it is important that a competent pavement engineer knows something of them.

In the well-known case of reinforced concrete, the reinforcement addresses the principal weakness of concrete, namely its relatively low tensile strength. In a similar way, in unbound materials,

Figure 7.3 Example rut predictions: load spread angle approach

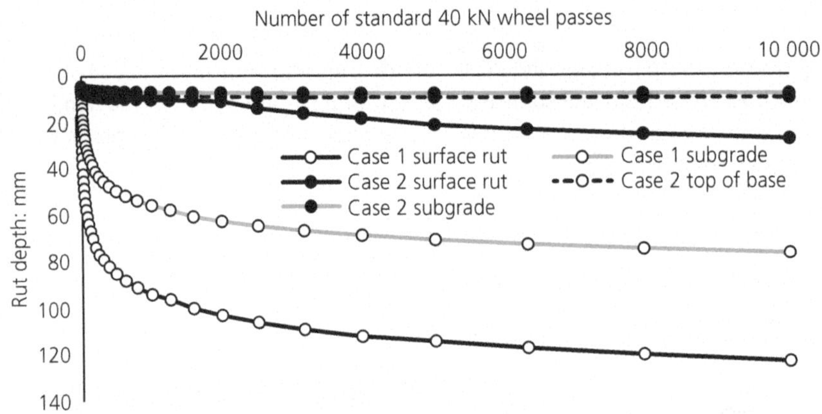

Case 1: 200 mm, $c = 10$ kPa, $\phi = 50°$/subgrade, $c = 35$ kPa, $\phi = 0°$
Case 2: 100 mm, $c = 80$ kPa, $\phi = 45°$/200 mm, $c = 10$ kPa, $\phi = 50°$/subgrade, $c = 35$ kPa, $\phi = 0°$

reinforcement needs to address their principal weakness, which is zero tensile strength. Even a small change here has the potential to make a large difference in overall behaviour.

7.4.1 Reinforcing mechanism

The key problem is how best to 'bond' a reinforcing element into an unbound material. By definition, this cannot be achieved directly, as there is no binding action available. This means that it is necessary to make use of the one mode in which an unbound material performs well, namely shear – achieved by means of the interlock between particles. The reinforcement must therefore achieve excellent interlock with individual particles. The critical particle size should logically be an important element in the design of reinforcement. In soils, this may only be a matter of micrometres (μm), which means that a membrane of some sort having an appropriate surface texture is likely to be effective; to reinforce an unbound base layer on the other hand, the 'texture' required would render a membrane impractical, and a grid becomes the logical choice. Figure 7.4 indicates the key features for efficient shear transfer.

The illustrations in Figure 7.4 imply that the particle size is very important and, ideally, a grid or membrane should be matched to the particular material to be reinforced. The 'texture' illustrated might be referred to as the 'macro-texture', generating interlock with individual particles. However, there will also be a 'micro-texture' effect, giving an angle of friction at slip between each particle and the material of the membrane or grid, and this will also affect the efficiency of the system.

7.4.2 Reinforcement effect

Unlike the surrounding material, reinforcement can withstand tension, and this means that any extensile strain in the plane of the reinforcement will be small and almost entirely recoverable. The effect will be to limit strain, particularly irrecoverable strain, in all directions in the region of the reinforcement. Logically, the effectiveness with which strain is limited will decrease with increasing distance away from the plane of the reinforcement. By these means, successful reinforcement can transform a layer of unbound material into one with much increased resistance to permanent

Figure 7.4 Reinforcement mechanism – unbound materials

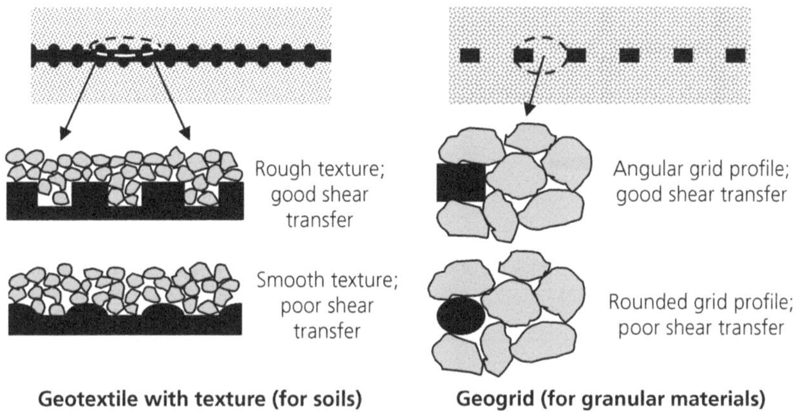

deformation and, sometimes, increased stiffness. The potential for reducing rutting in unbound pavement layers is clear.

7.4.3 Geotextiles

Geotextiles take the form of a continuous layer through which most particles of the surrounding material cannot pass. These can be further subdivided into 'woven' or 'non-woven', depending on the manufacturing process, and the material used can be polyester, polypropylene or glass fibre, as well as more traditional plant-based products. The point is that they should almost always be pervious, permitting the passage of water but not soil. Only in special cases is it likely that an impermeable membrane would be required.

In many cases, the primary reason for geotextile use is not reinforcement but separation of soil and overlying unbound material, preventing soil from rising into a granular matrix and clogging up the pore spaces, in which case the porosity and hydraulic conductivity become the principal design considerations. Rupture strength may also be important where large particles are involved or where construction activities might cause damage. Nevertheless, reinforcement is often at least one of the hoped-for effects from a geotextile, in which case the properties of surface roughness described above become important. In-plane stiffness is obviously a key factor, although the stiffness of the individual strands of fibre that interlock with soil particles may be equally important. It is at this scale that the true effectiveness of geotextile–soil shear transfer is determined.

7.4.4 Geogrids

The term geogrid means a grid reinforcement product used in the ground – or in a pavement. Grids can be manufactured from steel, glass fibre, polyester or polypropylene, as well as natural organic products such as bamboo. Figure 7.5 shows three widely used generic types. Clearly, these three types are very different in appearance and they also differ widely in the way in which they act – yet all are intended to perform a similar function. The truth is that they all have their strengths and weaknesses.

Needle-punched extruded polypropylene is likely to have the most advantageous (i.e. angular) strand profile and therefore the best interlock with surrounding stones. Bonded strips probably

Figure 7.5 Geogrid reinforcement types

Needle-punched extruded polypropylene sheet | Bonded polyester or glass fibre strips | Woven/twisted steel

have a price advantage and may also be stiffer longitudinally, but they will have a poorer strand profile and also a poorer out-of-plane bending stiffness. Steel gives the stiffest strands of all in bending, but the profile will usually be poor (i.e. rounded) and the stiffness at right angles to the main strands will be low due to the woven nature of the product. Unfortunately, these differences leave the user in a confused state. Venturing an opinion, however, the most important quality may be interlock, which demands a good strand profile and/or a stiff strand in bending.

7.4.5 Predicting effect on life: membrane effect

The load spread angle approach in Section 7.3.4 opens up a rational means of predicting the benefit to be derived from reinforcement. The first mechanism is commonly termed the membrane effect and it occurs due to the reinforcement being stretched as it is displaced vertically under the wheel-path but not either side. Calculation Sheet 7.3 goes through the force balance and the resulting spreading of the loaded area, resulting in lower stresses to underlying materials. Basically, the stiffer the reinforcement the stronger the membrane effect.

Calculation Sheet 7.3. Reinforcement: membrane effect

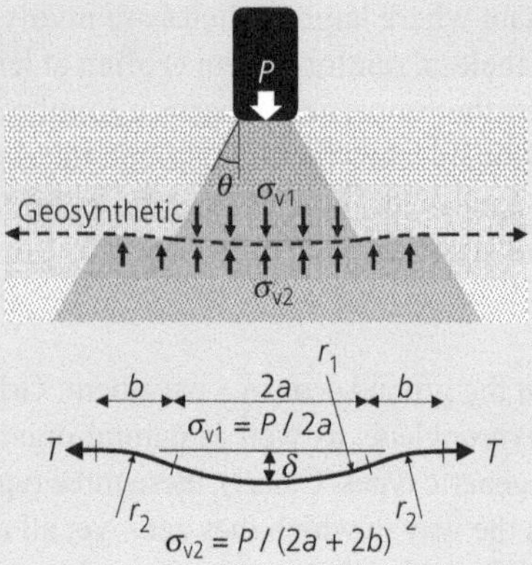

Assuming circular arcs:

Force balance, sides: $T = \sigma_{v2} r_2$

Force balance, centre: $T = (\sigma_{v1} - \sigma_{v2}) r_1$

Equating slopes: $a/r_1 = b/r_2$

Geometry: $\delta = a^2/2r_1 + b^2/2r_2$

Geosynthetic strain: $\varepsilon \approx \sqrt{\delta^2 + (a+b)^2} / (a+b) - 1$

Geosynthetic force: $T = \text{Stiffness}_{\text{geosynthetic}} \times \varepsilon$

Solve to find b and therefore σ_{v2}

7.4.6 Predicting the effect on life: confinement effect

The membrane effect can protect materials beneath the reinforcement but not above. However, the interaction of the reinforcement with unbound materials above and below, whether by friction or by interlocking with individual strands or ribs, inhibits their sideways movement under load. If the particles in contact with the reinforcement are restrained, then the next layer of particles will also be partially restrained due to the material's internal friction. But how strong will this restraint be? How far from the reinforcement will it be felt? And does it diminish with distance? Calculation Sheet 7.4 is presented as a logical – if idealised – way of answering these questions.

Calculation Sheet 7.4. Reinforcement: confining effect

As material above grid spreads horizontally:

Shear stress τ is induced $\tau = \sigma_{v1}\mu$

In the centre just above grid, this gives a maximum horizontal force of $\mu P/2$

The average horizontal force in the stressed zone is therefore $\mu P/4$

The average horizontal stress in the stressed zone is therefore $\mu P/(4a\tan\phi)$

Assume this dissipates to zero at a height of $a\tan\phi$ above the grid

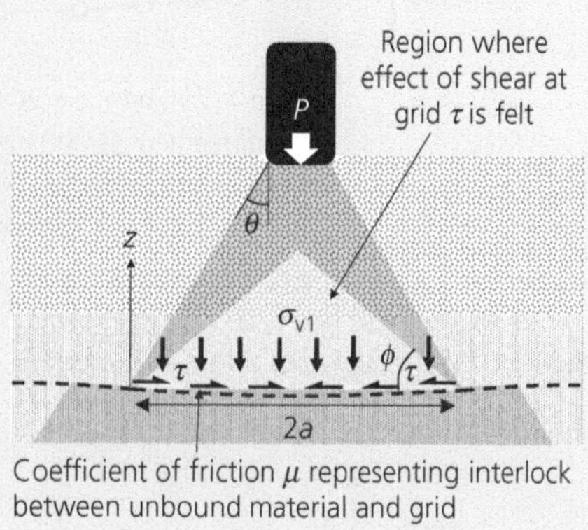

The average additional horizontal stress is therefore:

$$\Delta\sigma_h = [\mu P/(4a\tan\phi)] \times 2(a\tan\phi - z)/(a\tan\phi) = \mu P(a\tan\phi - z)/(2a^2\tan^2\phi)$$

A similar effect occurs below the grid

Once computed, the average confining stress in each sub-layer of the pavement can be added to the horizontal stress felt as a consequence of the overburden, thereby reducing the predicted rate of development of permanent strain. Figure 7.6 presents example predictions comparing reinforced and unreinforced unbound pavements.

7.5. Unbound surface courses

The problem with fully unbound pavements is that the stress conditions at the surface are particularly hard to withstand. Right at the edge of the wheel-load patch, the horizontal stress near the surface must (*a*) be sufficiently large to withstand the high vertical stress from the tyre without significant permanent strain but (*b*) not be so large that failure occurs outside the loaded area due to the zero overburden stress applying there. In an almost entirely frictional material such as a crushed rock, these twin requirements are virtually impossible to fulfil; the material needs a degree of *cohesion*. Where unbound pavements are to receive significant traffic loads, as is the case over

Figure 7.6 Example rut predictions: geosynthetic reinforced pavement

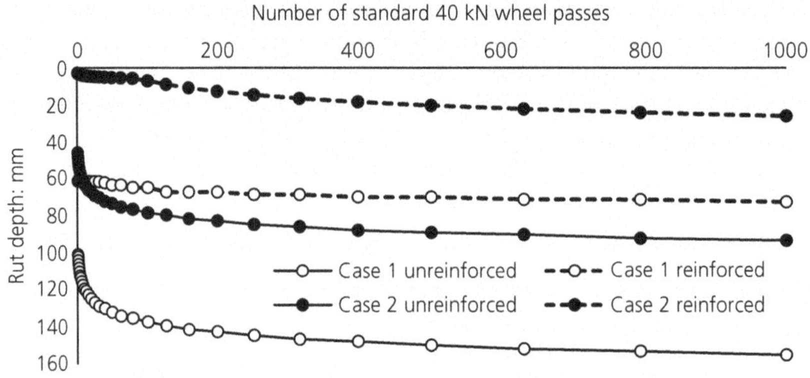

Case 1: 200 mm, c = 20 kPa, ϕ = 50°/ subgrade, c = 20 kPa, ϕ = 0°
Reinforcement @ 200 mm, 3MN/m, coefficient of friction 0.7
Case 2: 300 mm, c = 10 kPa, ϕ = 45°/ subgrade, c = 45 kPa, ϕ = 10°
Reinforcement @ 150 mm, 0.5 MN/m, coefficient of friction 1.0

Figure 7.7 Mohr circles of stress at surface

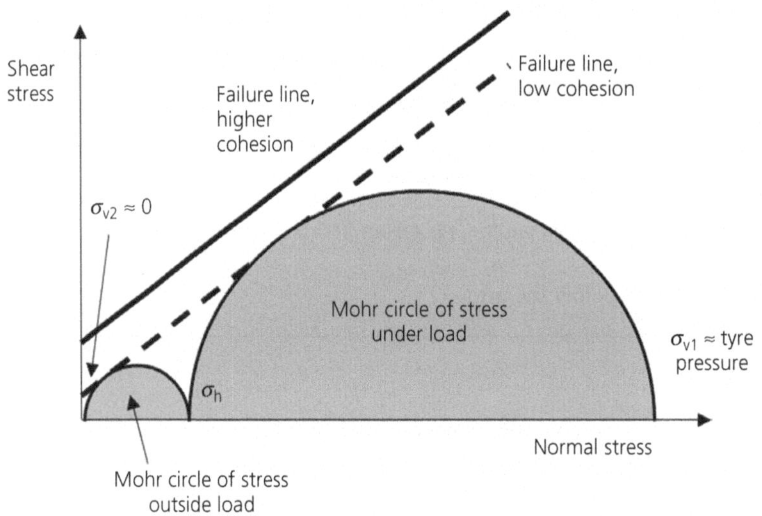

much of the world, it is therefore common to include clay fines (usually naturally occurring (e.g. in lateritic gravels)). Figure 7.7 illustrates the effect of cohesion in transforming a 'failure' state at the surface into one that can withstand the applied pressure.

However, clay brings its own problems. At exactly the right clay content, the beneficial properties (cohesion, very low permeability) are combined with the benefits of an aggregate skeleton, resulting in an excellent, stable and dust-free surface. The protruding aggregate particles can even give a reasonable level of skid resistance. On the other hand, if the clay content is too high, the result during wet weather will be a low-strength material with a dangerous low-friction surface. It is a difficult balance to achieve – but crucial to economic design. As a rough guideline, there should be between 25% and 45% retained on a 425 μm sieve, and the optimum plasticity index of material passing 425 μm would usually be between 4% and 9%. This advice follows the US Corps of Engineers recommendations for roads in wet climates (US CoE, 2006).

7.6. Effect of surface sealing

7.6.1 Surface dressing

Surface dressing, also known as *chip seal*, means the spraying of a layer of bitumen over the surface followed by the spreading of a layer of aggregate particles, usually 10–15 mm in size. Both bitumen spraying and aggregate spreading are normally carried out by means of automated pieces of equipment designed to give an even coverage of both; hand-controlled spraying and spreading is possible of course, but the finished product will never be as well controlled.

Aggregate selection depends primarily on the level of micro-texture required for the site in question. Size then depends on the degree of embedment likely into the existing surface. For example, surface dressing of an old concrete pavement demands a relatively small aggregate size, whereas an unbound pavement calls for a larger size. The rate of spread of particles is generally designed to give just about 100% coverage. As shown in Figure 7.8, it is possible to design for additional durability by including more than one aggregate spread operation. A 'racked-in' surface dressing includes a second application, of much smaller aggregate size, designed to fill the gaps between larger particles. One can go further and introduce a second binder spray operation and yet more aggregate, increasing quality – but at the same time increasing cost. In effect, multi-stage surface dressings resemble thin asphalt layers.

Bitumen selection depends on several factors, not least local economics. It is possible to use pure hot bitumen of low viscosity. More usually, an emulsion (see Chapter 13) will be used, allowing a harder grade of binder but ensuring that it can still be sprayed over the surface. The use of cut-back bitumen (bitumen dissolved in lighter oils) is reducing these days due to health and safety concerns. In all cases, polymer additives may be used. Bitumen selection will also depend on the ambient temperature, as this will affect the rate and degree to which the aggregate particles will develop

Figure 7.8 Surface dressing design

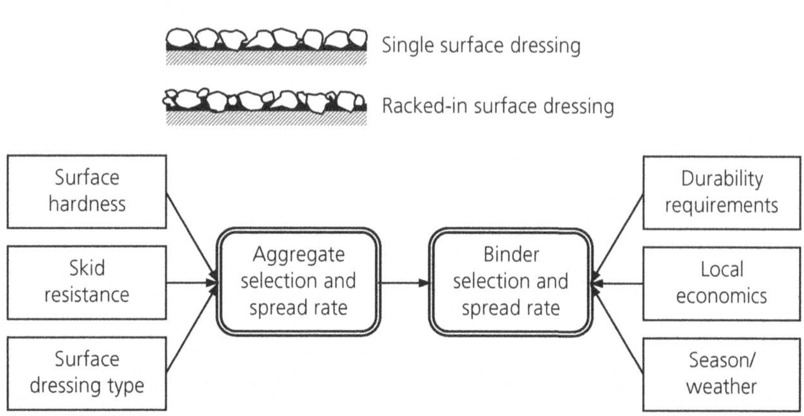

adhesion to the bitumen. Finally, the bitumen spread rate depends primarily on the aggregate size and spread rate to be used. Figure 7.8 summarises the decision-making process. Various documents from all over the world detail specific design advice; the UK Transport Research Laboratory Road Note 39 (Nicholls, 2002) gives a full set of design and construction guidelines.

In temperate northern climates, surface dressing is carried out during a limited season (e.g. March to October) in order to ensure that the adhesive bond between aggregate and bitumen is fully developed.

Surface dressing can give a long-life surfacing. It also gives a high skid resistance and therefore a safe surface, and has the very considerable advantage that it seals the pavement. It is also a logical choice for the first stage in the transition from an unbound pavement to a surfaced road. A high-quality surface dressing has sufficient flexibility and toughness to survive perfectly well when placed directly onto an unbound base, something that a genuine asphalt course may struggle to achieve because of its reduced flexibility under load. The point is that the curvature of the surface under wheel load will still be largely controlled by the base and, for a given curvature, a thicker layer will develop higher tensile strains. It is also likely (although hard to prove) that the fatigue characteristic of a surface-dressing layer is better than that of most asphalts because of its proportionally higher bitumen content.

7.6.2 Effect on pavement life prediction

It is often stated that surface dressing makes negligible structural contribution to a pavement other than its effect as a seal and therefore preventing excess water ingress – but this is not quite fair in the case of an otherwise unbound pavement. Surface dressing has a small but measurable resistance to bending, and the fact that it has to bend during deflection of a pavement under load means that a proportion of the wheel load is transferred outwards away from the contact area. Calculation Sheet 7.5 outlines a simplified way of taking account of this effect in the type of analysis introduced in Calculation Sheet 7.2.

Calculation Sheet 7.5. Load adjustment due to surface dressing

Assume surface profile is due to underlying materials, giving $y = y_0 r / (r + x)$

Curvature: $d^2y/dx^2 = 2y_0 r / (r + x)^3$

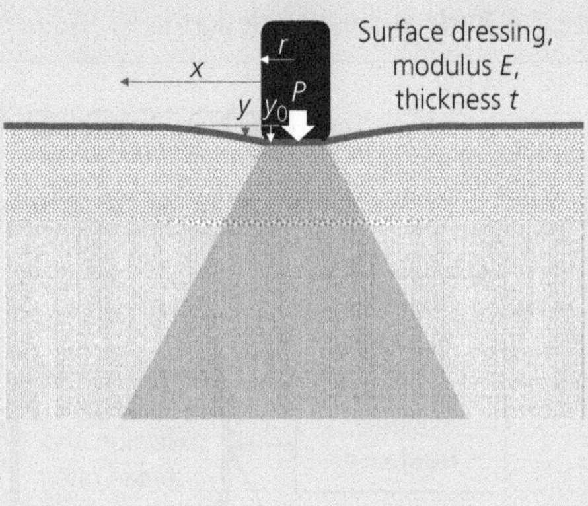

Curvature is related to bending moment and the rate of change of bending moment is shear force

Shear force summed at the edge of the tyre is the load supported by surface dressing (i.e. the reduction in load reaching the unbound layers, ΔP)

Adjusting for circumferential effects:

$$\Delta P = -\pi y_0 E t^3 / r^2$$

Naturally, plenty of assumptions have to be made. It isn't really possible to state that a surface dressing has a specific thickness but it is suggested that from 5 mm for a single surface dressing to 10 mm for a racked-in dressing represents a reasonable range. Similarly, it is very difficult to assign a modulus value, but in such a binder-rich material a lower value than a conventional asphalt mixture (see Chapter 9) is reasonable, perhaps 1000 MPa. While this approach at least allows a prediction of the effect of surface dressing to be made, theoretically it is rather small, probably dwarfed by the indirect effect of improving the average base and sub-base strength expected due to reduced average water content.

7.7. Block paving

Block paving is a widely used surfacing type, being a modern continuation of a millennia-old paving technique. The blocks can be anything from small elements of rock, known as 'setts', to relatively large concrete paving flags. However, a common standard size is a rectangular 200 mm × 100 mm concrete or baked clay element, with a usual thickness of 60 or 80 mm. Blocks are used for various reasons. In city centres, car parks and the like, the primary reason is often aesthetics; however, maintainability is an older and still applicable justification. In this section it is assumed that blocks overlie an unbound base and sub-base; Section 17.5 will cover their use as part of a heavy-duty pavement with a hydraulically bound base.

Blocks are like large nicely shaped rocks, forming a very well interlocked but still unbound pavement layer. While the blocks themselves are high-stiffness components, the effective layer stiffness is a function of the permissible rotation and shear at joints, which in turn depends on how well the joints are filled. Using a computational routine to be presented in Section 16.5.4 for a discontinuous layer, the effective stiffness generally comes out at 500–1000 MPa, and this range of stiffness has also been obtained in both field testing and laboratory investigations. In fact, it is common practice to assume a stiffness of 500 MPa for a combined block–bedding sand layer, and this can be used directly in a multi-layer linear elastic analysis.

If applying the sort of analysis presented in Calculation Sheet 7.3, the block layer can be considered as not being able to fail within itself; thus deformation calculation starts below the block layer where the wheel load is already spread by an assumed load spread angle – 35° as a default – and an overburden pressure already exists.

7.8. Maximising reliability

7.8.1 Specification

The principal form of failure in unbound pavements is deformation, whether in the form of ruts or corrugations (see Section 8.1.1), and the best way to minimise deformation in any given material is to ensure excellent compaction. Section 7.1.3 covered the concept of the optimum water content for compaction and the associated maximum dry density (MDD) and noted that it is common to write into specifications that density achieved on site should be a certain percentage of MDD. Basically, the higher the percentage the greater the resistance to permanent deformation. Typical values are 98% for a base layer and 95% for a sub-base. These are sensible levels and it is important to make sure that they are adhered to.

The other critical specification issue is the surface course. If no surface dressing or blocks are used then, as outlined in Section 7.5, the gradation of the surface course has to be just right, with just the right degree of plasticity to maximise suction in dry weather, minimise water ingress and avoid excessive softening in wet weather. The design has to be tailored to the climate, largely based on experience with available local materials, but the specification should make sure that the design is followed closely.

7.8.2 Road geometry

Appropriate road geometry can make the job of the surface course a bit easier. Drainage will be discussed in Chapter 24, but a key point here is that the steeper the cross-pavement gradient or 'crossfall' the more rapidly will surface water drain and so (*a*) the less time it has to soak into the pavement and (*b*) the less likely it will be that rutting or other forms of deformation will lead to water ponding.

7.9. Summary

Unbound pavements will continue to form an important component of transportation infrastructure where traffic is infrequent and funds are limited but the provision of a transport route is of vital commercial importance. Undoubtedly, the key to success with unbound pavements is maintaining a high level of suction within the pavement layers throughout most of the year and the efficient shedding of water, both surface and sub-surface, during periods when significant rainfall occurs. This then leads to the critical topic of pavement drainage, covered in Chapter 24.

REFERENCES

AASHTO (2021) M145-91: Standard specification for classification of soils and soil-aggregate mixtures for highway construction purposes. AASHTO, Washington, DC, USA.

ASTM (2016) D1556-16: Standard test method for density and unit weight of soil in place by sand-cone method. ASTM International, West Conshohocken, PA, USA.

ASTM (2020) D2487-17: Standard practice for classification of soils for engineering purposes (Unified Soil Classification System). ASTM International, West Conshohocken, PA, USA.

ASTM (2023) D6938-23: Standard test methods for in-place density and water content of soil and soil-aggregate by nuclear methods (shallow depth). ASTM International, West Conshohocken, PA, USA.

Austroads (2021) *Guide to Pavement Technology Part 2: Pavement Structural Design*. Austroads, Sydney, Australia.

BAA (British Airports Authority) (1993) *Pavement Design Guide for Heavy Aircraft Loading*. BAA, Group Technical Services Aircraft Pavements, London, UK.

Brown SF, Brunton JM and Stock AF (1985) The analytical design of bituminous pavements. *Proceedings of the Institution of Civil Engineers – Part 2* **79(1)**: 1–31, https://doi.org/10.1680/iicep.1985.1077.

BSI (1990) BS 1377-9: Methods for test for soils for civil engineering purposes. In-situ tests. BSI, London, UK.

BSI (2010) BS EN 13286-2:2010: Unbound and hydraulically bound mixtures. Test methods for laboratory reference density and water content. Proctor compaction. BSI, London, UK.

BSI (2021) BS EN 13286-4:2021: Unbound and hydraulically bound mixtures. Test methods for laboratory reference density and water content. Vibrating hammer. BSI, London, UK.

DoT (Department of Transport) (1996) TRH4: Structural design of flexible pavements for interurban and rural roads. DoT, Pretoria, South Africa.

NAASRA (National Association of Australian State Road Authorities) (1987) *Pavement Design: A Guide to the Structural Design of Road Pavements*. NAASRA, Milsons Port, NSW, Australia.

National Highways (2021) *Manual of Contract Documents for Highway works. Vol. 1: Specification for Highway Works. Series 800: Road Pavements – Unbound, Cement and Other Hydraulically Bound Mixtures*. Stationery Office, London, UK.

Nicholls JC (2002) *Design Guide for Road Surface Dressing*, 5th edn. Transport Research Laboratory, Crowthorne, UK, Road Note 39.

Powell WD, Potter JF, Mayhew HC and Nunn ME (1984) *The Structural Design of Bituminous Roads*. Transport and Road Research Laboratory, Crowthorne, UK, Report LR1132.

SANRAL (South African National Roads Agency SOC Ltd) (2014) Pavement design. In *South African Pavement Engineering Manual*. SANRAL, Pretoria, South Africa, Chapter 10.

TRB (Transportation Research Board) (1981) *General Soils Problems*. Transportation Research Board, Washington, DC, USA, TRR827.

US CoE (US Corps of Engineers) (2006) *United Facilities Guide Specifications: Aggregate Surface Course*. US CoE, Hyattsville, MD, USA, UFGS-321500.

Nick Thom
ISBN 978-1-83549-711-1
https://doi.org/10.1108/978-1-83549-710-420243009

Chapter 8
Maintenance of unsurfaced pavements

8.1. Patterns of deterioration

8.1.1 Deformation

All unbound materials deform under repeated load, as explained in Chapter 6 and applied in design in Chapter 7. Rutting is therefore expected. The greater the cohesion within the pavement layers – or if a surface dressing or block surface is present – the slower the rutting will be, but it is nevertheless expected and therefore has to be corrected.

However, in the specific case of unbound pavements, there is a further manifestation of deformation, namely *corrugations*. These are regular undulations along a road, which occur due to a resonance being set up between a vehicle's suspension system and longitudinal defects in the road surface. This results in slightly higher dynamic load being applied in depressed areas and the depressions are then magnified by the resulting permanent deformation. This is an expected occurrence on unsurfaced roads and does not necessarily imply any particular defect in design, material provision or construction. It does, however, present another maintenance need.

8.1.2 Surface problems

Unsurfaced pavements are likely to be exposed to tyre pressures that are just as high as anywhere else and yet the surface stones, those experiencing the direct forces from the tyre, are only held in place by other surrounding stones and a possibly slightly cohesive mortar of finer material. Loosening and then removal of stones under the combination of vertical and lateral forces is bound to occur.

The rate and nature of surface deterioration is weather dependent. In dry weather, the 'mortar' of fine material can develop a respectable suction and so grip the surface stones better, but if it dries out too much then it can be eroded in the form of fine dust. In prolonged wet weather, the suction effect can disappear, reducing the strength of the mortar and so making stone removal easier as well as contributing to more rapid deformation.

As with any other type of road, damage will not be even and when it starts in a particular location it is likely to accelerate there. Once sufficient deformation and surface wear has occurred, then the profile will no longer be able to shed water during rainfall events and ponded water will result in local weakening, which in turn will exacerbate the problem, leading – sometimes rapidly – to pothole development.

8.2. Maintenance processes

The problem with unsurfaced roads is that they are very maintenance intensive. When traffic is infrequent they may be economic, but under more intense traffic the required frequency of maintenance intervention can become unsustainable. In addition to local repairs – which rarely reach the same quality as the original – and ditch cleaning, the following sections introduce the standard treatment processes.

8.2.1 Regrading

Whatever the disadvantages of unbound materials in terms of their mechanical properties, they have the distinct advantage that they can be rearranged fairly easily. Once rutting and/or corrugations have become excessive, preferably before the water-ponding stage is reached, the surface can be reprofiled using a mechanical grader. In many parts of the world this will be carried out every few weeks on more heavily trafficked routes.

8.2.2 Regravelling

Stone loss rates on unsurfaced roads vary significantly (e.g. Pardeshi *et al.*, 2020), but eventually it becomes excessive and an additional layer of surface course material has to be added, graded and compacted. One problem, however, is that this does not remove any underlying variations in condition due, for example, to water ponding, and this means that the life of the resurfaced road may be less than the original.

8.3. Summary

While a concrete pavement can be built and then left to look after itself for decades, an unsurfaced pavement can only be left for weeks, or perhaps months. Maintenance costs and the availability of the required equipment have to be factored in as a major element in the whole-life cost of the infrastructure.

REFERENCE

Pardeshi V, Nimbalkar S and Khabbaz H (2020) Field assessment of gravel loss on unsealed roads in Australia. *Frontiers in Built Environment* **6**: 3.

Part 3

Hot- and warm-mix asphalt and flexible pavements

Nick Thom
ISBN 978-1-83549-711-1
https://doi.org/10.1108/978-1-83549-710-420243010

Chapter 9
Asphalt

An asphalt is basically an unbound material with an extra ingredient, bitumen, and a proper understanding of unbound material is therefore of considerable assistance when trying to make sense of asphalt. However, bitumen, the added ingredient, undeniably has a very important influence on behaviour, and so the first task is to get to grips with bitumen and the properties that make it such a useful binder in a pavement. This book will not go into the production of bitumen, which is basically a residue from the distillation of heavy crude oil, nor will it delve too deeply into the chemistry of bitumen, which is complex and still the subject of differing opinions regarding molecular interaction. However, the physical properties of bitumen are extremely important to a full understanding of asphalt.

9.1. Bitumen

Bitumen is a very different binder from cement – and the resulting mixture, asphalt, is quite different in its behaviour from concrete. Although not discussed separately in this book since economically viable equivalence to bitumen is yet to be achieved, the same principles apply to so-called bio-binders, which are alternatives to conventional bitumen derived from plant oils. Both bitumens and bio-binders remain liquid, even in service. This means that the primary mechanical property is viscosity and, as for any liquid, this is a function of temperature.

9.1.1 Viscosity

Viscosity can be measured directly in various ways. The most common technique nowadays is by means of a rotational viscometer, a cylinder rotated in a vessel of bitumen (e.g. AASHTO, 2022a; ASTM, 2012).

Figure 9.1 shows typical results for different so-called 'grades' of bitumen (see Section 9.1.2) and the dramatic effect of temperature is clear. In general, about 0.2 Pa.s is required for mixing, giving a temperature range of 140–180°C; 30 Pa.s is a maximum value for compaction, giving a minimum compaction temperature of 70–80°C. Notice the effect that warm-mix additives can have on mixing and compaction temperatures without impacting performance at in-service temperatures. There are a number of such additives on the market, typically waxes or fatty amides that melt at temperatures above about 100°C, giving much enhanced fluidity at mixing and compaction temperatures, or surfactants that reduce the binder surface tension and allow easier slippage between binder and aggregate at high temperatures. Such additives will typically bring down mixing and compaction temperatures by around 30°C, saving energy and potentially extending the usable working day.

A tremendous variation in in-service viscosity also occurs between a hot summer's day, when the pavement surface temperature may be well over 50°C, and the winter. It is therefore immediately obvious that asphalt performance will be quite different in summer compared with winter.

Figure 9.1 Bitumen viscosities

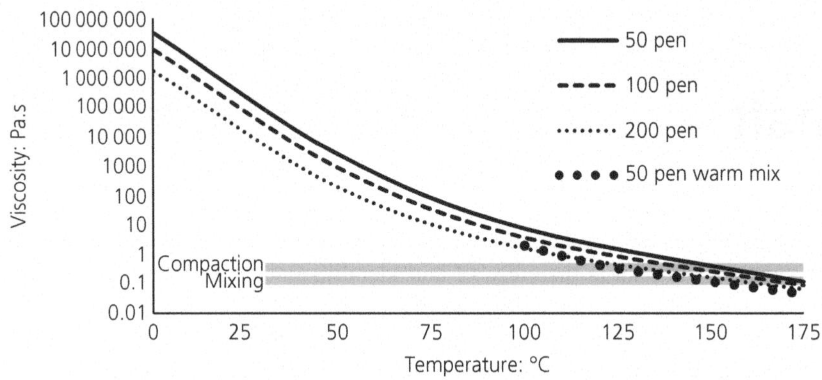

9.1.2 Penetration and softening point

Just as the liquid limit, plastic limit and the California bearing ratio are commonly used to characterise soils, so bitumen has its own index tests – equally non-fundamental and equally practical. The 'penetration' is inversely related to viscosity, as the test measures the viscous resistance to the penetration of a needle into a container of bitumen. The 'ring and ball softening point', on the other hand, is a temperature at which the bitumen has a certain viscosity (about 800 Pa.s), the viscosity being that at which a steel ball drops through a prepared disc of bitumen. Figure 9.2 shows both tests and also defines a quantity known as the penetration index (PI), which is related to the temperature susceptibility of the binder. For specification of the penetration test, see ASTM (2013) or BSI (2015a). For softening point determination, see ASTM (2014) or BSI (2015b).

Figure 9.2 Tests for penetration and softening point

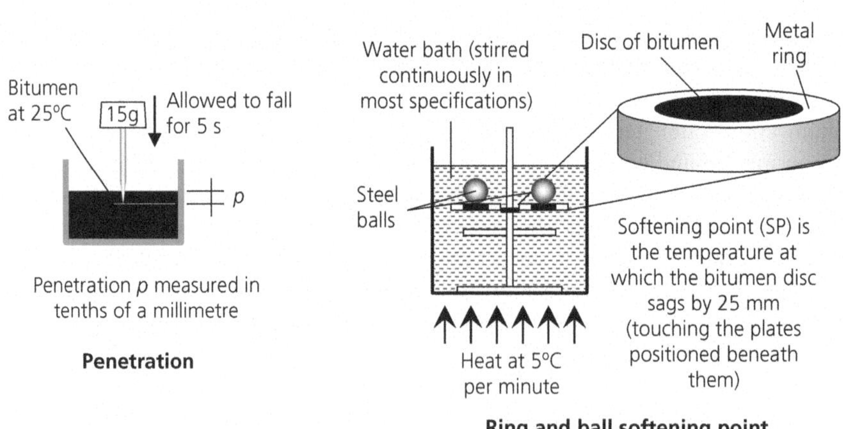

Penetration index (PI)

$(20 - \text{PI})/(10 - \text{PI}) = 50\,\{\log_{10}[100/\text{pen(at temperature } T)]\}/(\text{SP} - T)$

The most common way to describe a bitumen is still with reference to the penetration test. For example, a 40–60 penetration grade bitumen (colloquially known as 50 pen) would have a penetration of 4–6 mm in the penetration test.

9.1.3 Visco-elasticity

Although viscosity is the primary mechanical property, bitumen is actually a very unusual liquid because its viscosity is so high. This means that the elastic (i.e. recoverable) component of deformation becomes significant at in-service temperatures. Not only that, but there is actually a truly visco-elastic element; that is, a deformation that, though eventually recoverable, is time dependent. Figure 9.3 illustrates the full visco-elastic behaviour expected. It is undeniably complex and of keen interest to bitumen specialists. Many will fit parameters to a model such as the Burger's model shown in Figure 9.3. However, for the general pavement engineer it is sufficient simply to appreciate the different components without getting bogged down in the modelling details, particularly as the practical use of such modelling is slightly questionable.

Data on visco-elastic properties can be generated quite simply nowadays using a *dynamic shear rheometer* (DSR) (see Calculation Sheet 9.1), a device that applies repeated torsion to a small cylindrical sample of bitumen; as the test is straightforward, it is often carried out. However, interpretation is less straightforward. Figure 9.4 shows data for four quite different bitumens and, while the differences are obvious, the meaning (i.e. which will form the best binder) is not. In fact, it is still common for engineers to ignore detailed data in favour of the derivation of an 'equivalent penetration' value.

Figure 9.3 The visco-elastic nature of bitumen

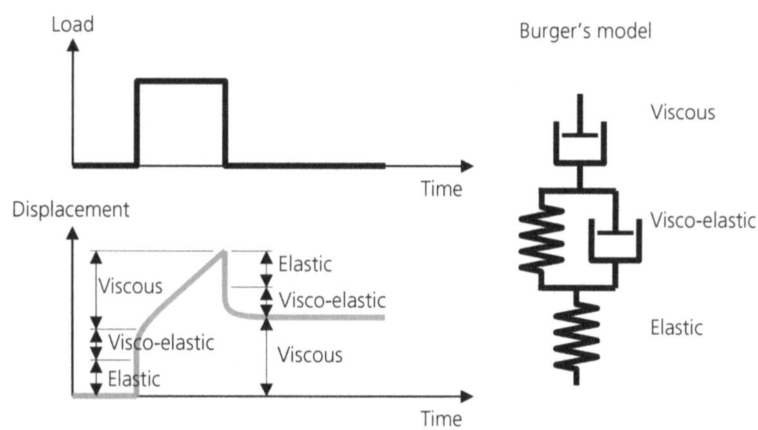

Calculation Sheet 9.1. The dynamic shear rheometer (DSR)

Shear strain at distance x from the centre of the bitumen specimen is given by:

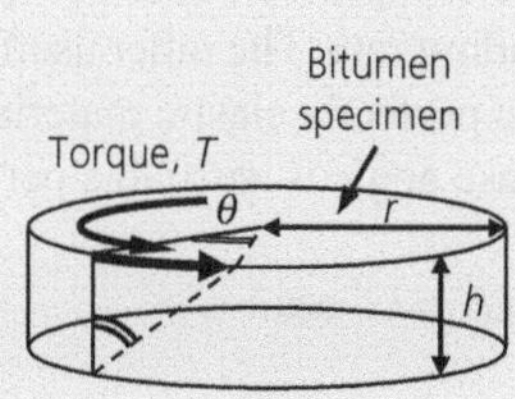

$\gamma = x\theta/h$

If G is the shear modulus:

Torque $T = \int G(x\theta/h).2\pi x.x \, dx = \int (2\pi G\theta/h).x^3 \, dx$

Integrating from 0 to r: $= \pi G\theta r^4/2h$

$\rightarrow G = 2Th/\pi\theta r^4$

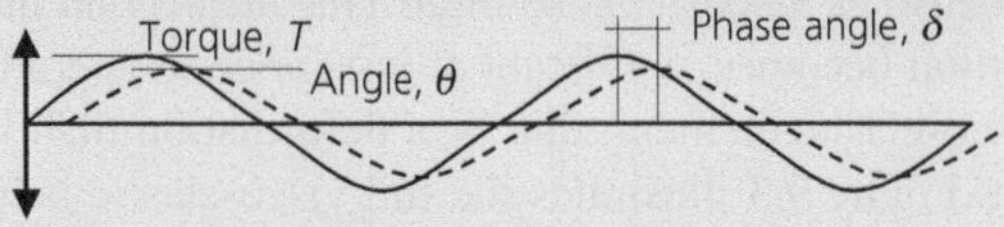

Testing is carried out sinusoidally:

Phase angle δ represents the relative viscous and elastic components of behaviour

A phase angle of zero implies fully elastic behaviour; 90° represents fully viscous behaviour

Figure 9.4 DSR data for different bitumens (Rahimzadeh, 2002)

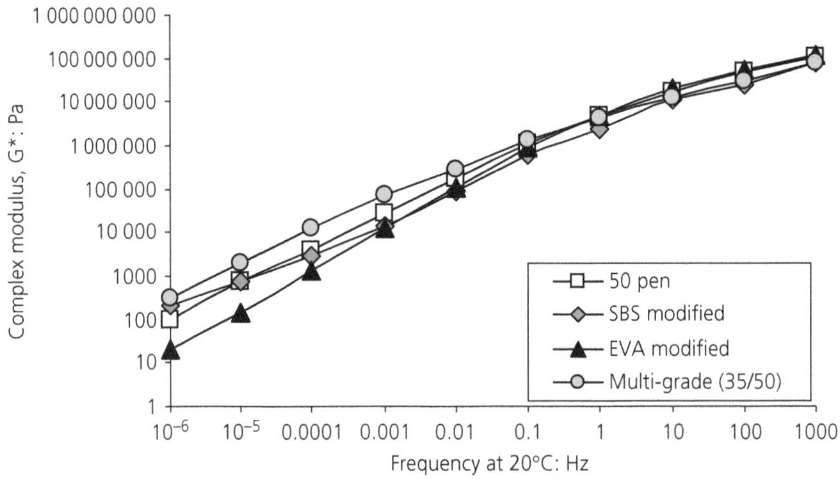

The quantity that is most usually quoted is known as the *complex modulus* (G^*) – complex because it includes both an elastic (G') and a viscous (G'') component. G' is known as the *storage modulus* and G'' as the *loss modulus*. Basically, G^* is the absolute magnitude of shear stress divided by the absolute magnitude of shear strain, ignoring any time lag between them due to viscous effects.

The complex modulus G^* gives an idea of the stiffness of a bitumen at a given temperature and loading rate. The other useful parameter is the phase angle difference (δ) between stress and strain. In a perfectly elastic material, stress and strain will be completely in phase – which means that the phase angle is zero; in a perfectly viscous material, they will be 90° out of phase.

$$G' = G^* \cos\delta \quad (9.1)$$

$$G'' = G^* \sin\delta \tag{9.2}$$

$$[G^*]^2 = [G']^2 + [G'']^2 \tag{9.3}$$

At 20°C and 10 Hz (a frequency typical of pavement loading), G^* is typically 10–30 MPa.

The US Superpave specification (Asphalt Institute, 2003) is, in terms of direct visco-elastic measurements, carried out in the DSR, and may therefore be considered much more scientifically based than specifications that use the penetration and softening point. Lower limits are placed on $G^*/\sin\delta$ (to protect against rutting) and $G^* \times \sin\delta$ (to protect against fatigue cracking). The specification also incorporates measurement of low-temperature visco-elastic properties using a *bending beam rheometer*, a device that induces flexure in a bitumen specimen of section 6.35 mm × 2.7 mm (Asphalt Institute, 2003).

9.1.4 Fracture and fatigue

This is an absolutely key area of bitumen performance. If stress is applied slowly enough, bitumen will simply deform viscously (with negligible elastic and visco-elastic components). However, if load is applied rapidly – or if the temperature is low – it can fracture. This behaviour is fundamental to a proper understanding of asphalt. A useful piece of equipment for investigating fracture is the direct tension test (DTT) (Asphalt Institute, 2003), illustrated in Figure 9.5, and it is usual to conduct tests at low temperature, recognising the tendency for cracks to appear at night in cold climates (thermal cracking; see Chapter 10). The best-known alternative low-temperature fracture test is the Fraass breaking point test (BSI, 2015c; Fraass, 1937), which consists of repeatedly flexing a thin strip of metal with 0.5 mm of bitumen applied to its surface while lowering the temperature. The temperature at which the bitumen fractures is the Fraass point, which gives an idea of the minimum serviceable temperature for that binder. Other impact-type tests for bitumen 'cohesion' are also in use, but none have achieved widespread recognition. Both the DTT and the Fraass test are intended to measure the susceptibility of a binder to fracture in cold weather.

Figure 9.5 DTT data – 50 pen grade bitumen (Thom *et al.*, 2006)

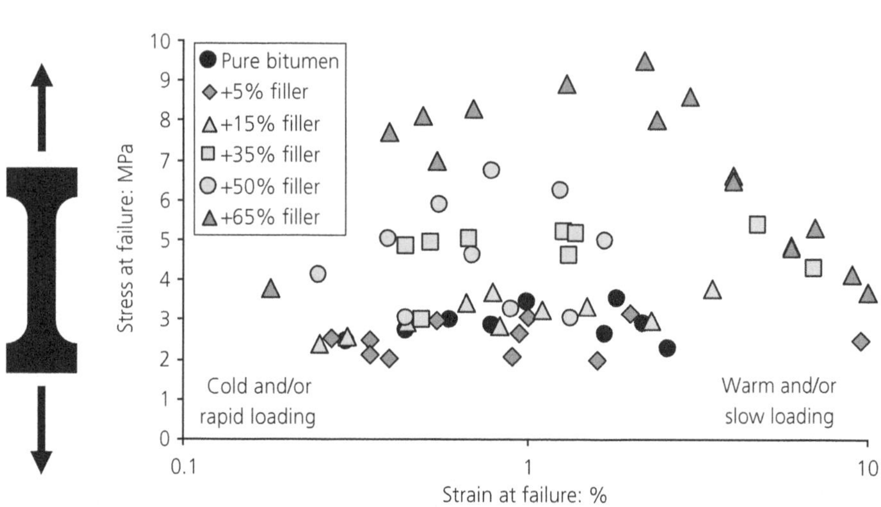

The DTT equipment can be used over a wide range of temperatures and loading rates, and Figure 9.5 shows such data for a particular 50 pen grade bitumen, including specimens with various filler contents (by mass) (see next section).

This is another highly complex aspect of behaviour, although – to a first approximation – bitumen can be seen to have a constant *fracture strength* (i.e. tensile stress at failure) but a highly variable fracture strain. In Figure 9.5, the fracture strength is around 3 MPa, becoming less at very low temperatures or high loading rates – a reason why low-temperature tests are appropriate. At the other end of the range, when the strain rate is slow enough or the temperature is high enough, fracture does not occur at all; it is replaced by steady-state creep.

Bitumen can also suffer *fatigue* (i.e. failure under repeated load at a stress less than the fracture strength) and a commonly used method to investigate this phenomenon is the linear amplitude sweep test (AASHTO, 2018), which is basically a repeated torsion test carried out in a DSR. Figure 9.6 shows a fatigue characteristic, in this case derived from repeated load axial testing of a small disc of bitumen, together with DTT strength data, and it is noticeable that the DTT fracture strength is close to the projected single-load application failure stress. In fact, just as bitumen can be considered to have a single fracture strength so, to a first approximation, it has a single fatigue characteristic when plotted against stress, independent of the loading rate and temperature – over a certain range.

9.1.5 The bitumen–filler mortar system

Filler is the term used for silt-size particles that are added to a bituminous mixture as a means of extending and enhancing the performance of the bitumen, and it is appropriate to consider the filler as a bitumen additive rather than as a component of the aggregate. It has the effect of

Figure 9.6 Bitumen and bitumen–filler mortar fatigue data (Thom *et al.*, 2006)

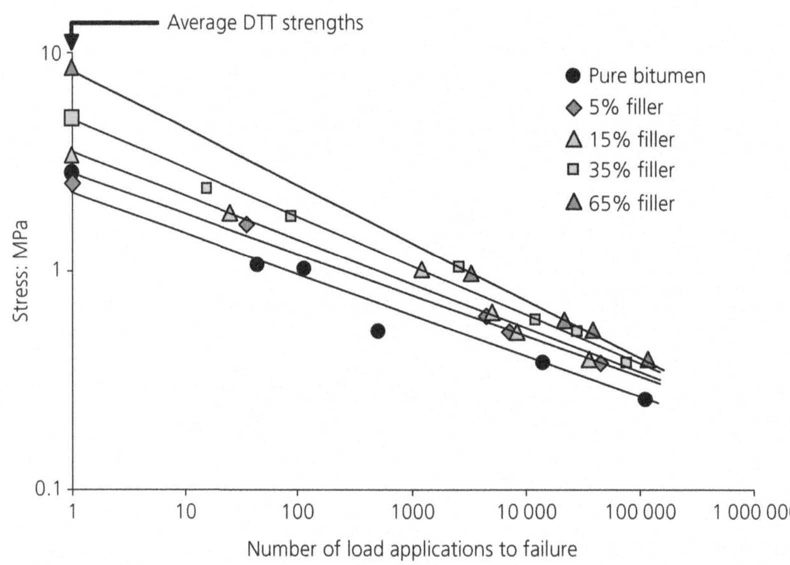

stiffening and strengthening the bitumen. The increased strength with filler content is illustrated in Figure 9.5. Figure 9.6 shows that a bitumen–filler mortar also has an improved fatigue characteristic in the sense that the life to failure is extended for a given level of applied stress. It can also be demonstrated that the stiffness of a bitumen–filler mortar is similarly increased. Calculation Sheet 9.2 outlines the theoretical effect of spherical filler particles. The actual effect is greater because of the irregular shape of real rock fines, and typically amounts to an increase in both strength and stiffness by a factor of 2–3.

Calculation Sheet 9.2. Bitumen–filler mastic

Stiffness

First approximation: assume filler particles are infinitely stiff and do not interact

$\rightarrow E_{mastic} = E_{bit}/(1 - V_f)$

where V_f is the filler volume fraction

$\rightarrow$ Typically, $V_f = 0.25$–0.5; enhances bitumen stiffness by $\times 1.3$ to $\times 2.0$

In reality, *particles interact*, particularly at V_f of 0.5 or more

$\rightarrow$ increased stiffness

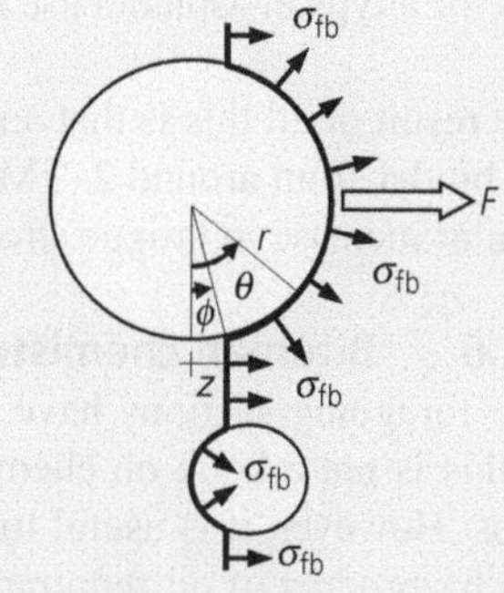

Strength

Assume fracture plane deviates around each particle

Integrate from ϕ to $\pi/2$: $F = \int \sigma_{fb} r \, d\theta \, 2\pi r \cos\theta / \sin\theta$

$$= 2\pi r^2 \sigma_{fb} \ln(r/z)$$

z has equal probability between 0 and r

$\therefore$ Integrate: average $F = \int 2\pi r^2 \sigma_{fb} \ln(r/z) \, dz/r$

$$= 2\pi r^2 \sigma_{fb}$$

Acting over average area of $\int \pi(r^2 - z^2) \, dz/r = 2\pi r^2/3$

$\rightarrow$ Average stress normal to fracture plane at particles of $2\pi r^2 \sigma_{fb} / 2\pi r^2/3 = 3\sigma_{fb}$

$\rightarrow$ Overall average stress normal to fracture plane of $3\sigma_{fb} V_f + \sigma_{fb}(1 - V_f) = \sigma_{fb}(1 + 2V_f)$

$\rightarrow$ Typically, $V_f = 0.25 - 0.50$; enhances bitumen strength by $\times 1.5$ to $\times 2.0$

In reality, *particles are non-spherical* $\rightarrow$ increased strength

Experience (and theory) indicates that it is beneficial to a mixture to include a certain proportion of filler, typically a slightly higher percentage by mass than that of bitumen, which equates to about a 2:1 bitumen/filler ratio by volume. If larger quantities are used, the filler begins to form part of the aggregate skeleton rather than acting as a bitumen enhancer. The principle governing the role of the filler is similar to that illustrated for unbound materials in Section 6.1.3. When the aggregate (including filler) gradation is compared with the appropriate Fuller gradation, a point of contact between the two can be found, defining a *critical particle size*. Larger sizes form the aggregate skeleton, while smaller sizes are 'lost' in the void spaces and would have no structural role at all were it not for the bitumen. Thus, there is an optimum filler content above which increased quantities no longer give improvement; in fact, excessive filler will result in poorer performance, as such small particles do not form an effective component in an aggregate skeleton. Figure 9.7 illustrates this for a typical asphalt base mixture.

The result of all this is that an appropriate volume of added filler will raise the tensile strength of the binder from around 2–3 MPa for pure bitumen to 5–8 MPa for a bitumen–filler mortar, with a similar increase in fatigue strength and stiffness.

9.1.6 Bitumen chemistry

The foregoing sections have treated bitumen as an inert fluid of indeterminate chemistry and, as this is not a book on chemistry, this is good enough for most pavement engineering applications. However, it is useful to have a broad appreciation of what actually goes into a bitumen. It is a by-product of oil refining and, because oils are complex and varied organic compounds, it is unsurprising that bitumens are similarly complex. The individual molecules in a bitumen are large and non-uniform, varying tremendously in molecular weight and therefore in resulting physical attributes. While bitumen chemists may criticise the following simplification, for engineers it is useful to discern four broad classes of molecules

Figure 9.7 Differentiating between aggregate skeleton and filler

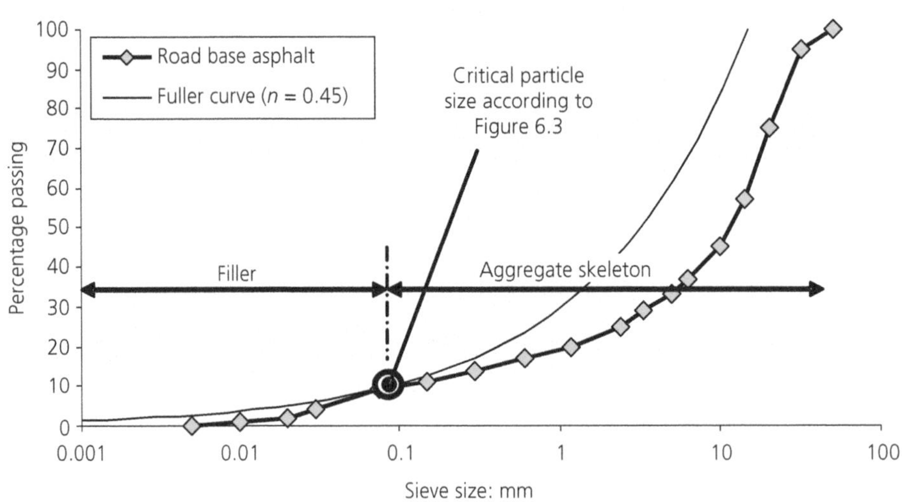

- asphaltenes
- resins
- aromatics
- saturates.

Asphaltenes are 'polar', which means the molecules carry positive and negative charge at different points, and therefore bonds can easily form between the molecules. Basically, the more asphaltenes there are in a bitumen, the harder and more viscous that bitumen is likely to be. At the other end of the scale, saturates are a lighter non-polar fraction, and they form a 'soup' surrounding the other types of molecules, giving fluidity to the overall compound. A bitumen with plenty of saturates will be softer and have a lower viscosity, and may also be more readily able to 'heal' (i.e. self-repair and re-form broken bonds).

The other molecule types, resins and aromatics, are intermediate components in the soup. Resins, like asphaltenes, are highly polar and they strongly affect the degree to which asphaltenes are dispersed within the bitumen. Aromatics make up the largest volume fraction; they are non-polar and act as a solvent to both asphaltenes and resins. A useful introduction to bitumen chemistry is given by Hunter (2014).

9.1.7 Bitumen ageing

It is an unfortunate fact that the chemistry of bitumen tends to change with time and this can result in loss of performance. This effect is termed 'ageing'. Ageing takes two main forms, as follows.

- *Oxidation.* Many of the polar molecules within a bitumen are readily able to combine with any free oxygen they find. The result is additional cross-linking between molecules, and a general stiffening and increase in viscosity. A less beneficial consequence is a reduction in fracture strain.
- *Absorption by aggregate.* The lighter saturate fraction of a bitumen, with smaller molecular size, has significant affinity for aggregate with the right absorptive quality. This is helpful in making sure that the bitumen adheres to the aggregate, but the long-term effect is that the molecular balance in the remaining bitumen is altered in favour of the harder and more brittle asphaltene fraction.

As well as the two ageing mechanisms, it is also usual to distinguish between two timeframes. Short-term ageing occurs during mixing, transporting, placing and compacting, while the bitumen is at an elevated temperature; it is responsible for a significant change in properties, amounting to a viscosity increase of over 50% (about 25% loss of initial binder penetration). Long-term ageing occurs gradually, such that bitumen becomes ever harder during its service life. To illustrate this point, data from three roads in the USA are presented in Figure 9.8. In hot climates, it is not uncommon to recover bitumen with a penetration of well under 10 (tenths of a millimetre), when the initial value would typically have been between 50 and 100.

The ageing rate is also strongly dependent on the air-voids content within a mixture, with higher air voids leading to more rapid ageing (Zhang *et al.*, 2022). There is also some evidence (Yin *et al.*, 2017) that warm-mix asphalts age more slowly than hot mixes.

Susceptibility to ageing by oxidation is most commonly measured by means of the 'rolling thin-film oven test' (see Figure 9.8) (AASHTO, 2023a; ASTM, 2022a; BSI, 2014), in which samples

Figure 9.8 Bitumen ageing data and the rolling thin-film oven test

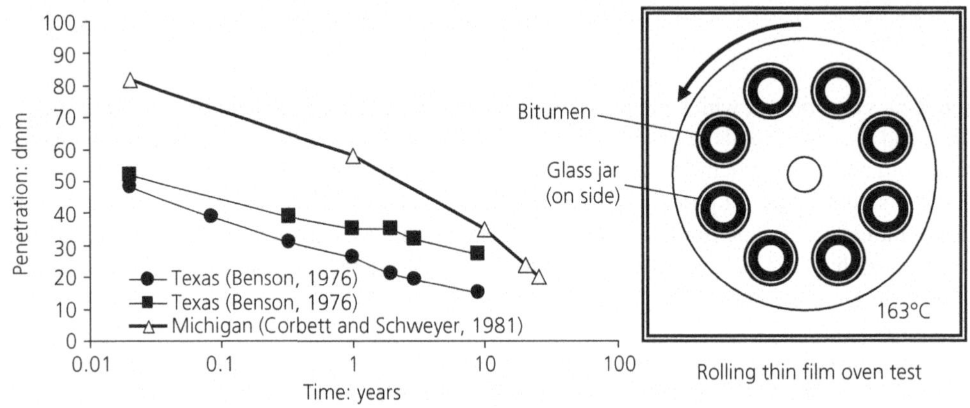

of bitumen are rotated and allowed to flow continuously around the inside of glass containers at a temperature of 163°C, while air is regularly blown into the containers. The properties of the bitumen are measured both before and after ageing. There is no current test to look specifically at ageing due to binder absorption by aggregate.

Because of the complexity of ageing and the large variation according to mixture type and climate, there is no reliable substitute for experience in predicting likely long-term binder properties.

9.1.8 Bitumen modification

A filler could be seen as a bitumen modifier since it effects a change in the properties of the 'binder', as illustrated above. And a fibre is, in reality, a long thin filler particle. Cellulose fibres have been used as additives for many years, their main purpose being to prevent so-called binder drainage (the migration of binder under gravity during asphalt transport at elevated temperature). However, there is evidence (Mohammed *et al.*, 2020) that fibres can also enhance binder strength and this may be partly due to changes in bitumen molecule arrangement induced immediately around the fibre surface, the point being that fibres have a lot of surface compared with their volume and so are very efficient at producing this change.

However, the term 'bitumen modifier' usually applies to a chemical additive, and this is a specialist area that is changing all the time as researchers continue to develop new additives. Nevertheless, it is important that pavement engineers have a broad understanding of the issues, possibilities and pitfalls. An unmodified bitumen has certain weaknesses: (*a*) it can 'flow' at high temperature, leading to pavement rutting, (*b*) it can fracture at low temperature, leading to pavement cracking and (*c*) the adhesion between bitumen and aggregate can break down under a combination of ageing and water attack. There are numerous products on the market that are claimed to improve performance in one or more of these areas, the most important of which are as follows.

- *Polymers*. Bitumen is already a complex group of varied polymers, but it is possible to design additives that emphasise beneficial properties while retarding others. Styrene–butadiene–styrene (SBS), styrene–butadiene rubber (SBR) and ethyl vinyl acetate (EVA) are three of the more commonly used, but polymers derived from waste plastics such as high- and low-density polyethylene (HDPE and LDPE) and polystyrene can also be effective. To varying degrees, all these products will increase the viscosity (and visco-elastic stiffness) of bitumen at high temperatures, but not at low temperatures. They may also give benefit in fatigue life, particularly in the case of SBS. However, polymer usage is a specialist field and not all bitumens are suited to blending with all modifiers. It is quite possible to blend an otherwise excellent additive with a given bitumen and achieve no benefit at all.

- *Simple chemicals*. Sulfur is the main chemical used. It is a liquid at >115°C, enhancing workability at high temperature, and a solid at in-service temperatures, when it acts as aggregate.
- *Rubbers*. Various forms of rubber can be added to bitumen in an attempt to enhance the 'resilience' (i.e. resistance to damage) of the binder. Much of the pressure for this comes from the need to find ways of recycling used vehicle tyres, and these can be added in two different ways: (*a*) a wet process in which small-size rubber crumb is blended into hot binder and (*b*) a dry process in which larger rubber crumb or shred is added as an aggregate replacement. The wet process is nowadays well accepted by many highway authorities and is specified in ASTM (2023b). The effect can be similar to that of polymer-modified bitumen blends. The dry process, however, presents a particular difficulty in that, over the course of weeks or months, the rubber absorbs some of the volatile components within the bitumen and the individual rubber particles swell to 2–3 times their original size. This significantly changes the volumetric proportions of the mixture and practitioners need to be aware of the possibility of long-term detrimental changes in properties, even where properties appear excellent in the short term.

One of the inevitable problems with bitumen modification is its complexity, and this means that it is difficult for engineers to specify with confidence. However, the performance grade approach taken in the US Superpave specification, in which upper and lower in-service pavement temperature limits are stated (e.g. PG 64-22, suitable from −22°C to +64°C), represents a sensible way forward. Superpave SP-1 (Asphalt Institute, 2003) describes the testing required to prove the applicable temperature range of a bitumen. The principal pieces of equipment used are the DSR for high and medium temperatures, and the bending beam rheometer for low temperatures. A particularly relevant test carried out in the DSR is the so-called multiple stress creep recovery (MSCR) test and the key measure is the degree to which strain is recovered after stress is removed; if strain is recovered, this equates to less rutting in the road and slower crack development. As illustrated in Figure 9.9, this test commonly shows the difference between modified and unmodified bitumens very clearly indeed, and allows selection of the optimum cost-effective modifier percentage.

9.2. The mechanics of asphalt behaviour

If a true understanding of asphalt is to be achieved it is necessary to think about the interaction between individual aggregate particles. An asphalt consists of an aggregate skeleton – as in the case of an unbound material – and a binder, which includes filler-size particles within it. There will also be a certain percentage of air voids. If strain is to take place, then this means that the aggregate skeleton has to deform and, as in the case of an unbound material, there are two basic mechanisms

- compression at particle contacts
- interparticle slip (combined with rotation and separation).

If no slip were to take place between particles, then the approximate stiffness of the material could be predicted from the stiffness modulus of the rock from which the aggregate is formed (say

Figure 9.9 MSCR test data on modified binders (Sudibyo, 2023)

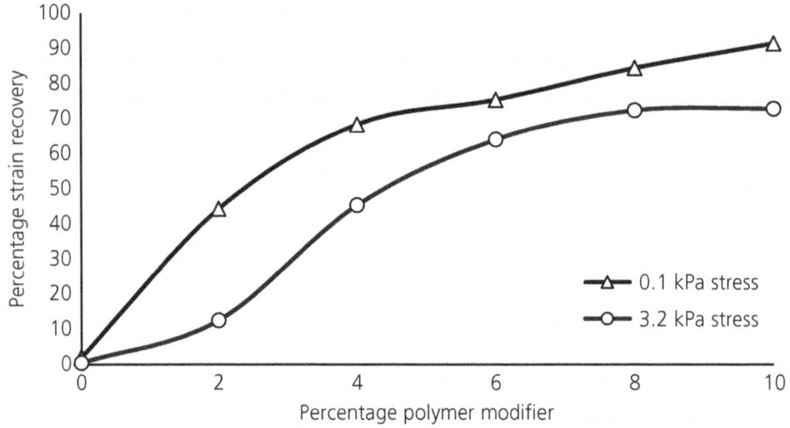

200 GPa), a contact law and a means of calculating the additional effect of the binder – in fact, exactly the approach proposed in Calculation Sheet 16.1 for hydraulically bound materials. In the case of an asphalt, the bitumen stiffness at 20°C and traffic loading frequency is typically between 20 MPa and 50 MPa, and bitumen–filler mortar stiffness is around 2.5 times this. According to Calculation Sheet 16.1, this would give a predicted mixture stiffness of 5300–7000 MPa, which is approximately the expected range for undamaged materials. The implication is that the bitumen initially allows very little interparticle slip or separation to occur; it is only as damage takes place that this situation changes.

9.2.1 Linear or non-linear?

Since unbound aggregate has been shown to be highly non-linear and bitumen is visco-elastic rather than elastic, one might expect the stiffness behaviour of an asphalt to be very far from linear – but this is not so. Why not? The point here is that the magnitude of strain taking place within the aggregate skeleton is small compared with that commonly encountered in an unbound material, and in the small-strain region even the stiffness of an unbound material is approximately constant. Furthermore, the loading rate is usually high enough to ensure that the bitumen is kept near the elastic end of its visco-elastic behaviour spectrum. The combination is sufficient to give an approximately linear stress–strain response, although the actual magnitude of the stiffness modulus will vary significantly with bitumen properties – which in turn depend on temperature and loading rate.

9.2.2 The micromechanics of asphalt damage

In a typical dense asphalt mixture, the aggregate occupies some 85% of the total volume (of which at least 5% acts as filler), with about 10% bitumen and 5% air voids. Thus, if the filler is considered as a binder additive, then the skeleton represents less than 80% of the volume and over three-quarters of the remaining space is filled with bitumen–filler mortar. Figure 9.10 illustrates the situation at a particle contact. If slip (or particle separation) is to take place, then this must impose a strain in the mortar, a strain that becomes ever larger the closer to the point of contact one looks. In fact, mathematically, the strain right at the contact point becomes *infinite* – and there is only one mechanism capable of producing infinite strain, and that is *fracture*. Damage to an asphalt, visible

Figure 9.10 Slip at particle contacts

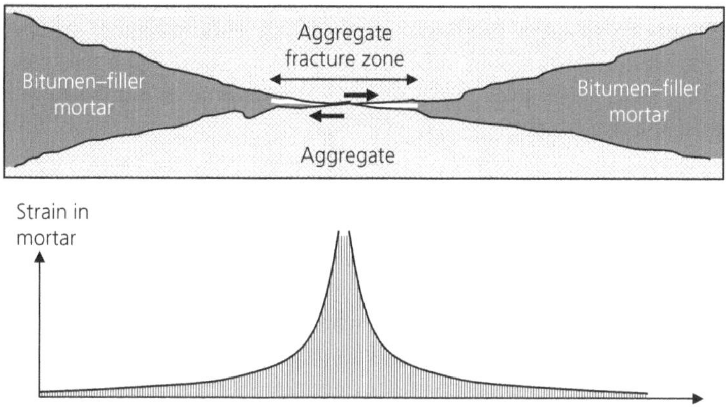

as a reduction in stiffness under repeated loading (e.g. comparing materials from trafficked and untrafficked areas in a highway), therefore takes the form of localised fractures at particle contacts, allowing slip and separation to occur, and experience is that such damage commences from the very first load application (see, for example, the data presented in Section 9.4.3).

Logically, where slip (± rotation or separation) takes place, a fracture zone will extend out from the contact point until it reaches a region where the strain is low enough for the mortar to remain intact. This then gives a measure of freedom to the two particles involved. They can slip (or separate) by a small amount each time a load is applied.

This is an area that is very imperfectly understood. Logically, these localised fracture zones (microcracks) cannot be static. Low temperature will lead to a stiff binder and restricted interparticle movement, which means additional stress taken through the binder, and therefore an increased chance of fractures occurring – and of existing fracture zones expanding. However, high temperature represents much more than the mere absence of high binder stress; it also allows the phenomenon known as *healing* to occur. Bitumen is a liquid, fractured or not, and this means that the molecules either side of a fracture can, over time, recombine to create continuous binder phase once more. This will happen most easily when binder viscosity is at its lowest, which means at high temperatures. Thus, there is a state of dynamic flux present. Winter leads to high stiffness but increased damage; summer leads to low stiffness but also healing. In theory, a state of dynamic equilibrium can be reached – but this, of course, depends just how much damage the traffic is doing and how quickly.

9.2.3 Bitumen adhesion

The micromechanical explanation given above makes one crucial assumption – namely a perfect bond (perfect adhesion) between the bitumen and the aggregates. If this is the case then there is no preferential fracture plane within the binder and the resistance to fracture at a particle contact will simply depend on the cohesive strength of the binder itself. However, if there is any lack of adhesion, then fracture will take place preferentially along one of the interfaces between an aggregate particle and the adjacent bituminous mortar, and the development of such a fracture will occur more easily than it otherwise might have. Adhesion is thus a very important property.

Unfortunately, adhesion is not a property that is easily measurable. The pneumatic adhesion tensile testing instrument (PATTI) (ASTM, 2022b) is among several tests that have been used to measure bitumen adhesion, but is certainly not an industry standard. However, the effects of lack of adhesion can certainly be observed indirectly from mixture tests and, more dramatically, in failed asphalt pavements. Mixture tests would indicate low stiffness, low fatigue resistance and high permanent deformation under multi-cyclic load, as all three properties are strongly related to the ease with which internal fracture occurs.

So the question arises: what causes poor adhesion? One important part of the answer appears to be the *acidity* or *alkalinity* of the aggregate. Limestone aggregate generally makes for high-quality asphalt, whereas many igneous rocks have proved to be less trouble-free, as have many natural gravels (although a rounded particle shape would cause poor performance even with good adhesion). Collop *et al.* (2004) are among those who have reported a significantly enhanced performance with limestone aggregate compared with granite. Current research is concentrating on the subject of 'surface energy' – a property of the aggregate surface similar in concept to the well-known 'surface tension' effect at a liquid surface. A particular issue is that most aggregates would attract water, if it were available, in preference to bitumen – hence the need to ensure that particles are absolutely dry before mixing with bitumen.

However, all is not lost. If acidity is the problem, then a logical move is to include an additive in the mixture, often *hydrated lime*, although other chemical additives are marketed and have also proved effective (ethylenediamine, formaldehyde, heptaldehyde and phenol have all been found to enhance bitumen–aggregate adhesion in certain cases). This can have a dramatic effect and, while there are some who doubt that the influence of such an additive is permanent, it certainly seems to be long-lasting enough for practical and economic usage.

9.3. Asphalt stiffness

The most basic property of an asphalt for pavement design is its stiffness modulus. However, from the foregoing discussion, it is clear that an asphalt stiffness modulus, although representing an approximately linear relationship between stress and strain, does not have a unique value; it is both temperature and loading-rate dependent because it depends on the efficiency with which the binder restricts interparticle movement in the aggregate, and this in turn depends on binder stiffness. It is also, unsurprisingly, dependent on the mixture proportions used.

9.3.1 Predicting asphalt stiffness

The first requirement is to predict binder stiffness, itself a complex visco-elastic quantity. For this, many equations and procedures have been devised, but the most widely adopted is undoubtedly that known as van der Poel's nomograph (van der Poel, 1954), a chart from which binder stiffness (E_{binder}) is derived from temperature (T) relative to softening point (SP), the penetration index (PI) of the binder (see Section 9.1.2 for details) and load pulse duration (t). The following approximate formula (Ullidtz, 1979) matches the predictions from the nomograph over a restricted range of input parameters.

$$E_{binder} = 1.157 \times 10^{-7} \times t^{-0.368} \times 2.718^{-PI} \times (SP - T)^5 \text{ (MPa)} \tag{9.4}$$

The parameter that investigators have found to be the most useful in describing mixture proportions is *voids in mixed aggregate* (VMA), basically the percentage of a mixture that is not aggregate or, to put it another way, the binder volume plus the volume of air voids. Here too, many different equations have been proposed, of which the following is one (Brown, 1978)

$$E_{\text{mixture}} = E_{\text{binder}} \times \{1 + (257.5 - 2.5\ \text{VMA})/[n(\text{VMA} - 3)]\}^{n} \tag{9.5}$$

where $n = 0.83 \log_{10}(4 \times 10^4/E_{\text{binder}})$ and VMA is a percentage.

Note that binder content is not taken into account, except as a component of VMA. This is logical, as it will only be the relatively small amount of binder that sits in the immediate vicinity of a particle contact that significantly influences the mixture stiffness. VMA itself acts as a broad measure of the efficiency with which aggregate particles are packed together.

In reality, such predictions can be no more than approximate. VMA is a very crude measure of packing when the infinite number of possible aggregate gradations is considered. Furthermore, as interparticle slip is an element in asphalt strain, this means that the frictional properties at contacts should also be relevant. In fact, it may well be that the commonly observed difference in stiffness between limestone and granite mixtures (see Section 9.2.3) is at least partly due to a difference in frictional properties, something that often leads to a similar proportional difference in unbound mixtures.

9.3.2 Measuring asphalt stiffness

The basic range of possible laboratory tests for asphalt stiffness is shown in Figure 9.11, namely tension–compression, indirect tension and flexure. Specifications for stiffness modulus tests on asphalt include tension–compression (AASHTO, 2015, 2022b), indirect tensile (ASTM, 2020a; BSI, 2018a) and four-point bending (AASHTO, 2022c; ASTM, 2021a).

Of these, the tension–compression test has the least complex stress conditions and is therefore favoured by many researchers; however, it is the least practical to carry out. The problem is the need to glue the end platens. Even if the glue is effective, there will be a concentration of stress around the perimeter at the glue–specimen interface, potentially causing premature specimen

Figure 9.11 Asphalt stiffness tests

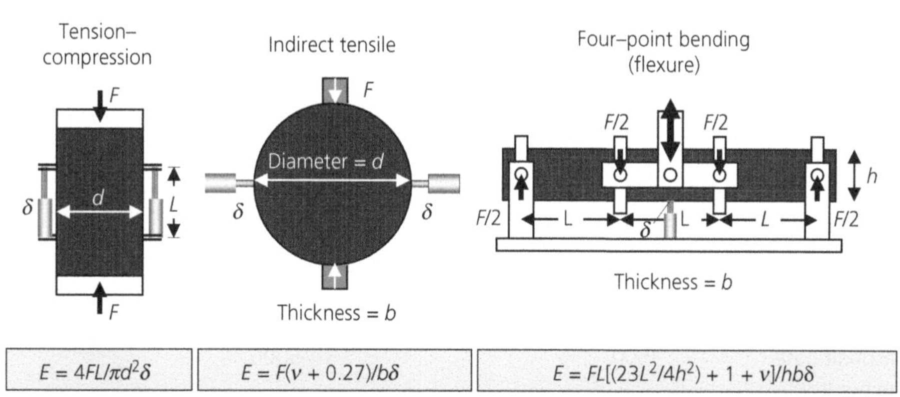

failure. The only way to overcome this is to use a necked specimen, greatly adding to the complexity of specimen preparation. In contrast, the indirect tensile test is very quick and easy to conduct, and is therefore favoured by industry, but it suffers from having highly complex stress conditions, particularly near to the loading points, and the interpretation is therefore less certain. Another problem is that, in the formula quoted in Figure 9.11 for stiffness derived from the indirect tensile test, the result is dependent on the value assumed for Poisson's ratio (ν), and this can only be estimated. The four-point bending test would not usually be carried out except when testing for fatigue, but it also gives a very satisfactory stiffness measurement. One of the key drivers is to be able to test a specimen that can be prepared from a core taken from the pavement – which means that the indirect tensile test scores highly.

It is worth making the point here that stiffness measured in any of these test modes cannot be used directly in pavement analysis and design without correction. The test temperature is, of course, important, and testing should preferably be carried out in a temperature-controlled environment, the specimen having also been stored at the test temperature for several hours prior to testing. However, the loading rate used is also critical. A pulse of load from a vehicle on a road may typically take about 10–15 ms to reach peak stress, whereas many tests are carried out at a considerably slower rate. Thus, for example, in the commonly used pneumatically powered Nottingham asphalt tester (NAT) (Brown *et al.*, 1994) and its subsequent developments, with 124 ms to peak, the stiffness measured would typically be only around 70% of that applying under fast-moving traffic. This is important since the NAT and its several commercial descendants has become one of the most commonly used pieces of asphalt testing equipment worldwide, providing an industry-friendly indirect tensile test, as well as offering both fatigue and permanent deformation options.

9.3.3 Typical stiffness values

As with similar sections in this book, the information given here is intended to give the reader a general idea of what to expect from various generic material types, but should definitely not be taken as applying in every case. Table 9.1 compares typical stiffness moduli at 20°C, while Figure 9.12 presents data on different asphalt types, together with a possible stiffness conversion equation.

Note that the values in Table 9.1 are for new asphalt, which means asphalt that is only a matter of months old. As already noted, bitumen ages – which means that the stiffness of a mixture will tend to increase throughout its life as the viscosity of the binder increases. For example, if a new dense asphalt base with 50 pen binder has a stiffness in the road of about 5000 MPa at 20°C at the

Table 9.1 Typical stiffness modulus data – new asphalt

Material	Stiffness modulus at 20°C: MPa	
	In the laboratory (125 ms to peak load)	In the pavement (10 ms to peak load)
Dense asphalt base (50 pen binder)	5000	7000
Dense asphalt base (100 pen binder)	3500	5000
Surface course	2000	3000

Figure 9.12 The effect of temperature on asphalt stiffness. Data from Thom *et al.* (1997) and Halle *et al.* (2012). NH equation from UK National Highways design standard CD229 (National Highways, 2020a)

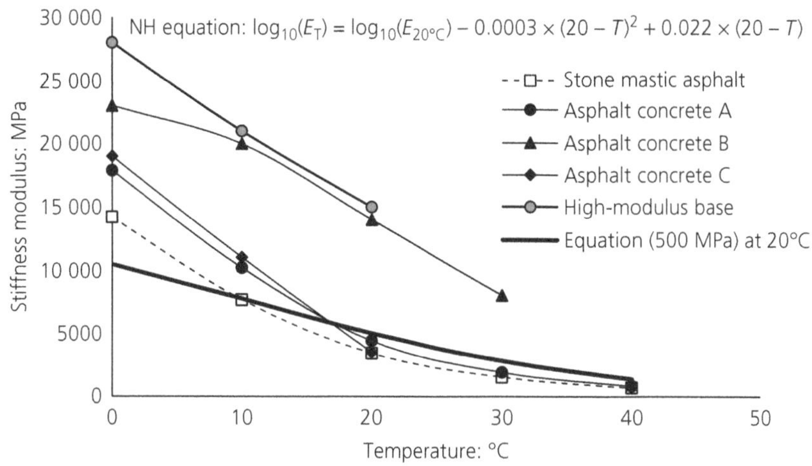

start of its life (by which time the penetration of the bitumen has already decreased to around 35 purely as a result of mixing and laying), then this will probably have increased to around 7000 MPa after 10 years in a climate such as that of the UK, with much more rapid stiffness increase in hotter climates. This would, of course, be offset by any stiffness decrease caused by microcrack damage.

Finally, it should be noted that a value of 0.3 or 0.35 is commonly adopted for Poisson's ratio of asphalt.

9.4. Fracture and fatigue of asphalt

One of the key design issues for an asphalt pavement is cracking. Cracks may arise for a number of reasons – as described in Chapter 10 – but they are always a problem, leading to loss of ride quality and also allowing water to enter the pavement foundation. They should, therefore, be properly understood.

9.4.1 Low-temperature fracture

This is a significant problem in continental climates. In fact, the label 'low-temperature' is slightly misleading, as the primary driving force is that of a large day–night temperature difference. The point is that asphalt, like any other material, expands and contracts with temperature changes. Bitumen itself is particularly susceptible, giving asphalt a typical thermal expansion coefficient of around 2×10^{-5}/°C, which is significantly higher than most engineering materials. This means that night-time cooling imposes a large volumetric strain. If the bitumen is soft, the asphalt can readily accommodate this by adjusting itself such that all the strain occurs either vertically or laterally, rather than longitudinally (i.e. along the road). However, if the bitumen is hard then such self-adjustment induces high internal stress and, in the limit, the stress level will be high enough to cause fracture. Chapter 10 covers the typical form of damage that results.

In climates such as those in North America, Northern Europe and Central Asia, the critical point occurs when the night-time temperature drops well below zero; in climates such as the deserts of

Arabia and the Sahara, it is still the night-time minimum that is the problem. Although the temperature might seem to be high enough for safety (often still above freezing), the problem is the rapid ageing of bitumens in desert climates, making them brittle enough to fracture.

Tests for susceptibility to low-temperature fracture are empirical. The primary tests are those already mentioned for the bitumen itself (see Section 9.1.4), but it is also not uncommon to carry out fracture strength tests on the mixture. BSI (2020a) and AASHTO (1993) both describe low-temperature fracture tests, including the thermal stress restrained specimen tensile strength (TSRST) test, in which a specimen is held at a constant length while the temperature is reduced. ASTM (2021b) describes a more general test for fracture energy of asphalt. Interpretation is based on experience.

9.4.2 Fatigue damage

There is much misunderstanding of this subject among pavement engineers. Part of the problem is that engineers are often taught about fatigue in the context of metals, in which a crack first has to 'initiate' at some defect within the material and then 'propagate'. This is an excellent representation of what happens in a relatively uniform material such as a metal, and expressions such as the well-known Paris law have been derived to describe this behaviour. The Paris law (Paris and Erdogan, 1963) relates the rate of crack propagation to a 'stress intensity factor' at the crack tip – a measure of stress concentration that takes account of the sharpness of the crack. However, asphalt is quite different from metals.

Asphalt consists of an aggregate skeleton surrounded by binder, with air voids scattered around. The key stress concentration points are at particle–particle contacts and, as described above, local fracture has to occur if damage is to take place. Fatigue therefore consists of the enlargement of these fracture zones at contacts and the development of new fractures (Maillard *et al.* (2004) and Nguyen (2020) describe tests that involve exactly this type of interparticle fracture). The effect is primarily controlled by movement within the aggregate skeleton (i.e. strain within the mixture); large movement will rapidly expand fracture zones and initiate new ones. Thus, it is not surprising to find that fatigue life is primarily a function of the 'elastic' strain under load within the mixture.

9.4.3 Tests for the fatigue of asphalt

Many of the same test configurations that have been introduced already are also suited to fatigue testing. Tension–compression, indirect tensile and flexural tests have all been successfully employed, as well as more complex configurations such as semi-circular bending. Tension–compression induces the simplest stress conditions and is used extensively by researchers, but it is not as convenient as other configurations. The trapezoidal shape included in Figure 9.13 may seem awkward, but it has the advantage that it produces a maximum tensile stress well away from any stressed boundary (typically at about one third of the specimen height). Figure 9.13 presents typical test data, expressed as loss of stiffness as a function of number of load cycles (relative to the number of cycles to specimen failure). Specifications for fatigue testing include ASTM (2021a) and AASHTO (2022c) for four-point bending and BSI (2019) and ASTM (2023a) for semi-circular bending.

A very important distinction exists between tests that are load controlled, in which strain increases during the test, accelerating towards failure at the end of the test, and those that are displacement controlled, in which stress reduces during the test as damage occurs and stiffness reduces. These two types of test give quite different fatigue lives, even with identical stress and strain at the first load cycle. In the case of displacement control, there is often no actual failure of the specimen and no visible damage at all until the stiffness has reduced to a small fraction of its original value.

Figure 9.13 Asphalt fatigue tests. Data from Choi *et al.* (2005)

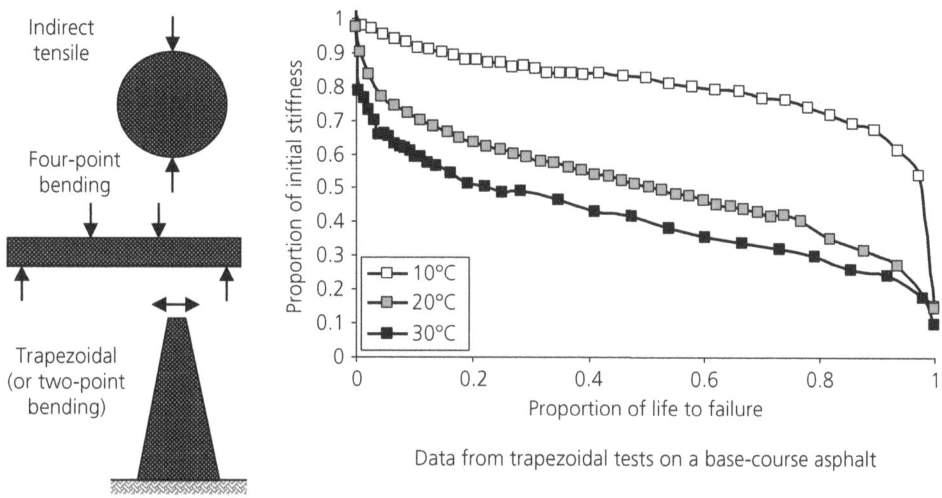

Specialists argue as to which type of test is the more appropriate for pavements, generally coming to the conclusion that displacement control is appropriate for thin asphalt layers, where strains are principally influenced by other layers, whereas load control is more appropriate for thicker layers, where the asphalt is the dominant material. The truth is, of course, that neither is 100% realistic. Furthermore, laboratory loading can never be realistic because it is impractical to duplicate the sequences of multiple loads followed by at least a second or two of 'rest' induced by real vehicle loading. The result is that those who wish to apply laboratory fatigue data in practice commonly have to treat the test as an indicator and then apply a considerable adjustment factor to the data in order to arrive at a usable result.

Key practical test requirements are as follows.

- The specimen size should not be so small in any of its dimensions that the result can be greatly influenced by local aggregate orientation.
- The test should be simple enough for practical application.

It is also desirable to avoid excessive permanent deformation during a test, this being a significant criticism of the indirect tensile and semi-circular bending tests. It means that, for some materials, particularly low-stiffness materials, the results may understate the true fatigue resistance.

9.4.4 Development of a fatigue characteristic

Whichever test is used, it is necessary to express the data in a meaningful way. The first task is to select a failure criterion. If the test is load controlled, as in the example in Figure 9.13, this is no problem; the specimen falls apart at the end of the test. However, if displacement control is used, then it is usual to work to 50% stiffness loss in place of total failure. This is an arbitrary value, but is justified as commonly representing the point where damage begins to accelerate after a relatively stable phase. Note that the data in Figure 9.13 suggest that, for any given material, this assumption is only valid over a certain temperature range.

Next, a series of tests has to be carried out at different levels of load (and therefore stress) or displacement (and therefore strain), generally at least six tests and preferably ten. The results can then be plotted as shown in Figure 9.14. It is usual to plot against strain (strain at the first load cycle in the case of a load-controlled test) rather than stress, as experience has shown that the resulting 'characteristic' is approximately independent of temperature and loading rate, at least within a broad range, whereas this is not the case in a plot against stress. This is not entirely unexpected because, as was argued in Section 9.4.2, damage is primarily related to interparticle movement, that is, strain.

Here, a warning must be given against assuming that the test temperature will never affect the fatigue life as long as it is expressed against strain – something that should be obvious from a glance at the different forms of behaviour shown in the example in Figure 9.13. Experience is that life will be longer (i.e. the fatigue characteristic in Figure 9.14 will be further to the right) if the temperature exceeds a certain value. This value will generally be over 30°C, but not in all cases. This means that it is *not* automatically acceptable to adjust the test temperature upwards in order to allow a target strain to be reached at a lower load level. In fact, there is some evidence (Chen *et al.*, 2021) that there exists a pessimum temperature, somewhere between 5°C and 20°C, at which the fatigue characteristic is farthest to the left, improving at both higher and lower temperatures.

The use of a fatigue characteristic in pavement analysis and design is explained in Chapter 10.

9.4.5 Healing

This phenomenon has been referred to already in relation to asphalt stiffness, the point having been made that bitumen fracture at particle contacts tends to occur in cold weather, but that binder continuity can be restored in warm weather due to viscous flow and molecular remixing. The healing effect is responsible for the improved fatigue lifetimes found in tests at high temperature; it also means that there will be differences between tests depending on the details of load pulse frequency and any 'rest periods' allowed between load pulses. This is illustrated in Figure 9.15. The data plotted (Oliveira, 2006) were actually obtained on a grouted macadam mixture (Section 9.7) with

Figure 9.14 An asphalt fatigue characteristic

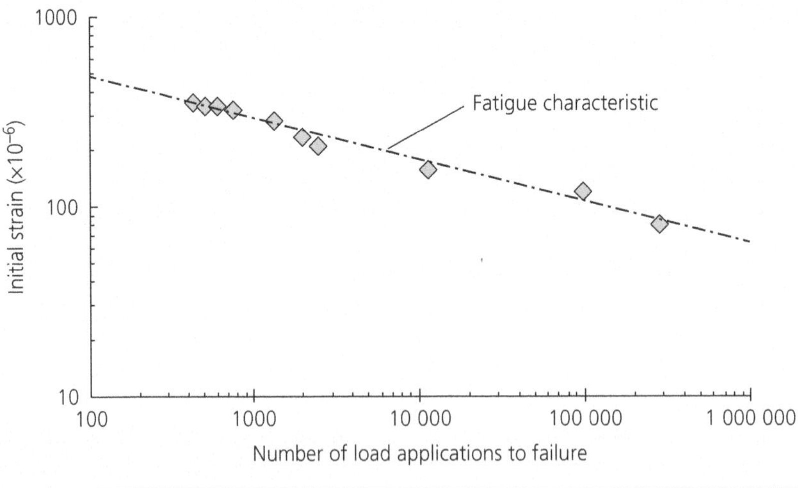

Figure 9.15 The influence of rest periods on fatigue behaviour. Data from Oliveira (2006)

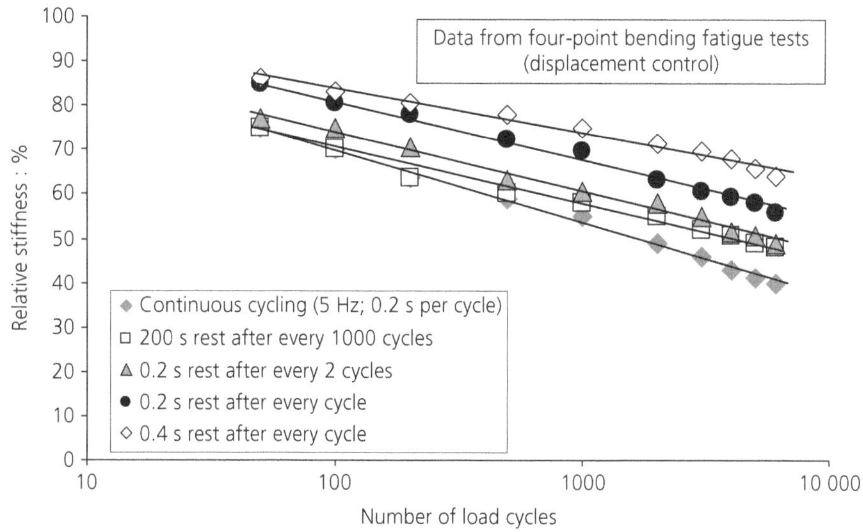

200 pen (i.e. very soft) binder. The use of this binder will have accentuated the healing effect, but the trends shown also apply to conventional asphalt mixtures to some extent.

However, it has to be admitted that understanding of healing is still incomplete. It is partly because of healing that it is not possible to apply a laboratory fatigue characteristic directly in pavement performance prediction, because the degree of healing in the pavement will be quite different from that in the laboratory test; a 'shift factor' (or 'transfer function') will be required (see Chapter 10 for details).

9.5. Permanent deformation

As in the cases of stiffness and fatigue, it is necessary to remember the fundamental structure of asphalt when considering permanent deformation. Asphalt can only deform if the aggregate skeleton deforms; thus, the resistance of an asphalt to deformation is intimately tied in with the stability of the aggregate skeleton. This means that the same properties that lead to low plastic strain in an unbound material (angular particles, broad gradation, large particle size) also give stability to an asphalt. Thus, although binder technology may help, it is impossible to produce a deformation-resistant mixture without an appropriate aggregate skeleton.

9.5.1 The influence of binder

Despite the secondary nature of the role of the binder, it is still extremely important. Plastic strain of the aggregate skeleton is, of course, greatly inhibited by the binder. In fact, throughout most of the year in a temperate climate, when pavement temperature is cool or moderate, the plastic strain will be negligible. This is because the binder takes much of the stress away from the particle contacts. However, as the temperature increases, the stress taken across particle contacts increases, leading to an increased danger of interparticle movement – and some interparticle movements will be irreversible. The binder may self-heal, but the aggregate skeleton will be permanently deformed.

Thus, a hard binder means that the temperature of the pavement has to climb higher before the onset of significant plastic deformation. In practical terms, experience is that it is dangerous to let the pavement temperature exceed the ring and ball softening point of the bitumen.

However, despite its beneficial binding action, bitumen also carries with it a latent danger. Surprising though it may sound, bitumen can actually exhibit the same sort of *pore pressure effect* as exhibited by water in an unbound material. As the aggregate skeleton strains elastically under load, some of the intervening voids will naturally decrease in size and some will increase, purely as a result of the aggregate geometry. If there is plenty of air in the system this is no problem, as the air will readily expand and contract. However, a void that is full of binder cannot change its volume, which means that the binder may have to carry more load than is good for it. Extra compressive stress in the binder means less normal stress at aggregate contacts, which means easier slip and, therefore, more plastic strain. This is important. Many pavements develop rutting simply because there is not enough air in the system to prevent this pore pressure effect. Most specifications guard against this by insisting that a minimum of 2–3% air voids is built into the mixture, but the engineer should be aware that further compaction under traffic is quite possible, and that an air-voids content under 2% will almost certainly lead to deformation (i.e. rutting).

9.5.2 Measurement of permanent deformation

There is no single property defining resistance to permanent deformation, which means that testing can only be indicative. The most usual forms of test are as follows.

- A repeated load axial test, whereby a cylinder of material is loaded repeatedly in compression (BSI, 1996). A similar test can also be carried out in the US asphalt mixture performance tester described in TP79 (AASHTO, 2015). Both generate indicative numbers that are usually interpreted based on empirical experience.
- A repeated load triaxial test – simply an axial test with the addition of confining stress. Confinement can be applied by means of an external pressure within a surrounding cell or by means of a (partial) vacuum applied to the specimen when sealed by a flexible membrane.
- A wheel-tracking test.

The axial test is the most convenient because the test arrangement is simple and the specimen can be obtained from a core through a pavement. The test is often criticised as being overly severe on base materials, where a considerable degree of confinement would in reality be present; however, as it is not intended to be more than an index test, this is probably not a fair criticism. In fact, the severe nature of the test means that intrinsically poor materials show up very clearly indeed, giving the designer useful information when considering whether, and at what level in the pavement, such materials may be used. Figure 9.16 presents datasets for materials of differing quality. As a rule of thumb, under the test conditions used for the data in Figure 9.16, 1% strain represents a safe limit, 2% spells danger and 3% means that deformation is practically certain if the material is placed within about 100 mm of the surface. Note that, in this case, tests were carried out both dry and following 'moisture conditioning' (see Section 9.6.2), as many materials are found to perform worse when water is present.

The triaxial test undoubtedly allows a more accurate representation of the real stress regimen applying in a pavement and also, in theory, allows the depth at which the material is to be used to be simulated by setting the level of confinement appropriately. However, the test is more complex than the axial set-up, and the 'appropriate' level of confinement is open to debate. As it is still only

Figure 9.16 Repeated load axial test data – 100 kPa, 40°C (Airey *et al.*, 2004)

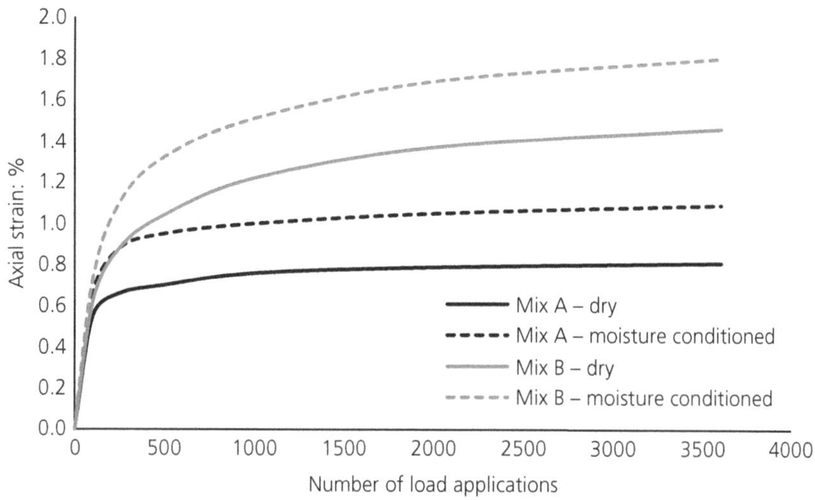

an index test, it is not clear that the triaxial test has sufficient advantages to warrant its general use, although it is undeniable that it gives a much better indication of the performance of low-stability materials such as porous asphalt (see Section 9.7).

The Superpave shear tester can be used to provide input into the Superpave asphalt mixture design method (see Section 9.7.2), and measures elastic, viscous and visco-elastic strain in shearing mode. Among other things, therefore, it gives a measure of resistance to permanent deformation.

Many mixture design specifications, however, insist on a wheel-tracking test, at least for surface course materials. All are relatively small-scale tests such that the device can be contained within a temperature-controlled cabinet. Details of the load used and test temperature vary from specification to specification, but it is usual to use a high temperature (45–60°C) in order to give a relatively rapid result (see BSI (2020b) for a range of testing options). The asphalt pavement analyser (AASHTO, 2010) is a popular and highly flexible US wheel-tracking system capable of applying very high stress over a range of temperatures. The close simulation of reality given by wheel-tracking means that the results, although still index numbers, can be treated as directly related to deformation susceptibility in the pavement.

9.6. Durability

Durability is a catch-all word, and it actually includes several different phenomena, the common theme being that they all take time to develop. Unfortunately, the differences between the different aspects of durability are not always recognised, which means that it is all too easy to allow substandard materials to 'slip through the net'. This section will consider each aspect of durability in turn.

9.6.1 Ageing

This subject has been identified already (see Section 9.1.7) as affecting all bitumens. Chemical changes over time mean that bitumen becomes stiffer – a good thing – and more brittle – not so good. Thus, a fatigue characteristic, plotted against applied strain as in Figure 9.14, should tend to shift slightly to the left as ageing takes hold. However, experience suggests that this shift is rarely significant in standard tests carried out at standard temperature (often 20°C), which unfortunately suggests to some that ageing is not a problem at all. The trouble is that standard tests are always rapid, producing failure in a matter of hours rather than years, and this means that the vital component of asphalt behaviour described in Section 9.4.5, namely healing, is not properly taken into account; and an aged bitumen will heal much less readily than an unaged bitumen. To put it another way, the temperature threshold above which healing is able to occur will be higher for an aged binder. Either way, the result is that, in the pavement, asphalt is often observed to reach a state of ageing where distress begins to accelerate.

It is important to understand the factors that affect ageing. These are: *temperature, aggregate absorption* (of binder) and the presence of *oxygen*. Ageing will therefore occur most severely in hot climates, near the road surface, in voided mixtures and where high-absorption aggregates are used. In a dense base, in a temperate climate, using a standard aggregate, ageing should be extremely slow, giving many decades of trouble-free life. Here, then, is another excellent reason for achieving good compaction during construction.

9.6.2 Water damage

The point has been made that it is essential to dry aggregate thoroughly during asphalt production in order to ensure good bitumen adhesion, but even then it is sometimes necessary to add lime or some other adhesion agent in order to facilitate the development of adhesion. The implication is that the bitumen–aggregate interface is potentially vulnerable to water. Once again the key is high density and low permeability. If no water ever reaches these vulnerable interfaces, there will never be any water damage – although the slow process of water diffusion, even through bitumen, will always result in some water molecules making it to the danger zone.

There is no agreement as yet on the most appropriate tests to determine susceptibility to water damage. Permeability can be measured directly in a high-pressure permeameter, although this is not a standard tool. Voids content can, of course, also be determined by comparing the density (mass/volume; usually determined from the weight of water displaced by the specimen – wrapped in foil or film) of a compacted mixture with that of the component pieces (the so-called ‘rice density’). An approximation of the density of the component pieces can be found by weighing the specimen in water unwrapped, allowing water to penetrate the voids, although it should be appreciated that enclosed voids will remain unfilled and that the estimate of voids content that results will be lower than is actually the case. However it is determined, experience suggests that a voids content of about 5% or less almost certainly means a sufficiently low permeability for water damage to be unlikely.

Direct testing can also be conducted by soaking specimens in water, usually at an elevated temperature, and then measuring the ‘retained’ stiffness (or strength); that is, a percentage value in comparison with the unsoaked condition (see e.g. AASHTO, 2022d; BSI, 2018b). Typically, a value of 80% or more is taken to represent non-water-susceptible material. Figure 9.17 illustrates some of the ‘tools of the trade’. Note that the saturation ageing tensile stiffness (SATS) test (Collop *et al.*, 2004; National Highways, 2020b), also known as the immersion ageing test, allows simultaneous application of water and accelerated ageing due to a raised temperature and pressurised air.

Figure 9.17 Approaches to water-susceptibility testing

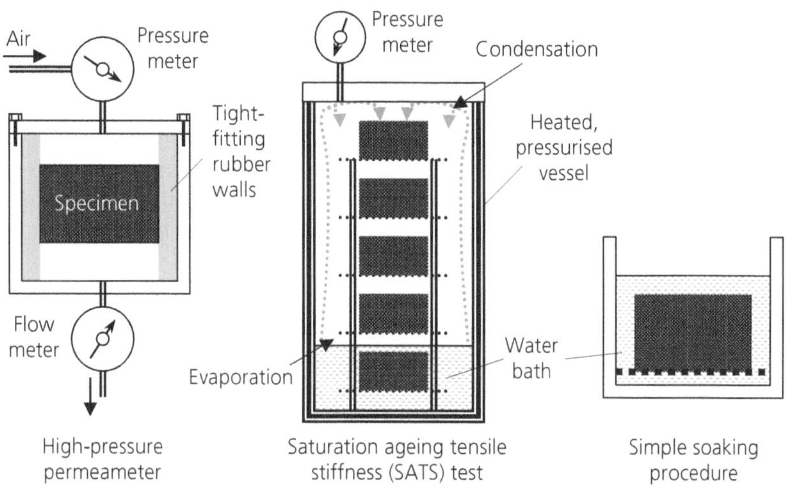

Even with these different tools and methods, the issue being addressed by no means covers the whole subject of water damage. A key element that is missing is the dynamic movement of water within and between void spaces under the action of traffic. For the specific case of degradation of a surface course, the Hamburg wheel tracker (AASHTO, 2023b; BSI, 2020b) allows trafficking of an immersed surface, and is particularly useful to determine the likelihood of surface ravelling (or fretting) (i.e. the loss of stones from the asphalt). Other tests are, at this stage, purely for research purposes (see Gubler *et al.*, 2004). The effect of water is most likely to be felt at the base of a layer, where the number of voids tends to be highest, and where water may become trapped due to high density at the top of the layer beneath. It is also likely where pavements suffer from flooding. However, the subject is so complex that it is hard to simulate correctly through direct testing. It must therefore simply be remembered that, if water is allowed to penetrate and accumulate within an asphalt, there will be trouble! Wider considerations of pavement drainage are presented in Chapter 24.

On a related note, there are certain aggregates that can be damaged by chemical attack. Specifications guard against such materials in asphalt by insisting on a level of particle hardness, determined indirectly from tests such as the Los Angeles abrasion test (AASHTO, 2019; ASTM, 2016, 2020b; BSI, 2020c) or the micro-Deval test (AASHTO, 2022e; ASTM, 2017; BSI, 2011), or by insisting on soundness against certain chemicals (AASHTO, 2022f; BSI, 2009).

9.6.3 Frost damage

This topic will be considered in relation to pothole formation in Section 12.1.5.1. Here it will simply be noted that in most climates, aggregates in asphalt have to be frost-resistant and that the same tests as listed in the previous section for hardness and chemical soundness are typically used as a surrogate for frost resistance.

In general, though, the secret of protection against frost is more or less the same as that of protection against water. If the asphalt is dense ($\leq$5% voids), then water will not find it easy to penetrate and ice formation will therefore not be a problem. If water is allowed to penetrate, expect problems.

9.6.4 Biodegradation

This is a topic that very few pavement engineers take seriously, yet it is a fact that certain bacteria can digest bitumen (e.g. Pendrys, 1989; Phillips and Traxler, 1963). This may help with eco-friendly credentials, but it is no help at all in ensuring a long-life road. On many occasions, pavement cores have revealed an almost totally disintegrated asphalt base and, while it is difficult to prove the cause, the apparent lack of visible binder strongly suggests a biological agent.

Unfortunately, this is not an area that is even being seriously researched yet, so it is certainly not possible to comment on which brands of asphalt are the tastiest! But, bearing in mind the current research drive to find bio-alternatives to bitumen, it is a topic that may yet reappear.

9.7. Mixture types and design

There are an infinite number of possible combinations of particle size distribution and binder content, but certain generic types of mixture can be identified as follows.

- *Asphalt concrete* (or *dense bitumen macadam*) has a broadly graded aggregate designed to give maximum aggregate packing density, which leaves limited void space and therefore demands a minimised binder content. Cost is therefore also minimised, making asphalt concrete the mixture type of choice for most applications.
- *Stone mastic asphalt* is a specialist surface course material. It is 'gap-graded'; that is to say it has a skeleton of coarse aggregate to give stability to the mixture and just enough fines and binder to nearly fill up the void spaces. The binder content is a little higher than for asphalt concrete.
- *Hot-rolled asphalt* (HRA) is also most commonly used as a surface course. It has a reduced coarse aggregate content, a high content of sand-sized material and a high binder content. This gives HRA excellent resistance to cracking but a relatively high cost. As a surface course, it requires the addition of bitumen-coated chippings to give skid resistance.
- *Porous asphalt* is deliberately designed with a high voids content (around 20%) in order to act as a drainage material. It therefore consists largely of coarse aggregate with only a limited amount of fines and a relatively low binder content. In order to stop the binder from draining out at elevated temperatures (i.e. during transport), cellulose fibres are added.
- *Grouted macadam* is basically a porous asphalt with voids subsequently filled using a strong cementitious grout. As it sets, the grout will fragment to form additional pieces of aggregate that interlock almost perfectly with the stones of the porous asphalt. The binder content will be very low and the deformation resistance (and the cost) very high.

However, these are broad descriptions only. There are a lot of potential pitfalls in designing an optimised bituminous mixture, and a lot of variables to be juggled. The only sensible route forward, therefore, is to tackle the problem one step at a time.

9.7.1 Aggregate particle size distribution

Here the choice will be based primarily on the role of the material in the pavement. Base layers will usually tend to contain larger sized particles because they are placed in reasonably thick layers, and this also allows the content of (relatively expensive) bitumen to be minimised. Surface courses will have a smaller maximum particle size, partly to give optimum ride quality, but also because such layers are usually thin; they also typically require a relatively high binder content to ensure low voids and high durability – the exception being *porous asphalt*.

Having selected a maximum particle size, the next issue is the gradation. Here, the approach taken by the US Superpave specification (Asphalt Institute, 2001), in which a comparison has to be made to an idealised Fuller curve, is believed to be fundamentally sound. Fuller curves, as introduced previously, represent particle size distributions that obey the formula

$$\text{Percent passing} = (d/d_{max})^n \tag{9.6}$$

The advice given in the Superpave specification is that $n = 0.45$ represents a 'perfect' gradation, in which there is just enough room between particles at every size to accommodate all the smaller-size fractions in the mix. This implies that a mix that follows the $n = 0.45$ Fuller curve will develop excellent interlock between particles and will require minimum binder content to 'glue' the particles together because the void space has been minimised. It will not be possible to pick out a single critical particle size; all size fractions will be 'pulling their weight' equally. This is the material known as asphalt concrete or dense bitumen macadam. Figure 9.18 illustrates.

Now, consider what happens if the gradation deviates from the perfect Fuller curve. A less steep gradation curve means that there are always more fine particles at every stage than are needed to perfectly fill the spaces between larger particles. The consequence is that the larger particles will be pushed apart and will 'float' in a sea of smaller material. The dominant particle size will be small (but not infinitely small because the curve cannot in reality follow such a path for ever at the small-size end of the spectrum) and the binder demand will increase. The result will tend to be a less stable mix because of the dependence on small-size particles in the aggregate skeleton and therefore a greater susceptibility to permanent deformation; however, it will be a relatively easy mix to work with and compact, and the final voids content should be very low indeed. Stiffness will be slightly lower than for the 'perfect' gradation, but fatigue resistance will be high because of the higher binder content. The traditional UK material known as hot-rolled asphalt has this sort of gradation and, if Figure 9.18 is a fair description, the *critical particle size* (where the grading touches the Fuller curve) will be 1–2 mm. This means that great reliance is placed on the properties (principally shape) of the sand-size fraction (i.e. the critical size). The German material known as *Gussasphalt*, commonly applied as bridge-deck surfacing because of its impressive durability, is an even more extreme example with a very small critical particle size indeed. The specification for this material is DIN 18317 (DIN, 2019).

Figure 9.18 Aggregate gradations in asphalt mixtures

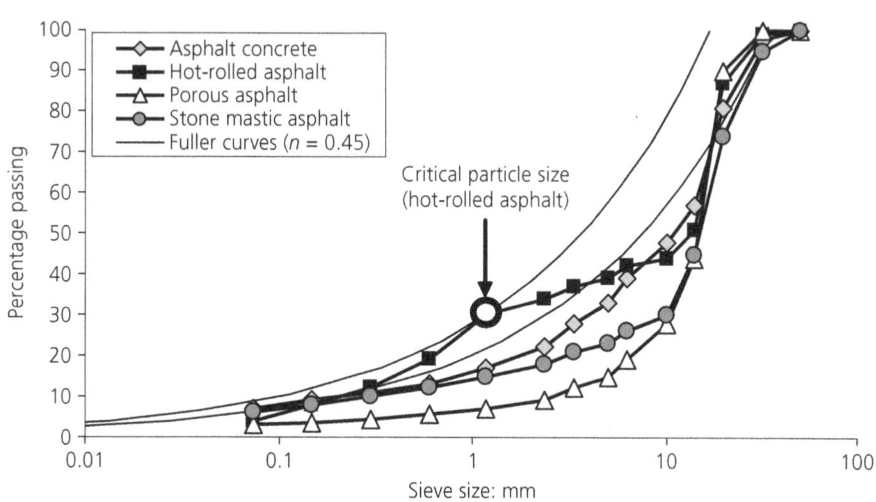

However, what happens if the gradation is steeper than the $n = 0.45$ Fuller curve? In this case, there are never enough smaller particles to fill the voids between larger stones. The result will be an aggregate skeleton comprising the larger stones, with the remainder of the material loosely packed into the spaces between. This is a recipe for high stability; it is also a recipe for high voids, low stiffness, high susceptibility to water damage and low fatigue life. In short, it is not desirable unless there is an overwhelming reason to adopt such a mix and the only reason likely to be powerful enough is the desirability of *porous asphalt* surfacing for drainage of surface water and low tyre noise (see Chapter 11).

Of course, most practical mixtures are compromises. Stone mastic asphalt follows a steep gradation initially, leading to a skeleton of large stones, but then flattens such that there are nearly enough fine particles to fill the voids, overcoming the potential problems of porous asphalt. However, this means that it is a slightly 'dangerous' mixture choice as it is very sensitive to small changes in gradation. A little too many fines and the stability provided by the large aggregate skeleton will be lost; a little too few and the problems of porosity, permeability, water damage and so on may reappear.

The key for the engineer is to be aware of the particular characteristics that apply to each generic mixture type. One should never take the view that all asphalt mixtures are equivalent.

9.7.2 Binder content

Having established the gradation to be used, the next question is how much binder to add. At this stage it is worth stating the obvious point that binder content as a percentage by mass is quite different from its percentage by volume – because the specific gravity of bitumen is so much less than that of rock. The volume percentage determines mixture properties and is typically in the range 8–12%; however, it is the mass percentage (typically 4–6%) that is most easily measured, and this is therefore the figure specified and 'understood' by practical engineers. Calculation Sheet 9.3 goes through the conversion between the two.

Calculation Sheet 9.3. Calculating mixture volumetrics

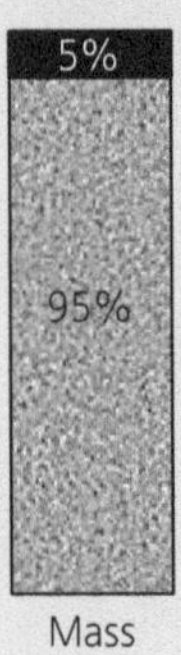

Mass

Say 1 kg of asphalt:
  5% binder content = 50 g
  95% aggregate = 950 g

Say the specific gravity of the binder is 1.03:
  So, 50 g binder → 48.5 cm^3

Say the specific gravity of the aggregate is 2.70:
  So, 50 g binder → 351.9 cm^3

Total volume = 400.4 cm^3

If 6% air voids → 94% binder + aggregate

Binder volume = 94 × 48.5/400.4 = 11.4%

Aggregate volume = 94 × 351.9/400.4 = 82.6%

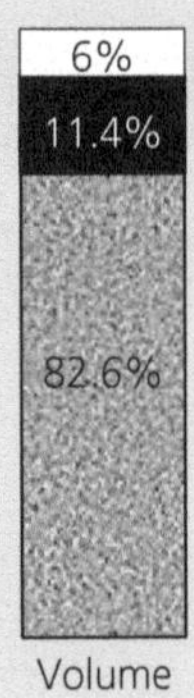

Volume

Anyway, back to the optimum binder content. Chapter 6 looked at the role of water in unbound material compaction and found there was an optimum, giving the maximum dry density (i.e. the best aggregate packing). In a hot- or warm-mix asphalt, bitumen replaces water and, for a given mix, there will be a particular bitumen content that gives optimum compaction, which usually means optimum strength, stiffness and resistance to deformation. The following subsections present the two best-known methods for determining design bitumen content in use today.

9.7.2.1 The Marshall mix design method

This practical technique was developed in the 1950s and its longevity speaks volumes for its user-friendliness and general reliability, despite the severe criticisms of purists. The design steps are as follows.

- Select a gradation.
- Make up a series of mixes at different binder contents.
- Prepare specimens using a Marshall hammer for compaction (Figure 9.19).
- Measure achieved densities.
- Carry out Marshall tests (see Figure 9.19) at 60°C and derive 'stability' and 'flow' values.
- Determine a design binder content based on stability, flow and density.

There are numerous modifications to the procedure in different specifications around the world, for example making use of a quantity known as the Marshall quotient (stability divided by flow) or replacing overall density by aggregate packing density, but in essence the procedure is as outlined in Figure 9.19. The most widely used is probably that given in MS-2 (Asphalt Institute, 2014).

The Marshall method gives optima for stability (logically linked to permanent deformation resistance) and density (linked to stiffness and durability). The limits placed on flow also guard against both poor workability (low flow) and excessive deformability (high flow). The exact choice of binder content will be a compromise. The user should, however, be aware of the following limitations of this method.

Figure 9.19 The Marshall mix design procedure

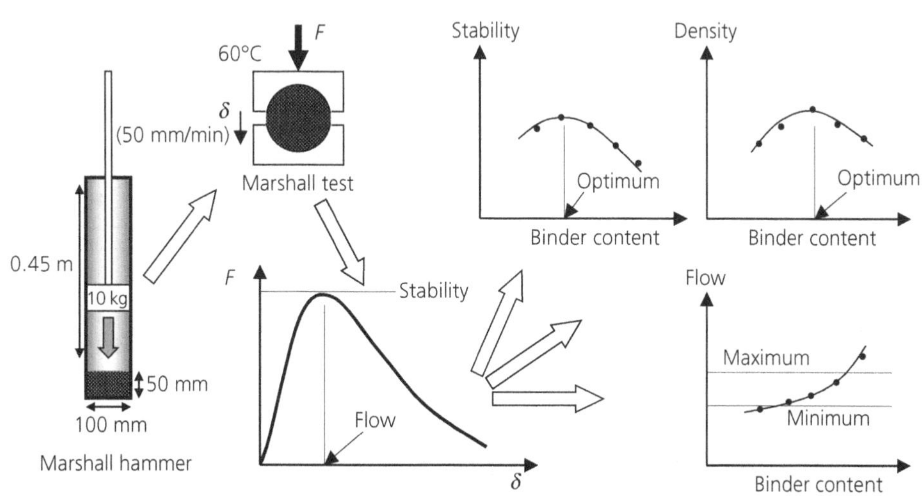

- While density may be linked to stiffness, it certainly is not a stiffness measure, and so gives no guarantee of a suitable value.
- There is no measure of fatigue resistance.
- It is unlikely that an adhesion problem would show up at 60°C.

In short, the Marshall method is a practical way of choosing the most suitable binder content, but on its own it does not ensure adequate performance in the pavement. It should always be used in conjunction with other checks.

9.7.2.2 The Superpave mix design method

This design approach grew out of research in the USA in the 1990s and is principally concerned with optimising mixture volumetrics (Asphalt Institute, 2003). The key steps are as follows.

- Select a gradation (only broadly graded mixtures covered; filler–binder ratio by mass between 0.6 and 1.2).
- Make up a series of mixes at different binder contents.
- Prepare specimens using a gyratory compactor (see Figure 9.20).
- Measure achieved densities.
- The optimum binder content is the one that gives a voids content of 4%.
- Prepare further specimens at the optimum binder content.
- Check voids at light and heavy compaction.

The additional checks at the end are to ensure that compaction does not occur too easily, an indicator of poor aggregate interlock, and that the voids content will never fall below 2%, even under heavy trafficking. Both checks are intended to avoid the danger of rutting.

Some of the same criticisms often levelled at the Marshall method could also be applied to the Superpave approach. It is another highly practical method for optimising volumetrics, but it does not in itself guarantee performance because there are no directly performance-related tests involved.

Figure 9.20 The Superpave mix design procedure

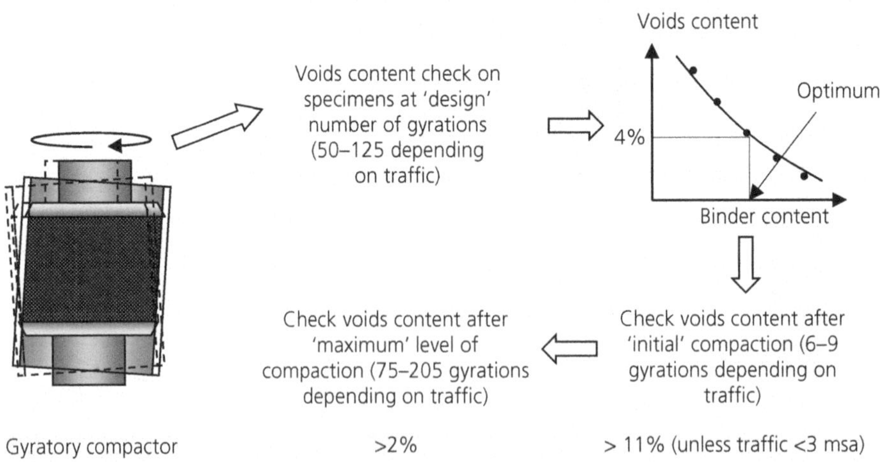

9.7.3 Binder grade

This depends largely on climate, explicitly so in the Superpave design approach and implicitly so elsewhere. The key point is that the binder should be able to perform satisfactorily over the full range of temperatures experienced in the pavement. Typically, these might be as shown in Table 9.2.

Low temperature presents the danger of fracture and fatigue, whereas high temperature can lead to rutting, and this means that there is a desirable working range of binder viscosity, approximately 5×10^3 to 10^7 Pa.s. Thus if, for example, the long-term binder penetration (of an unmodified bitumen) was expected to be about 50, this desirable viscosity range would equate to a temperature range of around −10°C to +45°C, and such a binder would be expected to perform well in a temperate climate but not in other climates. The US Superpave terminology designates binders to show their working range; thus a 64–28 binder is suited to temperatures between −28°C and +64°C.

In many climates it is just not possible to find a conventional binder that covers the expected temperature range satisfactorily; either the viscosity will be too low at high temperature or it will be too high at low temperature. In these cases, there are two options

- accept that damage will occur and plan accordingly
- pay extra and use a modified binder, extending the working temperature range.

Binder modifiers cannot work miracles. Styrene–butadiene–styrene (SBS) is one of the most effective polymers. When blended with the right bitumen, SBS has been reported to extend the working temperature range by up to about 50°C (approximately +30°C and −20°C) (Hunter, 2014). The highest Superpave grade listed is 76–40.

9.7.4 Filler

Filler, the proportion of aggregate passing a 75 μm sieve, is an extremely important component of the mixture. As any practical aggregate gradation will give a critical particle size of more than 75 μm, this means that filler will never form part of the asphalt skeleton; it simply loses itself between the larger aggregate particles, which puts it in the same ballpark as the bitumen. It is therefore rightly seen as a binder additive rather than as part of the aggregate.

Referring back to Section 9.1.5, filler can actually make a very effective binder additive indeed, multiplying the fracture and fatigue strength by a factor of up to about 3 (depending on exact filler

Table 9.2 Temperature ranges typical of different climates

Climate	Pavement surface temperature:°C			
	Winter		Summer	
	Low	Mean	Mean	High
Temperate (e.g. UK)	−10	+5	+20	+50
Cold continental (e.g. Canada)	−40	−15	+30	+55
Hot continental (e.g. India)	0	+15	+35	+60
Desert	−5	+15	+35	+65

content and type) and the stiffness by a similar factor. In fact, it appears from Section 9.1.5 that the more filler one mixes into a bitumen, the better the properties of the composite binder will become. Filler is, therefore, in principle a 'good thing' and it is critical to the properties of the resulting mixture. The successful use of a good-quality filler will

- enhance mixture stiffness and fatigue strength
- chemically assist in promoting aggregate–bitumen adhesion
- inhibit drainage of hot binder off the aggregate during transportation
- not prevent proper mixing
- not prevent proper bitumen–aggregate contact.

The last two points effectively put a limit on the filler content. Experience suggests that filler percentage by volume should always be less than bitumen percentage by volume in order to avoid mixing problems and to ensure that the large aggregate finds enough bitumen to coat it properly. In fact, mixtures are commonly found to work well when the filler percentage by mass is approximately equal to the bitumen percentage by mass – although of course this depends on the specific gravity and shape of the filler particles and the type of asphalt mix.

The second point in the above list is also important. Limestone and other alkaline fillers are known to be chemically supportive of adhesion development. Other filler types may perform perfectly well in most situations, but not with every aggregate type.

9.8. Summary

There are real differences between asphalts that perform well and last for decades and those that give endless maintenance headaches, and all the numerous subtleties of asphalt design described in this chapter are relevant. However, if one message is to be remembered it is this – make sure it can be compacted to a low voids content in the field.

Low voids mean high stiffness modulus, good rut resistance and, most importantly in most climates, protection against water damage. Other things are important (e.g. obtaining good adhesion between aggregate and bitumen), but achieving low voids is of over-riding importance – except of course in the special case of porous asphalt, which will be covered in Chapter 11.

REFERENCES

AASHTO (American Association of State Highway and Transportation Officials) (1993) TP10: Standard method of test for thermal stress restrained specimens tensile strength (TSRST). AASHTO, Washington, DC, USA.

AASHTO (2010) T340: Standard method of test for determining the rutting susceptibility of hot mix asphalt (HMA) using the asphalt pavement analyzer (APA). AASHTO, Washington, DC, USA.

AASHTO (2015) TP79: Standard method of test for the dynamic modulus and flow number of hot-mix asphalt (HMA) using the asphalt mixture performance tester (AMPT). AASHTO, Washington, DC, USA.

AASHTO (2018) TP 101-12: Estimating fatigue resistance of asphalt binders using the linear amplitude sweep. AASHTO, Washington, DC, USA.

AASHTO (2019) T96-02: Resistance to degradation of small-size coarse aggregate by abrasion and impact in the Los Angeles machine. AASHTO, Washington, DC, USA.

AASHTO (2022a) T316: Standard method of test for viscosity determination of asphalt using rotational viscometer. AASHTO, Washington, DC, USA.

AASHTO (2022b) T342: Standard method of test for dynamic modulus of hot-mix asphalt concrete mixtures. AASHTO, Washington, DC, USA.

AASHTO (2022c) T321: Standard method of test for determining the fatigue life of compacted asphalt mixtures subjected to repeated flexural bending. AASHTO, Washington, DC, USA.

AASHTO (2022d) T283-22: Standard method of test for resistance of compacted asphalt mixtures to moisture-induced damage. AASHTO, Washington, DC, USA.

AASHTO (2022e) T327-22: Standard method of test for resistance of coarse aggregate to degradation by abrasion in the micro-Deval apparatus. AASHTO, Washington, DC, USA.

AASHTO (2022f) T104: Standard method of test for soundness of aggregate by use of sodium sulfate or magnesium sulfate. AASHTO, Washington, DC, USA.

AASHTO (2023a) T240: Standard method of test for effect of heat and air on a moving film of asphalt (rolling thin-film oven test). AASHTO, Washington, DC, USA.

AASHTO (2023b) T324: Standard method of test for Hamburg wheel-track testing of compacted asphalt mixtures. AASHTO, Washington, DC, USA.

Airey GD, Collop AC, Thom NH and Zoorob SE (2004) Use of steel slag and recycled glass in bituminous mixtures. *Proceedings of the 3rd Eurasphalt and Eurobitumen Congress, Vienna, Austria*, pp. 1218–1232.

Asphalt Institute (2001) *Superpave Mix Design. Superpave Series No. 2 (SP-2)*, 3rd edn. Asphalt Institute, Lexington, KY, USA.

Asphalt Institute (2003) *Performance Graded Asphalt Binder Specification and Testing. Superpave Series No. 1 (SP-1)*, 3rd edn. Asphalt Institute, Lexington, KY, USA.

Asphalt Institute (2014) *Mix Design Methods for Asphaltic Concrete and other Hot Mix Types. Manual Series No. 2 (MS-2)*, 7th edn. Asphalt Institute, Lexington, KY, USA.

ASTM (2012) D4402-06: Standard test method for viscosity determination of asphalt at elevated temperatures using a rotational viscometer. ASTM International, West Conshohocken, PA, USA.

ASTM (2013) D5-06: Standard test method for penetration of bituminous materials. ASTM International, West Conshohocken, PA, USA.

ASTM (2014) D36/D36M-12: Standard test method for softening point of bitumen (ring-and-ball apparatus). ASTM International, West Conshohocken, PA, USA.

ASTM (2016) C535-16: Standard test method for resistance to degradation of large-size coarse aggregate by abrasion and impact in the Los Angeles machine. ASTM International, West Conshohocken, PA, USA.

ASTM (2017) D6928-17: Standard test methods for resistance of coarse aggregate to degradation by abrasion in the micro-Deval apparatus. ASTM International, West Conshohocken, PA, USA.

ASTM (2020a) D7369: Standard test methods for determining the resilient modulus of bituminous mixtures by indirect tension test. ASTM International, West Conshohocken, PA, USA.

ASTM (2020b) C131/C131M-14: Standard test method for resistance to degradation of small-size coarse aggregate by abrasion and impact in the Los Angeles machine. ASTM International, West Conshohocken, PA, USA.

ASTM (2021a) D8237-21: Standard test method for determining fatigue failure of asphalt-aggregate mixtures with the four-point beam fatigue device. ASTM International, West Conshohocken, PA, USA.

ASTM (2021b) D7313: Standard test method for determining fracture energy of asphalt–aggregate mixtures using the disk-shaped compact tension geometry. ASTM International, West Conshohocken, PA, USA.

ASTM (2022a) D2872-22: Standard test method for effect of heat and air on a moving film of asphalt (rotating thin-film oven test). ASTM International, West Conshohocken, PA, USA.

ASTM (2022b) D4541: Standard test method for pull-off strength of coatings using portable adhesion testers. ASTM International, West Conshohocken, PA, USA.

ASTM (2023a) D8044: Standard test method for evaluation of asphalt mixture cracking resistance using the semi-circular bend test (SCB) at intermediate temperatures. ASTM International, West Conshohocken, PA, USA.

ASTM (2023b) D6114: Standard specification for asphalt-rubber binder. ASTM International, West Conshohocken, PA, USA.

Benson PE (1976) *Low Temperature Transverse Cracking of Asphalt Concrete Pavements in Central and West Texas*. Texas Transportation Institute and State Department of Highways and Transportation, College Station, TX, USA, Cooperative Research Project Report 175-2F.

Brown SF (1978) Stiffness and fatigue requirements for structural performance of asphaltic mixes. *Proceedings of Eurobitumen Seminar, London, UK*, pp. 141–145.

Brown SF, Cooper KE, Gibb JM, Read JM and Scholz TV (1994) Practical tests for mechanical properties of hot-mix asphalt. *Proceedings of the 6th Conference on Asphalt Pavements for Southern Africa, Cape Town, South Africa*, vol. 2, pp. 29–45.

BSI (1996) DD 226:1996: Method for determining resistance to permanent deformation of bituminous mixtures subject to unconfined dynamic loading. BSI, London, UK.

BSI (2009) BS EN 1367-2:2009: Tests for thermal and weathering properties of aggregates. Magnesium sulfate test. BSI, London, UK.

BSI (2011) BS EN 1097-1:2011: Tests for mechanical and physical properties of aggregates. Determination of the resistance to wear (micro-Deval). BSI, London, UK.

BSI (2014) BS EN 12607-1:2014: Bitumen and bituminous binders. Determination of the resistance to hardening under influence of heat and air. RTFOT method. BSI, London, UK.

BSI (2015a) BS EN 1426:2015: Bitumen and bituminous binders. Determination of needle penetration. BSI, London, UK.

BSI (2015b) BS EN 1427:2007: Bitumen and bituminous binders. Determination of the softening point. Ring and ball method. BSI, London, UK.

BSI (2015c) BS EN 12593:2015: Bitumen and bituminous binders. Determination of the Fraass breaking point. BSI, London, UK.

BSI (2018a) BS EN 12697-24:2012: Bituminous mixtures. Test methods for hot mix asphalt. Resistance to fatigue. BSI, London, UK.

BSI (2018b) BS EN 12697-12:2018: Bituminous mixtures. Test methods. Determination of the water sensitivity of bituminous specimens. BSI, London, UK.

BSI (2019) BS EN 12697-44: Bituminous mixtures. Test methods. Crack propagation by semi-circular bending test. BSI, London, UK.

BSI (2020a) BS EN 12697-46:2020: Bituminous mixtures. Test methods for hot mix asphalt. Low temperature cracking and properties by uniaxial tension tests. BSI, London, UK.

BSI (2020b) BS EN 12697-22:2003: Bituminous mixtures. Test methods. Wheel tracking. BSI, London, UK.

BSI (2020c) BS EN 1097-2:2020: Tests for mechanical and physical properties of aggregates. Methods for the determination of resistance to fragmentation. BSI, London, UK.

Chen A, Airey GD, Thom NH, Litherland J and Adjetey Nii-Adjei R (2021) Modelling the stiffness development in asphalt concrete to obtain fatigue failure criteria. *Construction and Building Materials* **306**: 124837.

Choi YK, Thom NH and Collop AC (2005) Fatigue damage accumulation measured using different laboratory techniques. *Proceedings of the 4th International MAIRPAV Symposium, Belfast, Northern Ireland.*

Collop AC, Choi YK, Airey GD and Elliott RC (2004) Development of the saturation ageing tensile stiffness (SATS) test. *Proceedings of the Institution of Civil Engineers – Transport* **157(3)**: 163–171, https://doi.org/10.1680/tran.2004.157.3.163

Corbett LW and Schweyer HE (1981) Composition and rheology considerations in age hardening of bitumen. *Proceedings of the Association of Asphalt Paving Technologists* **50**: 571–582.

DIN (Deutsches Institüt für Normung) (2019) DIN 18317: German construction contract procedures (VOB). Part C: General technical specifications in construction contracts (ATV) – Road construction – Asphalt surfacing. DIN, Berlin, Germany.

Fraass A (1937) Test methods for bitumen and bituminous mixture with specific reference to low temperature. *Bitumen*, pp. 152–155.

Gubler R, Baida LG and Partl MN (2004) A new method to determine the influence of water on mechanical properties of asphalt concrete. *Road Materials and Pavement Design* **5(sup 1)**: 259–279.

Halle M, Rukavina T and Domitrovic J (2012) Influence of temperature on asphalt stiffness modulus. *Proceedings of 5th Eurasphalt & Eurobitume Congress, Istanbul, Turkey.*

Hunter (2014) *The Shell Bitumen Handbook*, 6th edn. ICE Publishing, London, UK.

Maillard S, de la Roche C, Hammoum F, Such C and Piau JM (2004) Bitumen healing investigation using a specific fracture test. *Road Materials and Pavement Design* **5(sup 1)**: 45–63.

Mohammed M, Parry T, Thom N and Grenfell J (2020) Microstructure and mechanical properties of fibre reinforced asphalt mixtures. *Construction and Building Materials* **240**: 117932.

National Highways (2020a) CD 229: data for pavement assessment. *Design Manual for Roads and Bridges. Pavement Inspection and Assessment.* Stationery Office, London, UK.

National Highways (2020b) *Manual of Contract Documents for Highway Works. Vol. 1: Specification for Highway Works.* Stationery Office, London, UK, vol. 1.

Nguyen VB (2020) *Design of Pavements Incorporating Polymer Modified Asphalt.* PhD thesis, University of Nottingham, Nottingham, UK.

Oliveira JRM (2006) *Grouted Macadam: Material Characterisation for Pavement Design.* PhD thesis, University of Nottingham, Nottingham, UK.

Paris PC and Erdogan FA (1963) A critical analysis of crack propagation laws. *Journal of Basic Engineering, ASME* **85**: 528–534.

Pendrys JP (1989) Biodegradation of asphalt cement-20 by aerobic bacteria. *Journal of Applied and Environmental Microbiology* **55(6)**: 1357–1362.

Phillips UA and Traxler RW (1963) Microbial degradation of asphalt. *Applied and Environmental Microbiology* **11(3)**: 235–238.

Rahimzadeh B (2002) *Linear and Non-linear Viscoelastic Behaviour of Binders and Asphalts.* PhD thesis, University of Nottingham, Nottingham, UK.

Sudibyo T (2023) *The Behaviour of LDPE Modified Asphalt Mixture in Pavement Application.* PhD thesis, University of Nottingham, Nottingham, UK.

Thom NH, Elliott RC and Cheung LW (1997) Comparisons between laboratory and in situ determined asphalt concrete moduli. *Proceedings of the 8th International Conference on Structural Design of Asphalt Pavements, Seattle, WA, USA*, vol. 2, pp. 1433–1442.

Thom NH, Osman SA, Collop AC and Airey GD (2006) Fracture and fatigue of binder and binder/filler mortar. *Proceedings of the 10th International Conference on Asphalt Pavements.* International Society of Asphalt Pavements, White Bear Lake, MN, USA, vol. 1, pp. 798–807.

Ullidtz P (1979) A fundamental method for the prediction of roughness, rutting and cracking in asphalt pavements. *Proceedings of the Association of Asphalt Paving Technologists* **48**: 557–586.

van der Poel C (1954) A general system describing the visco-elastic properties of bitumens and its relation to routine test data. *Journal of Applied Chemistry* **4**: 221–236.

Yin F, Arámbul-Mercado E, Epps Martin A, Newcombe D and Tran N (2017) Long-term ageing of asphalt mixtures. *Road Materials and Pavement Design* **18(S1)**: 2–27.

Zhang H, Soenen H, Carbonneau X, Lu X, Robertus C and Zhang Y (2022) Experimental and statistical analysis of bitumen's field ageing in asphalt pavements. *Transportation Research Record* **2676**: 495–511.

Nick Thom
ISBN 978-1-83549-711-1
https://doi.org/10.1108/978-1-83549-710-420243011

Chapter 10
Flexible pavement – structural design

Pavements with asphalt upper layers and unbound foundations are known as 'flexible'. Unlike concrete pavements (rigid) they can deform. If the subgrade wants to deform under applied load, none of the overlying layers are going to stop it; they will just follow it down. In one sense this is a bad thing as it means that pavement design has to guard against rutting, but on the other hand it means the pavement can accommodate a bit of ground movement without being damaged. Anyway, it is the task of this chapter to go through all the different aspects that have to be checked to come up with an optimised flexible pavement design.

10.1. Construction practicalities

Construction of unbound layers has already been covered in Chapter 7. This section will therefore concentrate on the defining material in flexible pavements, namely asphalt.

10.1.1 Asphalt production

The standard asphalt production process demands that hot binder is mixed with aggregate in a water-free environment, which effectively means that the stones have to be preheated to drive off any moisture present. The exact temperature required for mixing depends on the properties of the particular bitumen being used, but without additives the usual range is from about 140°C to 180°C, which means that asphalt production is a highly energy-intensive process.

There are two basic types of mixing plant: batch mixers and drum mixers. In both types, the aggregate has to be proportioned by weighing amounts from different storage hoppers and then feeding it into a drying chamber (which takes the form of an inclined rotating drum) where it is superheated to drive off all moisture. In batch mixers, the dried aggregates are fed through a set of sieves into hot-storage bins, from which they are dropped (in exactly measured proportions) into the mixing chamber, together with the appropriate volume of hot bitumen. The reason for the second stage of proportioning, from the hot-storage bins, is that the initial proportioning was of moist aggregate and may therefore not have been precisely correct. Once mixed, typically after about 30 s, the mixture is dropped into the back of a waiting truck and taken to site.

The other type of plant, the continuous drum mixer, has the advantage that higher productivity is possible. Drum mixers rely on accurate proportioning of the moist aggregate, with due allowance for the likely differences in moisture content of each size fraction, as the bitumen is fed directly into the drying chamber. The drying chamber therefore doubles as the mixing chamber, and the hot mixture is fed out continuously from the drum to a hot-storage hopper before being dropped into the back of a waiting truck. Figure 10.1 illustrates the principles of the two types of mixing plant.

10.1.1.1 Warm mixes

The use of warm-mix additives reduces the temperature range for mixing and compaction by about 30°C without any detriment to the material's performance. The potential savings in energy and

Figure 10.1 Batch- and drum-type asphalt plants

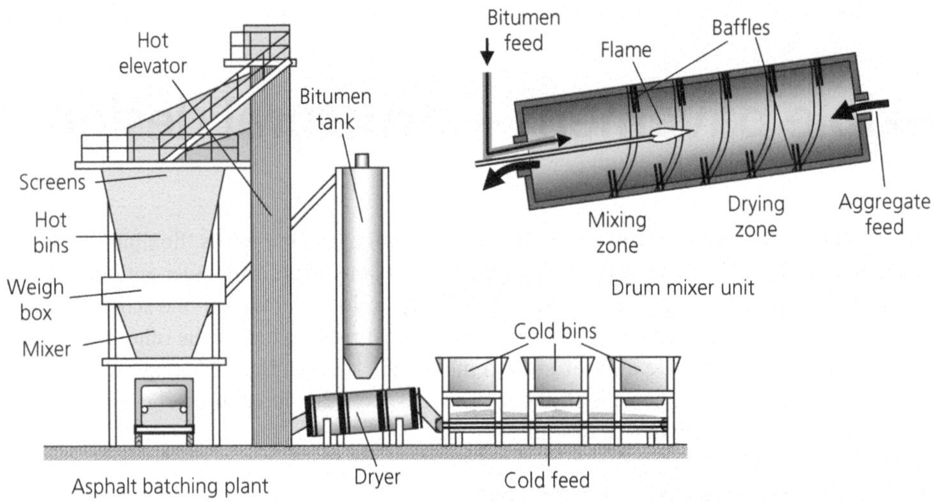

therefore cost are self-evident; a further key advantage for contractors is that the working time window is extended because of reduced cooling time. Thus, warm mixes have become the norm in many countries and this trend is only likely to continue.

Warm mixes do, of course, come with a cost, namely the cost of the additive itself. These come in different forms: waxes melt at intermediate temperatures, but are solid at in-service temperatures, thus giving no discernible loss in performance; chemical additives reduce the surface tension of the bitumen at elevated temperatures, allowing easier slippage between stone and bitumen. Both increase binder workability during mixing and compaction.

10.1.1.2 Inclusion of recycled asphalt planings (RAP)

Recycled asphalt planings (RAP) are a high-value resource, and the obvious way of maintaining their value is to re-incorporate RAP back into new asphalt. Naturally, this demands a degree of control over the source of RAP since binder content and binder hardness are likely to vary considerably from source to source but, if properly controlled, they can be fed directly into the mixer and combined with new binder and aggregate.

The first problem is heating the RAP without emitting toxic fumes and this is overcome by indirect heating once in contact with superheated aggregate, but this is a finite heat source which therefore limits the proportion of RAP that can be used; 20% is typical. If higher percentages are to be used then the plant has to be adapted to provide additional indirect heating to the RAP prior to being added to the mixer. Control of RAP gradation and binder content clearly also become increasingly critical the greater the percentage added.

The second problem is the blending of RAP binder with new binder, which will never be more than partial. Design methods (e.g. ASTM, 2016) allow selection of the grade of new binder such that, when combined with RAP binder, the appropriate grade is achieved; however, these methods implicitly assume perfect blending. The reality will always be that the resulting mix includes pockets of both binders with some blended zones in between; however, and perhaps surprisingly, experience is that this in no way compromises mixture performance.

10.1.2 Placement and compaction

The asphalt now has to be transported to the site, paved and compacted, for which the temperature should still be greater than about 120°C for a hot mix or about 90°C for a warm mix if proper compaction is to be achieved. This means that the following conditions must be met.

- Trucks for transporting hot- or warm-mix asphalt have to be well insulated and covered.
- The journey must not be too long to avoid excessive cooling.
- Binder drainage – the phenomenon of bitumen flowing to the bottom of the mixture under gravity – must be avoided; if necessary, cellulose fibres have to be included in the mixture.

The mixture is then paved by means of an asphalt paver (Figure 10.2). This results in an even layer of material placed to a controlled level and, in modern machines, with a significant degree of compaction already induced by the vibrating screed.

However, to achieve full compaction, the 'mat' of asphalt has to be rolled (before it has cooled excessively) and there are various different types of roller available.

- Vibratory steel-drum rollers induce large numbers of cyclic load applications into the mixture, encouraging densification but also resulting in some loosening of surface material behind the roller.
- Dead-weight pneumatic-tyred rollers impart many fewer load applications but at a higher load level; they avoid the problem of loosening of surface material.
- Dead-weight steel-drum rollers (can be vibratory rollers with the vibration mode switched off) are principally used to compact the surface layer, particularly as a finishing process after using vibratory compaction.

Figure 10.2 Asphalt paving

Compaction is critical to asphalt performance, just as it is for unbound materials. If the mat is too cool, it will tend to form surface cracks beside and behind the roller; if it is too hot, it will deform too much and it will be difficult to achieve the required level tolerances. A skilled roller operator is a very valuable member of staff indeed. However, no operator – however skilled – can achieve the impossible and the mathematics of asphalt mat cooling (see Calculation Sheet 10.1) demand that the material is not placed either too thinly or too thickly. It is no good if the operator has to wait so long for the main body of the mat to reach its ideal rolling temperature that the top and bottom have already cooled and can no longer be densified. In fact, the base of the layer represents a problem that simply has to be accepted, as the almost immediate heat transfer between the hot asphalt and the colder substrate inevitably results in very rapid local cooling. Figure 10.3 shows predictions of asphalt mat temperature against time for different mat thicknesses assuming a 110°C delivery temperature, using the methodology outlined in Calculation Sheet 10.1.

Calculation Sheet 10.1. Temperature loss from an asphalt mat

Specific heat capacity of asphalt = S (typically 920 J/kg.K)
Thermal conductivity of asphalt = C (typically 2.0 W/m.K)
Density of asphalt = ρ (typically 2400 kg/m^3)

Radiation equation: $$Q_{rad} = 5.67 \times 10^{-8} (T_{surface})^4 \quad (1)$$

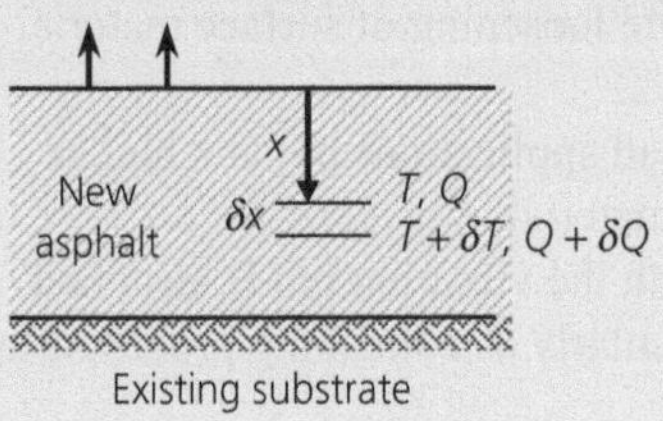

T = temperature (K)
Q = heat flow (W/m^2)
t = time (s)

(a) $\delta T/\delta t = -\delta Q/(S.\rho.\delta x) \rightarrow S.\rho(\mathrm{d}T/\mathrm{d}t) = -\mathrm{d}Q/\mathrm{d}x$

(b) $Q = C(\mathrm{d}T/\mathrm{d}x) \therefore \mathrm{d}Q/\mathrm{d}x = C(\mathrm{d}^2T/\mathrm{d}x^2)$

Combine: $$\mathrm{d}^2T/\mathrm{d}x^2 = -(S\rho/C)\,\mathrm{d}T/\mathrm{d}t \quad (2)$$

Equations 1 and 2 can now be applied on a spreadsheet, using discrete time steps

The practicalities of compaction mean that it is unwise to place either hot- or warm-mix asphalt in layers of over about 120 mm thickness. The lower limit on thickness is a function of aggregate particle size, and a limit of 2.5 times the maximum particle size is often used. Whatever the particle size, however, any layer under about 25 mm thickness is clearly going to be difficult to compact due to its rapid heat loss.

10.1.3 Interlayer bond

Asphalt pavement performance relies on there being a good bond between individual asphalt layers; if this is not the case, the upper layers can detach, crack and eventually give rise to potholes. It is therefore important to at least make sure that there is nothing present that might inhibit the development of such a bond, in particular water or dirt. However, to achieve a higher level of assurance, it is necessary to physically add a bonding agent (a 'tack coat' or 'bond coat'). This comprises a thin (0.15– 0.6 l/m^2 depending on the substrate) layer of bitumen sprayed onto the surface shortly

Figure 10.3 Temperature changes in an asphalt mat

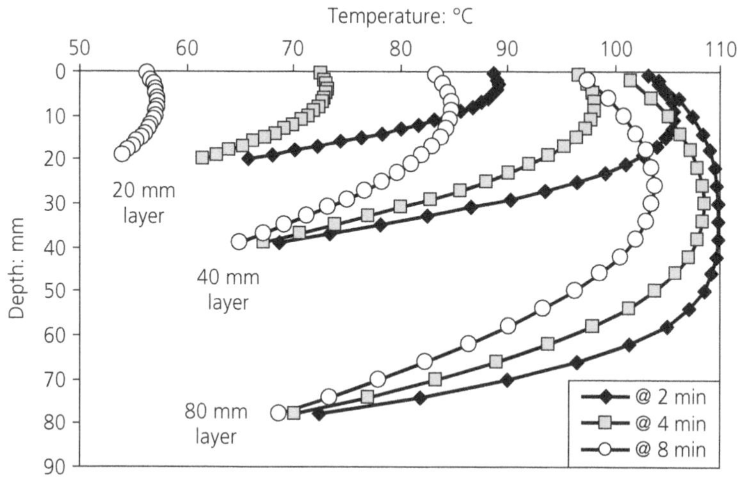

before placing the upper lift of asphalt. In the UK, the expression 'bond coat' is reserved for a modified bitumen emulsion used below a relatively thin surface course layer.

The need or otherwise for an interlayer bonding agent is still subject to debate and, in many specifications, reliance on surface cleanliness is considered sufficient. At the other extreme (Stöckert, 2001), bond strength is specified and measurements are carried out on specimens cored from the pavement. Bond can be measured using the Leutner test, a modified version of which is specified by National Highways (2020a), or a torque test (BBA, 2004). The principles are shown in Figure 10.4. Where bond strength is specified, it is typically around 1.5 MPa at the surface course–binder course interface and 1.0 MPa at lower interfaces.

Figure 10.4 Interlayer bond strength tests

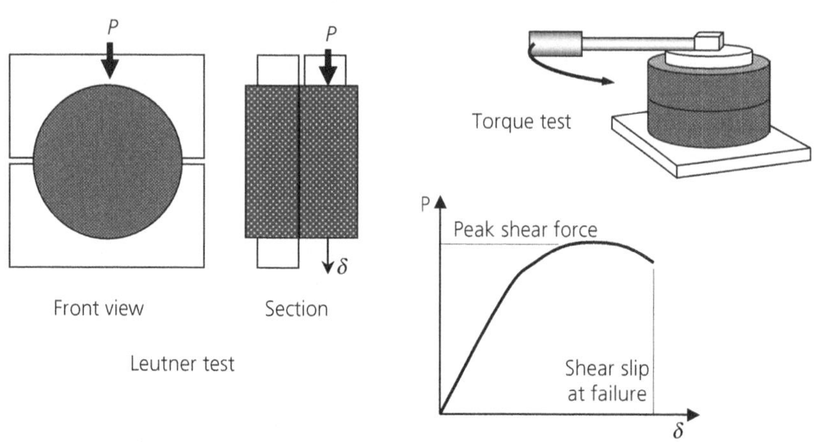

10.1.4 Joints

A short walk along most asphalt-surfaced roads will reveal straight lines – both longitudinal and transverse – of voided material, sometimes with stones missing and the beginnings of potholes. These are asphalt joints and they frequently give rise to significant weaknesses. Longitudinally, they represent the join between two lines of paving; transversely, they are formed between the end of one paving operation and the start of the next, for example on the following day. The problems are as follows.

- It is impossible to compact a stony material well against a hard edge.
- The edge will usually be at ambient temperature, causing an instantaneous local loss of temperature in the new material.

The result is a locally high voids content, potentially allowing water to enter and leading to the various durability problems outlined in Sections 9.6.2 and 9.6.3.

So what can be done about it? First, longitudinal joints should never be located in wheel-paths. Second, the faces of hardened asphalt should be covered with a generous coat of bitumen before paving the adjacent material. Third, hardened asphalt faces can be softened by heating using a hot-air lance immediately prior to paving. Fourth, joints should be sealed at the surface with further bitumen application. Fifth – and this is rarely possible – paving can be carried out in echelon; that is, two or more pavers paving adjacent strips of asphalt and following each other closely. This last technique effectively gets rid of longitudinal joints.

Joints are a problem that should be taken very seriously as many otherwise excellent road surfaces end up marred by joint damage.

10.1.5 Asphalt reinforcement

This short section is included simply to warn the reader that special care is needed where reinforcing products (geogrids, geotextiles) are used in asphalt. In theory, there should be no disruption to asphalt paving and no special measures should be needed. However, experience suggests that the fixing of reinforcement to the layer beneath, for which various different techniques have been devised, is critical. If it is well fixed, then no problems should occur; if not, then the reinforcement product will tend to move around under the action of the paver and the likelihood is that this will result in voiding at the base of the layer. If this happens, then any benefit that the reinforcement might have given is more than offset by the danger of deterioration due to a high voids content, particularly if water is able to accumulate at the level of the reinforcement.

10.1.6 Surface finish

In general, an asphalt surface is well suited to vehicle use. It absorbs more energy than concrete, which makes for a smoother ride, and modern pavers are quite capable of obtaining the required longitudinal profile. However, the requirement for skid resistance demands certain actions – depending on usage of course – and there are also desirable features, such as spray retarding and noise absorption, that also affect the choice of surfacing. The options available are illustrated conceptually in Figure 10.5.

The material labelled bituminous macadam in Figure 10.5 is illustrative of a wide range of different mixtures, all having the property that they need no further treatment in order to form the finished pavement surface. In fact, options range from dense mixtures (having broad aggregate gradation)

Figure 10.5 Asphalt surface options

Bituminous macadam	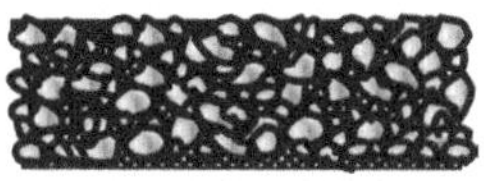	Aggregate is coated with bitumen, mixed, transported and paved. The grading chosen results in a suitably textured surface
Grooved surface	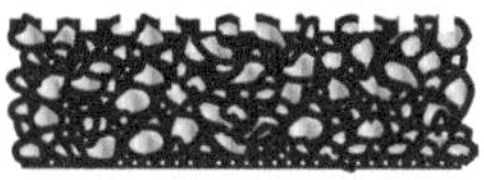	In cases where the natural texture is insufficient, grooving can be carried out in the same way as for concrete. This is common on airfields
Rolled-in chippings	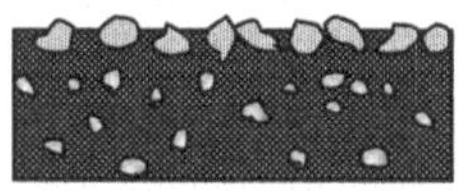	An alternative is to pave a layer of fines-rich asphalt and then spread chippings on the hot surface prior to rolling
Surface dressing		In 'surface dressing' bitumen is first sprayed onto the surface (usually as an emulsion) and chippings are then spread and lightly rolled

known as *asphalt concrete*, through *stone mastic asphalt* (with a slightly coarser gradation) to *porous asphalt* (almost single sized). Chapter 9 has described the materials and Chapter 11 will cover surface design in more detail.

10.2. Current design practice

Most pavement designs today start and end with reference to a particular document or computer package – and in some cases little real engineering knowledge is required. No criticism is levelled here at such an approach. An authority in charge of a network of pavements needs some assurance that those carrying out designs on its behalf are following the same broadly sensible standards, and it makes sense to codify designs that are known from past experience to have been successful. So, while most of this part of the book will look at the real engineering of a pavement in some detail, it is appreciated that for most day-to-day designs, for new pavements at least, this is not really necessary.

Nevertheless, it makes sense that a user of a particular design method should at least be aware of the basis and rationale of that method. The following paragraphs provide an overview of certain selected methods.

10.2.1 AASHTO *Guide for Design of Pavement Structures*

Although this method (AASHTO, 1993) has now been officially superseded in the USA by the AASHTO computer-based mechanistic-empirical pavement design guide (AASHTO, 2007) described in Section 10.2.2, it is still in common use around the world and probably will continue to be because it has the key advantage of simplicity. Here are some of its features related to flexible (and also composite) pavements.

- The strength of a pavement structure is expressed as a 'structural number' (SN).
- The variation in the subgrade throughout the year is explicitly taken into account by choosing a modulus for each month (following guidelines) and converting to a single equivalent foundation modulus.

- An equation relates the required SN to the equivalent subgrade modulus value, the design traffic (in equivalent standard axles – see Chapter 4), an appropriate design reliability and an acceptable end-of-life condition (the 'design serviceability loss').
- For each potential pavement layer, a structural coefficient has to be determined. For asphalt, this is based on the modulus. For unbound materials, it is based on one of several alternative measures including the California bearing ratio (CBR) and modulus.
- For unbound layers, a moisture coefficient also has to be determined based on an estimate of the likely effectiveness of drainage and the prevailing climate.
- Design then consists of selecting a combination of layers and thicknesses such that the sum of the products of thickness, structural coefficient and moisture coefficient for each layer adds up to the required SN.

This approach has several attractive elements. It forces the designer to consider seasonal effects, and excellent reason to treat variations in subgrade water content with respect was detailed in Chapter 6. Consideration of drainage is also explicit – another very positive feature. The inclusion of design reliability and end-of-life condition allow the method to be built into a highway management strategy. Finally, while the simple addition of effects from different layers is open to criticism, this approach lends itself well to the design of strengthening treatments, whether consisting of the simple addition of a new layer (an overlay), or the subtraction of part of an existing layer and its replacement by a new layer (an inlay).

The basis for the method is experience, initially from the so-called AASHTO road trials carried out in the late 1950s, but then supplemented by decades of subsequent information. It therefore remains a highly practical tool in the design and subsequent management of asphalt roads.

10.2.2 AASHTO *Guide for Mechanistic-Empirical Design of New and Rehabilitated Pavement Structures*

This method (AASHTO, 2007) represents the antithesis of the earlier AASHTO approach in that it is massively complex and only comes in the form of a large computer program (AASHTOWare). Key features are as follows.

- The program effectively sums damage – day and night, summer and winter – throughout the lifetime of the pavement. This requires that climatic data appropriate to the pavement's location are fed in and that asphalt and unbound layer properties are varied accordingly.
- Analytical computations are carried out to predict both the deformation and fatigue cracking damage within each layer as a function of time. This requires that fairly detailed material and traffic data are supplied by the user.
- The output from the program consists of predictions of the development of each form of distress (rutting, cracking of various forms, surface ravelling, loss of evenness etc.).

This design method is best described as all-encompassing and is now in common use by several highway authorities in the USA and around the world. However, despite the complexity, it is only as good as the algorithms it uses, some of which are excellent and well-founded, others of which are more questionable. One clear advantage is that sensible account can be taken of climatic factors, and so the method can be transposed from the USA to other countries with a reasonable chance of success. Furthermore, traffic details can be altered to reflect local conditions; for example, realistic levels of overloading. The disadvantage is that the method is, to some extent, a 'black box' and its complexity means that there is a danger of misuse by those without a full understanding of the required input. Those organisations that are using it successfully have tended to go through a process of calibration, ensuring that they are happy with the answers it gives for their particular circumstances, and have then continued to use it with their chosen inputs.

10.2.3 UK National Highways *Design Manual for Roads and Bridges*

Significant sections of this massive document relate to pavement design (National Highways, 2020b, 2020c, 2021a) and they build on several earlier design standards. The approach, which is based on empirical evidence and extended through analytical calculation, can be described as follows.

- The designer first has to select a 'foundation class' (based on effective modulus) and then designs the lower layers of the pavement to achieve it, using the charts or equations provided.
- These foundation designs are a function of subgrade CBR/modulus (an equation linking the two is assumed) and the thickness and moduli of one or more unbound or hydraulically bound layers.
- Upper pavement design is then derived from charts into which the design traffic (in equivalent standard axles), the selected road base material(s) and the selected foundation class are fed.

This is a practical tool and extremely easy to use. It allows the designer a certain amount of freedom, for example, in trading increased foundation class for reduced road-base thickness, while restricting choice to certain defined material options. It is, therefore, well suited to the particular types of pavement managed by UK National Highways. However, it cannot readily be applied elsewhere, as the user has no freedom to alter items such as design reliability or acceptable end-of-life condition. It also assumes the typical mix of soils and material sources present in the UK and the particular asphalt (and hydraulically bound) mixtures permitted by UK National Highways. It does not allow the user to extend it to cover other alternatives.

10.2.4 Federal Aviation Administration *Airfield Pavement Design and Evaluation*

In its most recent version (FAA, 2021), this method is contained in the computer program FAARFIELD. It is a user-friendly means of optimising the design thicknesses for airfield pavements under a selected mix of aircraft traffic and has the following features.

- The user has complete freedom to define a mix of aircraft from an extensive database and can also add other types not listed. The program will analyse this input and determine a design-load case.
- The list of possible material types is a little more restrictive (based on US standards), but even here the user can amend the modulus values to suit a particular case.
- Once the traffic and layer properties are defined, including the CBR or modulus of the subgrade, the program will iterate to an optimised thickness combination such that there is just enough overall pavement strength to protect the subgrade while ensuring that asphalt (or concrete) surface layers do not crack.

This design package is an excellent tool for airfield pavements. It can also be extended for non-standard use, for example, by inputting the wheel configuration and loads of a piece of equipment trafficking an industrial pavement. However, it should not be used for road traffic because this takes it far outside the range of cases for which it has been calibrated.

10.2.5 Defence Estates *Guide to Airfield Pavement Design and Evaluation*

This guide (Defence Estates, 2006) is widely used and is popular because of its flexibility. It is based on charts/equations and it is relatively easy for the user to see what the underlying principles are at each stage. The relevant points are as follows.

- The user has a certain amount of work to do in reducing actual aircraft traffic to an equivalent number of a selected design aircraft. Historically, an attractive feature for airport authorities has been that the internationally accepted Aircraft Classification Number (ACN) is used to describe damaging power. In this system, the so-called Pavement Classification Number (PCN) has to be greater than the ACN of the largest aircraft using that part of the airfield. However, the International Civil Aviation Authority (ICAO) has now replaced this with the ACR/PCR (Aircraft Classification Rating/Pavement Classification Rating) system, and there is no simple correlation between ACN and ACR.
- As well as standard designs for different base types, the document also facilitates rehabilitation/strengthening design by allowing the user to assign 'condition factors' to existing pavement layers.
- It is also permitted to transfer thickness from one layer to another using an equivalence principle similar to the structural factors used in the AASHTO (1993) method, a feature that is also particularly useful in rehabilitation design.

Basically, this guide is a good combination of simplicity, transparency and sound engineering. As with most practical design methods it is based on a combination of experience and pavement analysis. Although its simplicity means it is possible to make improper use of it, in the hands of a competent engineer it is an excellent tool.

10.2.6 Interpave *Structural Design of Heavy Duty Pavements for Ports and Other Industries*

This document (Knapton, 2007) was primarily written with port pavements in mind, although it can be applied more widely. The approach is as follows.

- Designs are carried out for individual load cases. In each case, the actual vehicle has to be reduced to a single equivalent wheel load, with an explicit requirement to take dynamic effects into account.
- Once the single equivalent wheel load is calculated, the required thickness of 10 MPa compressive strength concrete is read from a chart. This can then be converted using equivalence factors into whatever materials (e.g. asphalt) are actually to be used.

This is another relatively simple method, with designs being primarily based on empirical evidence. It is particularly useful when dealing with unusual load cases, including container feet. The most questionable element is the use of wide-ranging equivalence factors, which mean that the user should be slightly cautious in converting between material types.

10.2.7 Other methods

The six methods commented on in foregoing sections were selected from a very wide range on account of their international standing. However, mention should also be made of two other generic types of approach.

- *Catalogues* – numerous national pavement design standards use a catalogue approach (the French, German and South African national design standards are among those that do).

Basically, the user decides on a class of traffic and a class of subgrade, turns to the page in the catalogue that covers the desired type of pavement and simply reads off the design to be used. It is quick, easy and requires no understanding of pavements at all.

- *Analytically based charts* (e.g. Asphalt Institute, 2008) are based on asphalt properties and use so-called multi-layer linear elastic analysis (see Section 7.3.2); they only cover asphalt-base roads.

The one aspect that all the methods mentioned have in common is that the user does not need to carry out any pavement analysis because this has all been done already in the development of the method – and it is acknowledged that, for most purposes, this will be sufficient. Nevertheless, the use of standard design methods does not promote real understanding of what should go into pavement design and, for that reason, this part of the book will continue rather than closing here!

10.3. Subgrade deformation

In general, natural ground will not be strong enough to bear traffic load directly; it would deform and rut. The pavement must therefore be stiff enough and thick enough to distribute the load from vehicle wheels over a wide enough area not to cause excessive subgrade deformation.

The subgrade strain criterion was introduced in Section 7.3.1 and the logic for its use explained – and the logic is exactly the same for asphalt-surfaced pavements as for unbound. Reference may therefore be made back to Chapter 7 for the equations relating subgrade strain to the number of load applications permitted. The recommended means of calculating the subgrade strain under the application of a design wheel load is multi-layer linear elastic analysis, explained in Section 7.3.2.

This approach is, of necessity, highly approximate since every subgrade is unique. Nevertheless, bearing in mind the tremendous uncertainty present at the design stage regarding the state of the actual subgrade and the way it is likely to vary through the year due to weather events, it represents a sensible simplification, so long as an appropriately conservative assumption is made for long-term subgrade properties.

Also covered in Chapter 7 was the concept of closeness to failure and the use of an equation to calculate the build-up of strain within an unbound material. This approach is not recommended for asphalt-surfaced pavements because it does not take into account the detailed interaction between deformation in a pavement foundation and bending in the overlying asphalt.

10.4. Asphalt deformation

Chapter 9 made the point that asphalt deforms in a very complex way and, for that reason, the usual procedure is not to even try to estimate the rate of rut development in an asphalt layer, but to concentrate on mixture design and to aim for a notionally deformation-resistant mix, the principles of which were outlined in Chapter 9. On the other hand, asphalt frequently forms ruts, and it would be useful to be able to make some sort of a prediction in advance based on measurable material properties. The following subsections outline one such approximate technique.

10.4.1 Viscous analysis

Just as multi-layer linear elastic analysis can be used to calculate elastic strains within a pavement, *multi-layer linear viscous* analysis can be used to calculate viscous strains. In fact, the same programs can be used, replacing the elastic modulus with a viscosity; computed strain is replaced by strain rate and deformation by deformation rate. The computation can readily be carried out by engineers – as long as they can assign appropriate viscosities. But is such an approach valid?

Figure 10.6 shows average rut depth accumulation data from many hundreds of kilometres of major road in one UK county over a number of years. The striking thing is how linear it is, and near-linear development implies that it is not unreasonable to assume a constant mixture viscosity – but on what basis can values of viscosity be chosen?

There are two main approaches. The first is to carry out static creep tests – the standard type of test for viscous materials – but the problem is that the resulting strain is often far from linear with time. The second approach, which is definitely recommended here, is to carry out repeated load tests such as the repeated load axial test (RLAT) introduced in Chapter 9. The growth of strain is not linear in this test either, particularly at the beginning of the test, but it becomes much closer to being linear during later stages. The recommended approach is, therefore, to use the slope of the plot of strain against number of load applications towards the end of the test; in UK practice, this means over the last few hundred load applications out of a total of 3600, each of which lasts 1 s. Figure 10.7 shows three sets of data and the viscosities determined from them.

The problem being evaluated here is not a whole-pavement problem; it is restricted to the asphalt. Therefore, the procedure is to run a multi-layer linear viscous analysis by inputting data for the asphalt layers, substituting viscosity for elastic modulus and assigning a very high viscosity to the infinite layer beneath the asphalt. The design wheel load is applied and the deformation (actually deformation rate) is calculated under the centre of the load. For example, if the three materials shown in Figure 10.7 are used, with 40 mm of mixture A in the surface course, 60 mm of mixture B in the binder course and 180 mm of mixture C in the base, and a 40 kN standard wheel load is applied over a circular area of radius 150 mm, then the following are obtained.

- Surface deformation rate of 1.809×10^{-7} m/s.
- Deformation rate at top of binder course of 1.454×10^{-7} m/s.
- Deformation rate at top of base of 0.880×10^{-7} m/s.

Figure 10.6 Average trend for rut depth increase (Abed *et al.*, 2023)

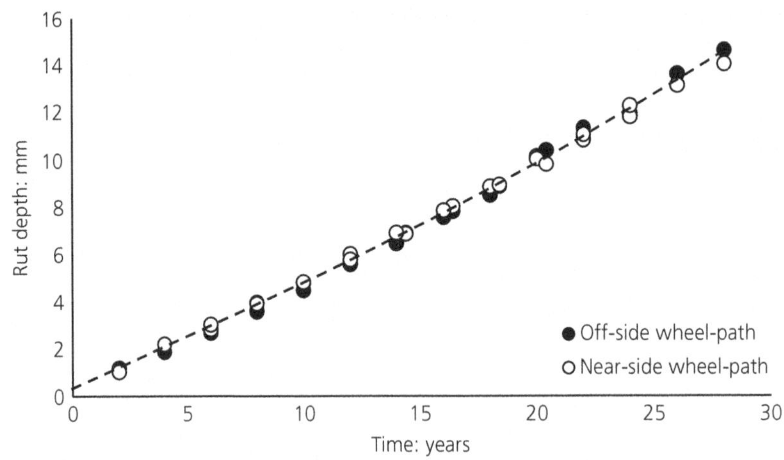

Figure 10.7 Derivation of mixture viscosity from RLAT, 100 kPa at 40°C

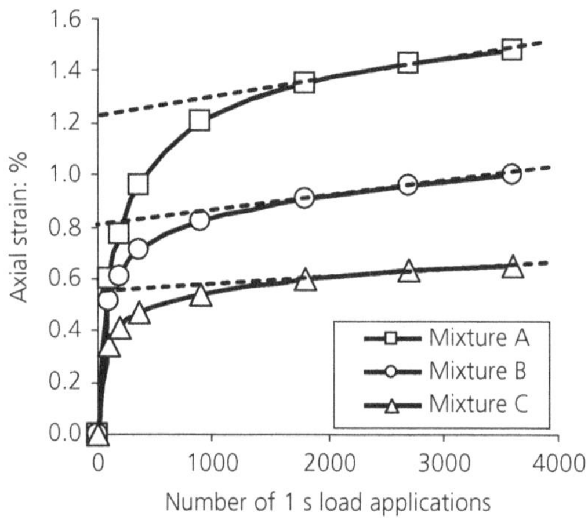

Mixture A

Slope ≈ 0.26% per hour

= $0.72\eta\varepsilon$/s (under 100 kPa)

Viscosity = 100/0.72 = 139 GPa.s

Mixture B

Slope ≈ 0.20% per hour

= $0.56\eta\varepsilon$/s (under 100 kPa)

Viscosity = 100/0.56 = 179 GPa.s

Mixture C

Slope ≈ 0.11% per hour

= $0.31\eta\varepsilon$/s (under 100 kPa)

Viscosity = 100/0.31 = 323 GPa.s

Thus, it may be observed that, in this case, the largest component of deformation per millimetre of thickness occurs within the binder course. This is actually quite common and reflects the fact that, on roads, the largest shear stresses commonly occur at a depth of around 50 mm.

The next step is to convert from seconds to numbers of load applications. This is achieved by working out the length of time taken for a vehicle travelling at design speed to cover a distance slightly greater than the length of a single tyre contact patch. For example, for a heavy goods vehicle the distance is about 400 mm and, at 60 mph (about 25 m/s), this will be covered in 16 ms. Thus, one second of load in the calculation is equivalent to about 60 load applications; if a limiting rut value of 10 mm is adopted for design, then this implies that the pavement is suitable for about 4.0×10^6 load applications – equivalent to around two million goods vehicles.

10.4.2 Accounting for temperature variation

The final problem is that the RLAT data in Figure 10.7 were obtained at a temperature of 40°C and asphalt viscosity is strongly dependent on temperature. A reasonable assumption for this dependency (based on the author's experience rather than any fundamental analysis) is that a 10°C decrease in temperature causes a *three-fold* increase in viscosity, and this provides a basis for converting a total annual traffic load into an equivalent number of vehicles at 40°C. The three-fold increase in viscosity is, of course, a highly approximate figure – but fit for the purpose for which it is used here. Pure bitumens would tend to have a rather higher change in viscosity (around four- to five-fold per 10°C), but the influence of the aggregate reduces this for mixtures. Calculation Sheet 10.2 presents such a conversion, using temperature data obtained from the surface course of a pavement in the UK. In this case, the result is a factor of 6.4 between total traffic numbers and equivalent 40°C load applications, and the pavement analysed in the previous section can therefore take rather more traffic than at first appeared – about 25 million standard 40 kN wheel load applications.

The reader is cautioned against taking the absolute magnitude of the prediction too literally; however, the method is a relatively easy one to use and the RLAT is a simple enough test.

Calculation Sheet 10.2. Converting traffic for pavement temperature

If a 10°C temperature change produces a three-fold viscosity change:

$$\eta_T = \eta_{40^\circ C} \times 3^{(4-0.1T)}$$

Suggest use average noon pavement temperatures since most traffic occurs during the day, giving the following monthly factors based on pavement temperatures in Nottingham, UK

Month	Pavement temperature	η factor	Strain rate factor
January	+3°C	×58.3	÷0.017
February	+7°C	×37.5	÷0.027
March	+13°C	×19.4	÷0.052
April	+21°C	×8.1	÷0.123
May	+25°C	×5.2	÷0.192
June	+30°C	×3.0	÷0.333
July	+32°C	×2.4	÷0.417
August	+33°C	×2.2	÷0.455
September	+24°C	×5.8	÷0.172
October	+14°C	×17.4	÷0.057
November	+5°C	×46.8	÷0.021
December	+3°C	×58.3	÷0.017

Average through the year

→ ÷0.157 strain rate factor

→ × 6.4 viscosity factor

i.e. $\eta_{design} = \eta_{40^\circ C} \times 6.4$

10.4.3 Other factors affecting rut development

Ageing and binder saturation also have a significant bearing on rutting.

10.4.3.1 Ageing

Unless RLAT data have been obtained from specimens recovered from an existing road, it is likely that a calculation such as that given above relates to new, unaged material. However, during the course of its life, bitumen will age significantly, resulting in an increase in viscosity. This increase may reach 400% in a dense mix in a climate such as the UK, but could be more like 5000% (a factor of 50, which represents a drop in penetration from 100 to 10 or slightly less) in a tropical or desert climate. In rough terms, this viscosity increase may be translated directly to a decrease in rutting rate. In the UK, therefore, this will make a significant difference, and in many climates ageing demands a large decrease in predicted asphalt rutting.

10.4.3.2 Binder saturation

Chapter 9 made the point that mixes that are overfilled with binder are some of the poorest performing because the excess binder acts to reduce particle–particle contact forces and so renders interparticle slip much easier. However, this effect only becomes apparent under high stress, and especially under high loading rate, and these are not the conditions present in a RLAT, nor in any other commonly used creep or repeated load creep test. Laboratory testing cannot, therefore, be expected to reveal this problem fully (although some increased deformation is usually observed). Furthermore, this is a problem that may not be apparent in early life but which becomes evident after trafficking has compacted the asphalt, reducing the void space available for the binder. The conclusion, therefore, is that no amount of simulative testing and analysis removes the need for sensible mix design, ensuring that the voids content is kept well above 2%.

10.5. Low-temperature cracking

This phenomenon manifests itself in the form of transverse cracks across the full width of a pavement and, as stated in Chapter 9, susceptibility to this type of cracking is generally related to empirical tests on bitumen; analysis is rarely carried out. In point of fact, such analysis is not difficult (Calculation Sheet 10.3 outlines what is involved), but the problem – as with rut prediction – is that one of the key inputs to the analysis, asphalt viscosity, is not easy to estimate accurately. However, taking as an example a viscosity at 0°C of 30 000 GPa.s (binder penetration around 50) and assuming a factor of 3 increase for every 10°C temperature drop (as suggested in Section 10.4), then, with an asphalt fracture strength of 5 MPa (see Section 9.1.4), the following are examples of daily surface temperature variations (assumed to be sinusoidal) that would be predicted to cause fracture.

- –10°C to –28°C
- –15°C to –30°C
- –20°C to –31°C.

However, low-temperature cracking is actually a repeated loading phenomenon, and the fracture strength should therefore be adjusted using a fatigue characteristic. For example, if the slope of the fatigue characteristic is –4 (or –0.25) on a log(strain)–log(load cycles) scale, the daily temperature ranges to cause failure after 50 cycles become

- zero to –20°C
- –10°C to –23°C
- –15°C to –25°C.

Therefore, supposing that 50 cycles represents a winter season, low-temperature cracking could be expected to occur, in this example, in a climate where pavement temperatures commonly fall below –20°C. (It is assumed that summer temperatures allow partially fatigued asphalt to make a full recovery.)

Clearly, viscosity is the key parameter. In the example in Calculation Sheet 10.3, a factor of 2 decrease in viscosity would allow around 5°C lower temperatures before the onset of cracking; hence the use of low-viscosity binder in cold climates. In a desert climate, where binder penetration frequently reaches single figures, viscosity may be at least ten times that of a new 50 penetration grade mixture, in which case it is predicted that 50 cycles between + 20°C and 0°C would be sufficient to cause fracture.

Calculation Sheet 10.3. Low-temperature cracking

Assumption: Uniform temperature within the asphalt

As the asphalt cools at night, the contractive strain due to temperature loss ΔT is matched by an extensive strain due to tensile stress σ and viscosity η

Thermal shrinkage: $\varepsilon_{\text{contraction}} = \alpha \Delta T$ **(1)**

Viscous flow: $d\varepsilon_{\text{extension}}/dt = \sigma/\eta$ **(2)**

where $\varepsilon_{\text{contraction}} = \varepsilon_{\text{extension}} = \varepsilon$

Differentiate Equation 1: $d\varepsilon/dt = \alpha dT/dt = \sigma/\eta \rightarrow \eta = \sigma/(\alpha dT/dt)$

Example:

$\alpha = 1.8 \times 10^{-5}$; $dT/dt = 1°C/h$; $\sigma_{\text{fracture}} = 5\,MPa \rightarrow \eta_{\text{fracture}} = 10^{15}\,Pa.s$

If $\eta = 3 \times 10^{13}\,Pa.s$ at 0°C, increasing by × 3 every 10°C:

η reaches $10^{15}\,Pa.s$ at −32°C

As viscosity is the critical parameter and bitumen is the viscous element in asphalt, design against low-temperature cracking can only be achieved by appropriate *binder selection*. In moderate climates this is not difficult, and standard penetration grades are suitable, but in climates with very low winter temperatures or in hot climates with large day–night temperature variation, where binders are likely to age rapidly, modified binder may be the only answer – if indeed an answer can be found at all. However, even advanced binder technology cannot work miracles and, in certain climates, low-temperature cracking simply has to be accepted as a fact of life and annual maintenance budgeted accordingly.

10.6. Wheel-path cracking

It is, of course, commonly observed that cracks occur predominantly in wheel-paths, which means they are clearly caused by repeated wheel loading – perhaps assisted by climatic factors such as low-temperature stresses. Chapter 9 presented the fact that a relationship is generally found between tensile strain under load in a fatigue test and the number of load applications until failure occurs; it is therefore quite logical that tensile strain within the asphalt layers in a pavement under a wheel load should also be related to cracking. Figure 10.8 illustrates the situation conceptually.

Cracking can occur anywhere where tensile strain is present; Figure 10.8 indicates three such areas

- A – at the bottom of the asphalt immediately under the load
- B – near the surface just outside the loaded area
- C – at the surface in the tyre tread contact zone.

The first of these is generally (but sometimes incorrectly) assumed to be dominant. Numerous so-called 'analytical pavement design' methods all over the world are based on this assumption, and it is fair to say that this approach has stood the test of time. It has the advantage of simplicity

Figure 10.8 Tensile strains in asphalt under wheel loading

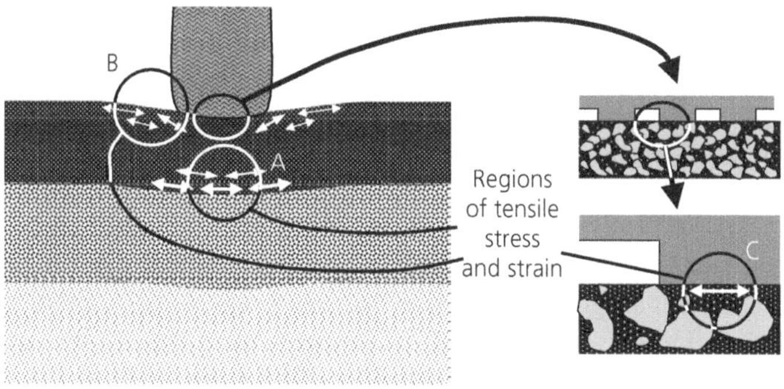

because the strain at the bottom of the asphalt can be calculated unambiguously using multi-layer linear elastic analysis.

However, engineers can't help noticing that many, perhaps the majority, of the cracks they see have clearly originated at the surface. It is common, particularly in thick asphalt pavements, to find that a wide and dangerous-looking crack in a wheel-path actually only extends a centimetre or two in depth, and it is therefore natural to want to explain this and to build a top-down cracking prediction into pavement design. Unfortunately, there is still considerable disagreement as to the cause of top-down cracking. Analysis clearly predicts that the immediate contact zone beneath the tyre tread will be extended horizontally by the mechanics of the tyre rubber, since each individual element of the tyre tread will compress on loading, giving a Poisson's ratio effect horizontally. This is restrained by shear from the surface, generating a theoretical strain high enough for early crack formation – but only extending to a depth of a few millimetres. See De Beer (2006), De Beer *et al.* (1997) and Liang *et al.* (2021) for measurement of the complex stress pattern under a tyre and Casey *et al.* (2016) for an analysis of its impact. Also, immediately to the side of the tyre is a deeper reaching region of high shear, leading to a tensile strain in a direction at an angle to the horizontal. The magnitude is not usually calculated to be as high as at the bottom of the asphalt, but this is compensated by the fact that bitumen commonly ages more at the surface than at the base of the layer.

Conceptually, therefore, top-down cracking is not difficult to explain – but it is far from easy to predict. So it is unsurprising that the great majority of analytical pavement design methods still concentrate on the strain at the *bottom* of the asphalt layer. The next few subsections describe this commonly used and enormously important approach to design.

10.6.1 Design temperature

The point has been made many times that asphalt is a temperature-sensitive material. Its stiffness changes by a factor of about 3 for every 10°C temperature change, and stiffness is a key input to calculation of tensile strain. The truth is that temperature also affects fatigue behaviour. Perhaps fortunately, this effect is masked by the accelerated way in which laboratory fatigue testing is carried out, so test data are not usually particularly sensitive to temperature; but in real pavements fatigue cracking certainly is. Experience in temperate climates is that most wheel-path cracking damage takes place in winter, when the stiffness is highest and the computed strain is lowest, and some design procedures recognise this by building in a stiffness-dependent element to the fatigue characteristic; that is, the predicted fatigue life (at a given strain level) reduces as stiffness increases.

So what are the implications? A rigorous analysis would have to take account of the contributions to cracking at all temperatures and then sum them over a year, but this is hard to do with confidence. The US mechanistic-empirical design system (AASHTO, 2007) does actually take this computationally intensive approach, using Equation 10.2 (see next section) for the rate of fatigue damage that depends on asphalt stiffness and, therefore, temperature. Nevertheless, when the AASHTO (2007) fatigue assumption is applied to a typical asphalt pavement on a major road, it still turns out that the most rapid cracking is predicted to occur at the highest temperatures, something that is not borne out by observation.

For most designers, however, the only practical approach is to admit that the problem is far too complex to analyse in such a rigorous way. They therefore select a design temperature and carry out the analysis with stiffnesses appropriate to that temperature. In the UK, the temperature usually selected is 20°C; in France it is 15°C. In reality, fatigue damage probably occurs at lower temperatures in both countries. But this is not really relevant. It is accepted that the computation carried out will only be indicative; there will have to be a significant *calibration factor* between any prediction of life based on laboratory fatigue data and life expected in reality.

10.6.2 Traditional analytical pavement design

The principal enabling task that has to be undertaken is the development of a *fatigue characteristic* for use in design, and this can only be achieved by calibrating against observed pavement performance. Any relationship between strain and asphalt fatigue life found from laboratory testing will require calibration before it can be trusted for real, partly due to the lack of realism present in all laboratory test set-ups and procedures, but also a function of the lack of realism inherent in traditional pavement analysis (relying on calculation of the tensile strain at the bottom of the asphalt layer). Such calibrations have been carried out by a number of organisations since the 1970s, notably the Shell oil company (Shell, 1978), the Asphalt Institute in the USA (Asphalt Institute, 2008) and the Transport and Road Research Laboratory in the UK (Powell *et al.*, 1984).

For example, the main Transport and Road Research Laboratory equation for the number of load applications to fatigue failure of asphalt (N_f) as a function of tensile strain (ε_t), derived from experience of dense asphalt mixtures on typical UK roads, can be expressed as

$$N_f = 4.17 \times 10^{-10} \varepsilon_t^{-4.16} \tag{10.1}$$

In the USA, the AASHTO mechanistic-empirical pavement design procedure (AASHTO, 2007) makes use of the stiffness-dependent formula

$$N_f = 0.00432 k_1' C \varepsilon_t^{-3.9492} E^{-1.281} \tag{10.2}$$

where $k_1' = f(h)$, h is the asphalt thickness, $C = f$(mixture volumetrics) and E is the asphalt stiffness modulus.

An approach that is both sensible and practical is to adjust the actual equation applying in a particular case by treating laboratory testing as a purely comparative exercise. Thus, any proposed new asphalt material could be tested using a selected fatigue test and the results compared with a dataset determined for a standard material with known performance. This would give a shift factor to be applied to the standard fatigue characteristic (i.e. a multiplication factor on predicted pavement life).

Figure 10.9 An example of the traditional analytical design process

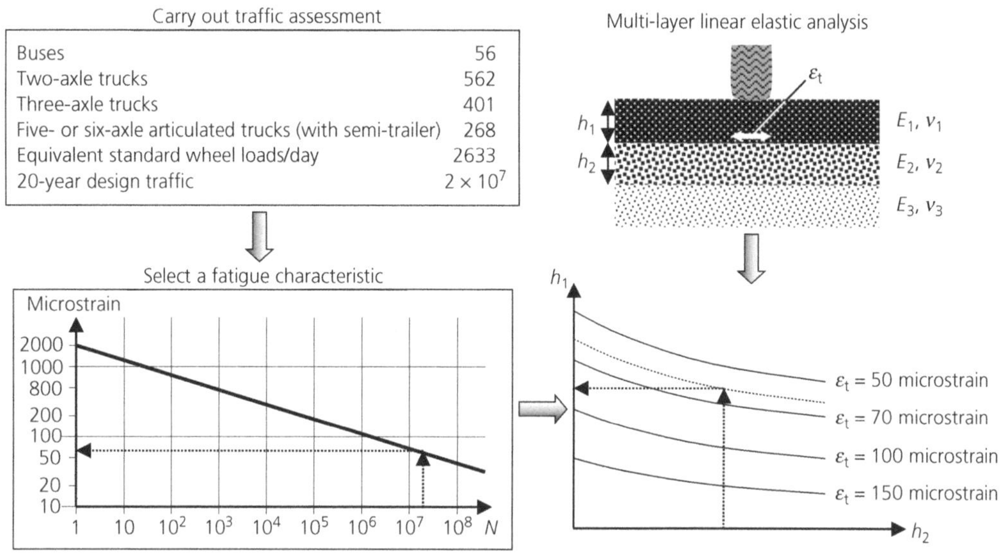

Once an appropriate fatigue characteristic has been selected, design is fairly straightforward. Figure 10.9 illustrates the procedure, which involves calculation of the tensile strain at the base of the combined asphalt layer using a multi-layer linear elastic analysis program (see Section 7.3.2).

Use of this type of design approach is widespread and, in the main, successful. It is often criticised for the underlying assumption that cracking starts at the bottom of the asphalt, but this is a flawed argument because these methods do not rely on fundamentally correct modelling but on calibration against reality. However, there are genuine limitations of which users should be aware. Most importantly, the calibrated fatigue characteristic cannot be assumed to apply outside the climatic region for which it was developed. Nor can it be applied with any real confidence to significantly non-standard asphalts, design loads, traffic numbers, pavement foundation materials and so on. In such cases, the message is – handle with care!

10.6.3 Pavements with a thin asphalt layer

A particular problem with this type of analysis occurs with thin asphalt layers, as illustrated in Figure 10.10. The problem is that deflection (and therefore surface curvature) of pavements with thin asphalt layers is dominated by the stiffness of the underlying support, which means that it only increases relatively slightly as asphalt thickness reduces; and strain due to flexure, which is proportional to curvature but inversely proportional to thickness, actually starts to reduce as thickness reaches very low values. However, while this may be theoretically true, it results in a highly unrealistic prediction of pavement life. A practical approach (shown in Figure 10.10) is simply to fit a characteristic curve to points with greater thickness and to project this characteristic into the low-thickness region. Also shown in the inset in Figure 10.10 is a comparison of pavement life determined in this way with a prediction using the much more realistic (but also much more complex) crack propagation modelling approach outlined in the next two sections, and the match is seen to be reasonably good down to a thickness of 30 mm. So, although this construction cannot easily be justified theoretically, it appears to generate realistic solutions.

Figure 10.10 Adapting conventional analytical design to thin pavements

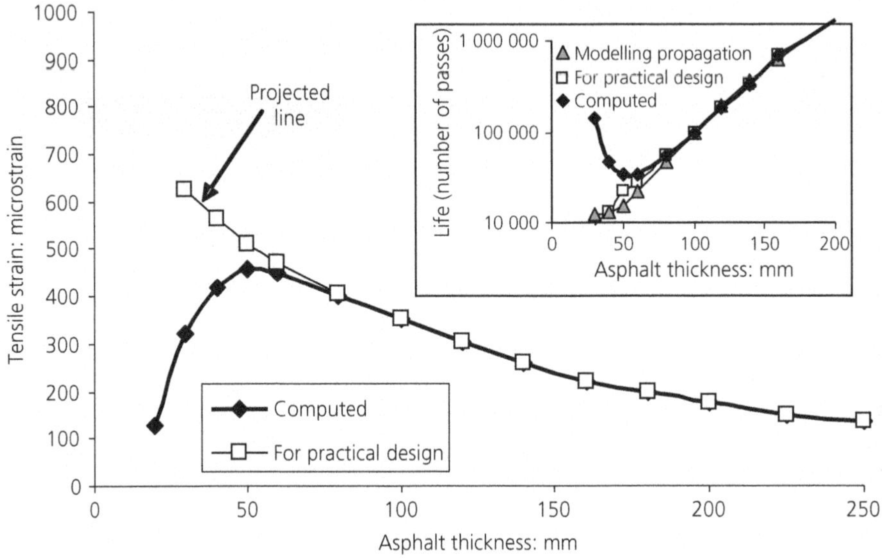

10.6.4 Predicting crack propagation

Most traditional design approaches make no attempt to model the progression of a crack. The result is simply the number of load applications until the 'state of cracking' has reached some defined condition, such as the first appearance of cracking on the surface. This has the clear advantage of simplicity, but does not really have logic on its side. For instance, two pavements may have the same calculated tensile strain, but one has twice the asphalt thickness of the other (due to dramatically different foundation stiffnesses, for example). With traditional design, the predicted lives would be the same, despite cracks having to propagate through twice the distance in one case compared with the other. The AASHTO (2007) mechanistic-empirical pavement design method is an exception in that a constant (k_1') dependent on thickness (h in inches) is included.

$$k_1' = 1/[0.000398 + 0.003602/(1 + e^{(11.02-3.49h)})] \qquad (10.3)$$

for bottom-up cracking; slightly different for top-down.

This constant is actually an empirical means of addressing the 'thin asphalt' problem.

An alternative way to tackle this anomaly, and potentially develop an improved predictive technique, is to interpret fatigue test data somewhat differently and to postulate a *crack propagation law*. Perhaps the obvious choice here would be the so-called Paris law (Paris and Erdogan, 1963), developed to model crack growth in metals, which relies on the calculation of a 'stress intensity factor' at the tip of a crack, and there are researchers who do indeed make use of this model for asphalt also. However, the differences between asphalt and metal are highly significant, and Chapter 9 has made the point that cracking in asphalt actually consists of the progressive breakage of bonds between individual aggregate particles rather than the propagation of a discrete crack.

For this reason, the Paris law is not recommended here. However, there is no reason why the observed relationship (from laboratory tests) between strain and number of load applications to failure cannot be re-expressed as a relationship between strain and crack propagation rate, always accepting that 'damage progression' may be a better description in reality than crack propagation. The suggested equation is of the form

$$\mathrm{d}c/\mathrm{d}N = A\varepsilon_{\text{crack}}^{n} \quad (10.4)$$

where c is crack length, and A and n are constants. Or, integrating over the propagation length c

$$\int \varepsilon_{\text{crack}}^{-n}\,\mathrm{d}c = AN \quad (10.5)$$

The parameters A and n have to be determined from laboratory testing, and Calculation Sheet 10.4 suggests how this might be achieved. The details are open to debate, and Calculation Sheet 10.4 certainly represents a highly simplified understanding of the stress distributions applying under conditions of tension and flexure.

Calculation Sheet 10.4. Crack propagation parameters

Tensile tests (including indirect tensile)

Assume cracks develop as shown, each crack a minimum distance from its neighbours. As they grow, dominant cracks progress; lesser cracks stop growing. At each stage the picture is assumed still to be as in the diagram. The specimen fails rapidly once the crack length c reaches about 40% of the specimen dimension

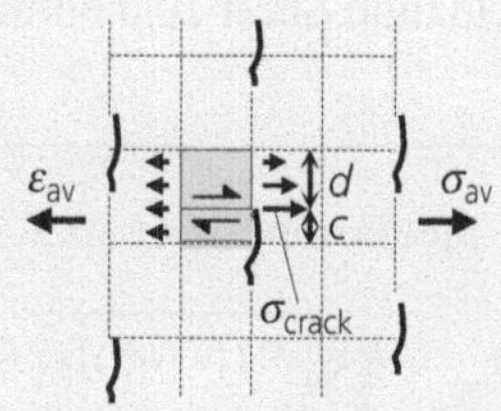

Approximate analysis of shaded area:

$$\sigma_{\text{crack}} \approx 2.63\sigma_{\text{av}} \;\therefore\; \varepsilon_{\text{crack}} \approx 2.63\varepsilon_{\text{av}}$$

$$\int \varepsilon_{\text{crack}}^{-n}\,\mathrm{d}c = AN \;\therefore 2.63^{-n}\varepsilon_{\text{av}}^{-n}c = AN$$

Derive A and n from plot of ε_{av} against N

Flexural tests

Assume uncracked thickness $(h - c)$ has normal beam stress distribution. The value of c depends on when the test is stopped. 50% apparent beam stiffness when $c \approx 0.33h$

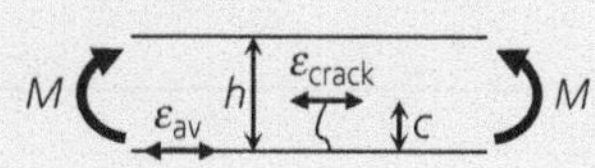

$$M = \sigma_{\text{av}}h^2/6 = E\varepsilon_{\text{av}}h^2/6 = E\varepsilon_{\text{crack}}(h-c)^2/6$$

$$\varepsilon_{\text{crack}} = \varepsilon_{\text{av}}h^2/(h-c)^2$$

$$\int \varepsilon_{\text{crack}}^{-n}\,\mathrm{d}c = AN \rightarrow \varepsilon_{\text{av}}^{-n}\int (h-c)^{2n}\,\mathrm{d}c/h^{2n} = AN$$

$$\varepsilon_{\text{av}}^{-n}[h^{2n+1} - (h-c)^{2n+1}]/(2n+1)h^{2n} = AN$$

Derive A and n from plot of ε_{av} against N

A particular issue is selection of an appropriate value for the final crack propagation length (c), representing the end of the test. Under load control, the test ends when the specimen fails. In the case of a tensile (or indirect tensile) test, Calculation Sheet 10.4 implies that when a crack extends across about 40% of the specimen width, then the assumed crack pattern breaks down and failure will occur rapidly; under flexure, with cracks developing from both top and bottom of the beam, rapid failure will occur (under load control) once c reaches approximately one-third of beam thickness (h).

Having determined parameters A and n, the crack propagation equation can then, in theory, be applied to a pavement, although it is necessary to develop a method of pavement analysis that allows calculation of the strain in the region of a crack tip. This could be a finite element analysis, but it could also be a simpler evaluation of asphalt bending (and shear), making an allowance for the reduction in effective stiffness as cracks develop. One such simplified form of analysis, suitable for spreadsheet work, is the so-called 'method of equivalent thicknesses' (Odemark, 1949). This method is based on the assumption that the stresses, strains and displacements beneath a surface layer (e.g. asphalt) are solely a function of the bending stiffness of that layer and the properties of the underlying material. Thus, a thickness h_1 of an upper layer of stiffness modulus E_1 and Poisson's ratio ν_1 is assumed to be equivalent to a layer of thickness h_{eq2} of modulus E_2 and Poisson's ratio ν_2, E_2 and ν_2 being the properties of the underlying material. Stresses, strains and displacements can then be calculated using Boussinesq's half-space equations (Boussinesq, 1883). Calculation Sheet 10.5 outlines the necessary steps.

Calculation Sheet 10.5. Method of equivalent thicknesses

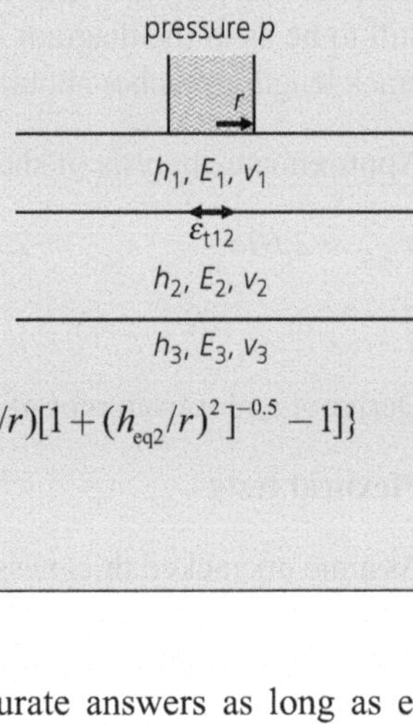

$$h_{eq2} = h_1 \sqrt[3]{[E_1(1-\nu_2^2)]/[E_2(1-\nu_1^2)]}$$

from Boussinesq:

$$\sigma_{v12} \approx p\{1-[1+(r/h_{eq2})^2]^{-1.5}\}$$

$$\varepsilon_{v12} \approx [p(1-\nu_2)/E_2]\times\{(h_{eq2}/r)[1+(h_{eq2}/r)^2]^{-1.5} - (1-2\nu_2)[(h_{eq2}/r)[1+(h_{eq2}/r)^2]^{-0.5} - 1]\}$$

$$\varepsilon_{t12} \approx \{[(1-\nu_2)/2\nu_2](\sigma_{v12} - E\varepsilon_{v12}) - \nu_2\sigma_{v12}\}/E_2$$

The method of equivalent thicknesses generates reasonably accurate answers as long as each layer is significantly stiffer than the one beneath. In the case of a partially cracked asphalt layer, the value of E_1 used should be reduced to take account of the cracking present (Figure 10.11); the computed strain is then converted to a bending moment, from which the strain in the region of the crack tip can be determined (following the principle shown in Calculation Sheet 10.4 for flexure). Note that the expression 'in the region of the crack tip' is appropriate here because the 'crack' is, in reality, not continuous, but represents a region of concentrated damage (i.e. broken bonds between particles).

Figure 10.11 Calculation of a crack propagation increment

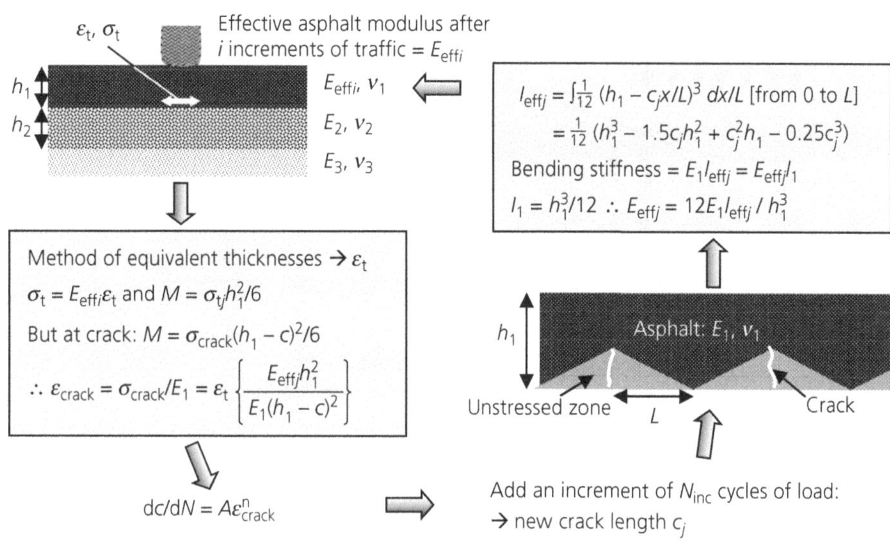

The procedure is incremental. For each increment of a certain number of load applications (logically related to a certain growth in crack length), a new effective asphalt layer modulus is calculated, from which, using the method of equivalent thicknesses, the strain in the region of the crack tip is determined. This leads to the calculation of a crack growth rate, followed by a revised crack length and so on.

A sensible procedure for determining an equivalent stiffness modulus for the combined asphalt layers is needed, and this might reasonably be taken as a weighted average, weighted according to the relative thickness of each sub-layer. It has to be remembered that this is an approximate calculation method, but it lends itself readily to spreadsheet work, which is an appreciable advantage. It also allows different asphalt layers of differing properties to be taken into consideration, for example, allowing the benefit from the use of a modified binder in one of the layers to be evaluated.

However, despite this approach taking crack propagation into account, *calibration* against known performance will still be required. This is partly due to the fact that different types of fatigue test tend to produce differing fatigue characteristics, but it is also due to the inevitable differences between laboratory and site in loading rate, pattern of loading and temperature.

10.6.5 Top-down cracking

The problem with this sort of realistic modelling of crack propagation is that the issue of top-down cracking raises its head once more. It is no good carefully predicting the course of cracks from the bottom of an asphalt layer if the true mode of deterioration is by surface-originating cracking! Logically, therefore, it is necessary to develop a technique for calculation of strain at the surface, and a way of achieving this is shown in Figure 10.12. Admittedly, this is an area where there is no consensus as yet, and the equation shown in Figure 10.12 is nothing more than a useful working

Figure 10.12 Surface strain computation

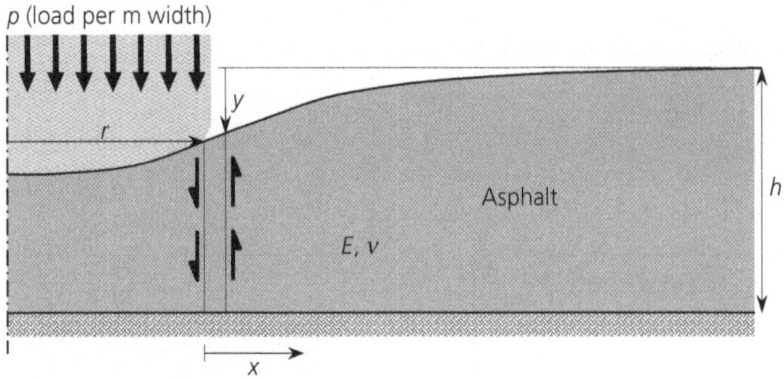

Suggested equation: $y = (ph/E)\exp\{-(x/h)\sqrt{[6(1 + \nu)]}\}$

$dy/dx = (p/E)\sqrt{[6(1 + \nu)]}\exp\{-(x/h).\sqrt{[6(1 + \nu)]}\}$

Maximum at x = 0: $= (p/E)\sqrt{[6(1 + \nu)]}$

$=$ shear strain γ

And assume that $\varepsilon_{tensile} \approx \gamma$

approximation; the key assumption made is that surface strain is primarily a function of shear adjacent to the loaded area – and different researchers and practitioners may have differing views as to the most appropriate assumptions to make.

When using this computational method for crack propagation prediction, the equation for shear strain should change as the crack progresses. The simplest approach is to assume that the pressure due to wheel loading dissipates with depth into the pavement; the value of p in Figure 10.12 should therefore be reduced as the crack grows, assuming a load spread angle (e.g. 35°) acting in the longitudinal direction (i.e. along the crack).

Note that the equation in Figure 10.12 is for shear strain and that the direction of maximum tensile strain is at 45° to the horizontal, which would logically produce an inclined crack. However, the *lateral distribution of tyre paths* across the pavement surface means that different load applications produce different maximum stress points and directions, and the effect may be for inclined increments of crack growth to cancel each other out, leading to a near vertical crack. In fact, an adjustment for the lateral distribution of wheel loads, often with a standard deviation of around ±0.15 m on highways but considerably more on most airfield pavements, can also be built into a spreadsheet-based prediction.

A point to emphasise is that the designer must never be afraid of approximation. The nature of pavement performance prediction will always be approximate because of the inherently uncertain materials involved and the unfathomable complexities of temperature effects. There is thus little real value in precise analysis – only in taking a fundamentally correct approach.

Figure 10.13 shows predictions for both bottom-up and top-down cracking, carried out using the simplified techniques described above (Thom, 2000), for two conventional asphalt pavements with identical unbound pavement foundations, but with different asphalt thicknesses. Whether accurate or not, the predictions clearly match the observed fact that top-down cracking tends to predominate on thick pavements but that bottom-up cracking is dominant for thin pavements. For the thick pavement, top-down cracking is predicted to occur relatively rapidly but then to slow at a depth of around 50 mm, which matches what is typically observed and gives a degree of confidence that the assumptions made are sensible. Note that further rapid crack propagation only occurs after the bottom-up crack finally starts to make significant progress.

10.6.6 Design with reinforced asphalt

The following comments reflect the author's own views based on experience, both in practice and in research (Thom, 2003), and it is hoped that they will assist the reader; however, they do not represent universally accepted truth.

Figure 10.13 Predictions of crack propagation in asphalt pavements

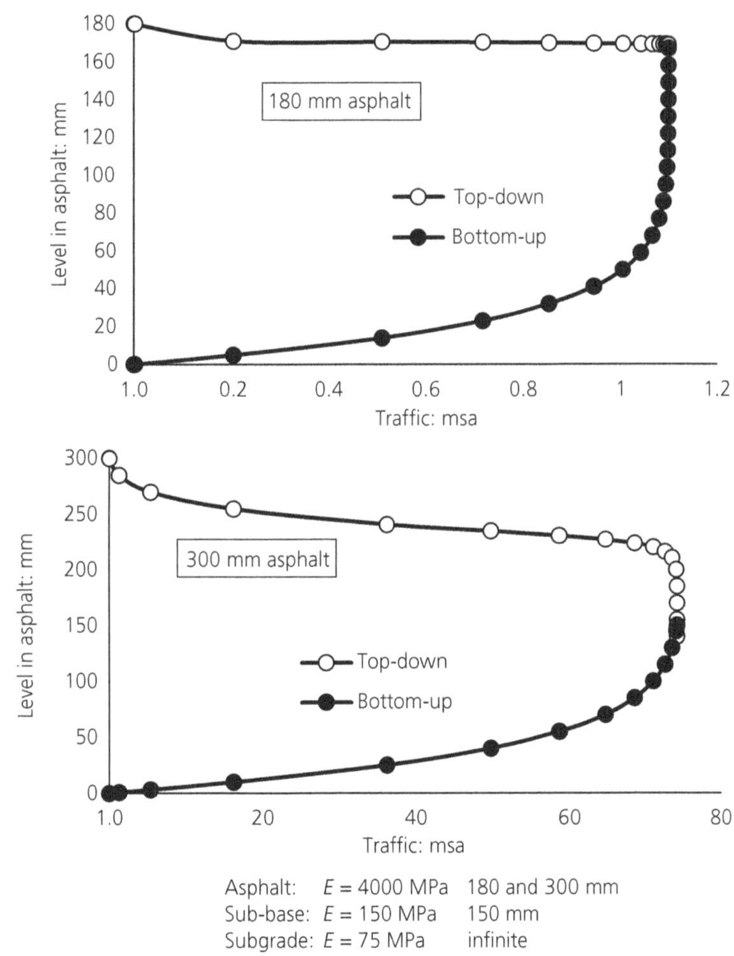

Chapter 9 introduced the concept that fatigue damage, at least in its early stages, consists of the enlargement of local fracture zones around particle contacts and the initiation of new ones. To allow this to happen, it follows that aggregate particle realignment is taking place since this is the only physical mechanism through which new fracture points can be formed; and 'realignment' really means localised permanent deformation. So, at a fundamental level, fatigue damage and permanent deformation are inextricably linked. Reinforcing products are usually made of polymers or bonded glass fibre and they are not stiff enough to affect an asphalt's elastic behaviour; but they are quite good at inhibiting permanent deformation, hence their use in unbound materials as presented in Chapter 7. And, if they can inhibit localised permanent deformation, that would explain how they might inhibit cracking – but by how much? Field evidence (e.g. Elsing and Sobolewski, 1998; Huhnholz, 1996) suggests a life enhancement factor of around 4 is often observed in practice, which means asphalt reinforcement is worth taking seriously, but what is the best way to take account of it in design?

The type of mechanistic analysis described in Sections 10.6.4–10.6.5 is well suited to non-standard designs, and the use of reinforcement certainly falls into this category. It is straightforward to include a slow-down in crack propagation rate above and below the reinforcement. It is also possible to take account of the effect of a reinforcing grid in 'stitching' a crack, effectively preventing it from opening at the location of each reinforcing strand – although not between strands. Figure 10.14 illustrates and presents an example prediction of pavement life with and without geogrid reinforcement, in this case showing an improvement factor of about 4.

It is simply not possible to generalise about the effect of reinforcement, as every design case is different, although predictions using the crack propagation approach described above often suggest that a layer of reinforcing grid may be equivalent to around 30–50 mm of asphalt in a conventional asphalt pavement in terms of its effect in delaying crack development. However, it should be recognised that there will be additional post-cracking benefit since reinforcement will inhibit narrow cracks from developing into larger failures, maintaining pavement serviceability for longer.

Figure 10.14 Prediction with geogrid reinforcement

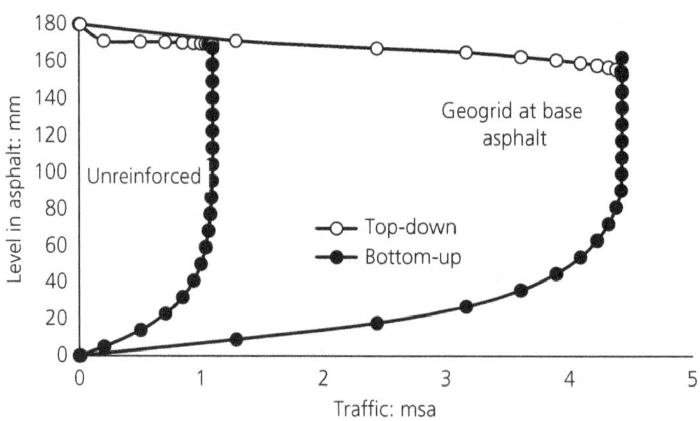

Asphalt: E = 4000 MPa 180 mm
Sub-base: E = 150 MPa 150 mm
Subgrade: E = 75 MPa infinite

However, one potential negative has to be taken into consideration, namely 'recyclability'. Some geogrids fragment under the action of a planer and so become part of the resulting RAP (recycled asphalt planings) – but this means the RAP probably cannot be reused in fresh hot- or warm-mix asphalt, although it is still likely to be suitable for cold mixes. Other geogrids are tough enough to cause problems to the planer and can be a real headache to remove from the pavement during a partial reconstruction operation. They can still be used but the designer has to be certain that they will remain in place for a long period.

10.6.7 Pavement edge effects

All the analysis presented thus far has assumed a continuous pavement of infinite extent all around the loaded area, which is obviously not the case, and clearly a load near to a pavement edge would produce an increased strain. Chapter 20 will present a method of analysis for concrete pavements, which typically finds edge load stresses to be up to 80% higher than those distant from an edge. Substituting a typical asphalt stiffness modulus, this reduces only slightly. Assuming an asphalt fatigue characteristic with a slope of −4 (or −0.25 depending on how the equation is formulated), this means that each edge load application induces about 10 times the damage caused by a normal wheel-path load application. Thus, if heavy vehicles keep well away from the edge 90% of the time or more, then edge loading is not critical; if, on the other hand, over 10% of passes are at or near the edge, then this becomes the critical design case (i.e. a reduced number of load applications but an increased asphalt tensile strain).

The question is: how far is 'well away' from the edge? Unfortunately, this question has no single answer. For thick pavements with wide deflection 'bowls', the author's experience with in situ deflection testing suggests perhaps a metre; for thinly surfaced pavements it may be no more than half a metre. Assuming a 2.0 m centre-to-centre distance between near-side and off-side wheel sets on heavy goods vehicles and that vehicles travel centrally along a traffic lane, this suggests edge lanes on major highways should have a width of around 4.0 m, whereas 2.8 m is sufficient on minor roads. In practice, the 4.0 m requirement would normally be replaced by a narrower lane and an adjacent hard strip, usually 1 m wide in the UK. Problems are most common on minor roads, with insufficient width (less than 5.6 m total) to accommodate two heavy goods vehicles without encroaching close to the edge.

10.7. Maximising reliability

Mathematical treatment of reliability was covered in Chapter 5. The questions here are: what can go wrong and how can the pitfalls be avoided? The following reflects the author's experiences, some direct, some based on the findings of others.

10.7.1 Materials specification

Materials may be selected based on sound engineering principles and the pavement design carried out diligently and correctly – but how does the designer make sure that it is constructed to an acceptable standard?

10.7.1.1 Recipe and method

This technique is that of a cookbook – list the ingredients and their proportions, specify the mixing method (temperature, time, equipment), give instructions as to how to produce the perfect cake! If the instructions are properly expressed, then anyone should be able to follow and achieve excellent results. It's simple and low risk, and many asphalt specifications fall into this category, for example most UK National Highways specifications (National Highways, 2021b). The specification will typically include

- aggregate description (e.g. hard, angular, percentage of crushed faces, flakiness limits)
- aggregate gradation limits
- binder grade – which is specified with limits on penetration and softening point
- binder content limits
- mixing temperature limits
- minimum mixing time
- paving temperature limits
- number of roller passes – which depends on roller type used.

Inevitably, the use of limits and descriptive words gives scope for quite a bit of variation but, in a given context (e.g. single country or state with a limited range of aggregate sources and a mature construction industry), this approach can lead to a suitably reliable product.

10.7.1.2 End product

But what happens when we want to do something non-standard, such as incorporate a polymer modifier into the bitumen or include RAP into the mix? A desirable option would be to leave all these decisions up to the contractor and simply specify that the finished pavement has to pass some sort of post-construction test, or just insist that it fulfils minimum serviceability criteria, with appropriate penalty clauses in the event of failure. The problems are (*a*) there is no post-construction test that gives assurance of pavement life and (*b*) it's difficult to design contracts with the right degree of incentive and penalty. As the highway authority, the last thing you want is to over-penalise and see your contractors go bust; that way your quality assurance disappears with them. But you also don't really want to pay out more than you have to. It's a fascinating area and it's worth highway authorities taking the time and effort to get such contracts to work as well as they can – but regretfully this is outside the scope of this book.

So, back to post-construction testing: what can be achieved? Realistically, there are two options. The stiffness of the whole (or partially constructed) pavement can be measured by a falling weight deflectometer (see Chapter 25) or, for unbound layers, a light-weight deflectometer (Section 6.3.10), or cores can be extracted and tested in the laboratory. But in neither case is it straightforward to write requirements into a specification.

The problems with deflectometer testing are (*a*) the stiffness of the subgrade is outside of the contractor's control, (*b*) the properties of unbound materials change with the water state and therefore with the weather and (*c*) asphalt stiffness changes with temperature, requiring correction to the reading. The result is that sometimes, although the tests themselves may be specified, the results will be for information only (i.e. without limits imposed). It's still a useful thing to do; it's just the contractual side of things that is tricky.

Core testing is more straightforward. Specifications can be written for air-voids content, percentage refusal density (BSI, 1998) – similar to the percentage maximum dry density in unbound materials – stiffness, fracture strength, deformation resistance, so-called retained strength (ratio of soaked to unsoaked strengths), even for fatigue. Testing in this way gives a very direct measure of the quality of an asphalt layer and so a really high level of assurance that it will perform as designed. The contractual challenge is to determine the appropriate sanction when the specified end-product performance level is not reached. If the sanction is too great (rip the road up and start again!) then you will suddenly find the cost of all your contracts rising sharply due to excessive contractor risk.

In the end, the type of specification has to be appropriate for the type of pavement and its economic importance.

10.7.2 Interlayer bond

In previous sections, the assumption has been made that the combined asphalt layer, which may in reality consist of four or five individual courses, can be treated as a monolithic whole; indeed, the success of the design depends on it. But what happens if this is not achieved? Section 26.8.1 will illustrate and discuss more fully in the context of pavement evaluation, but here it should be appreciated that the apparent stiffness of the layer as a whole will be reduced, giving an increased (in most cases) strain within the asphalt. This effect is most certainly real and, unfortunately, common. Reduced layer stiffness is routinely found when carrying out in situ testing, as described in Chapter 25. Furthermore, a specific problem with debonding of the surface course is delamination leading to pothole development.

The conclusion here is that it is worth spending a little extra on the interlayer bond, as outlined in Section 10.1.3, to avoid the very real danger of a considerably reduced life. This might entail an increased coverage rate for the tack coat (or bond coat) used or the use of a polymer-modified bitumen; it might be ensuring that tack coating is not carried out in wet conditions or that the surface is properly cleaned before application of the tack coat.

10.7.3 Construction conditions

It is obviously easier to achieve good results in pleasant weather in daylight than in cold conditions at night. No amount of adhering to procedures can eliminate the human factor, and the human factor is most productive when feeling comfortable! Perhaps this oversimplifies the issue, but Figure 10.15 is taken from a recent study of surface deterioration of major UK highways (Khojinian, 2019), which found that the construction season was by far the biggest factor affecting longevity of the surface course, at least as far as ravelling is concerned.

The problem of course is that the solution to this issue is not really an engineering one. It demands that those looking after spending by highway authorities find a way of managing their budgets to concentrate asphalting in the most clement months of the year. Other construction activities can be carried out in the winter but, ideally, asphalting should not be.

Figure 10.15 Percentage ravelling of UK major highways according to construction season (Khojinian, 2019)

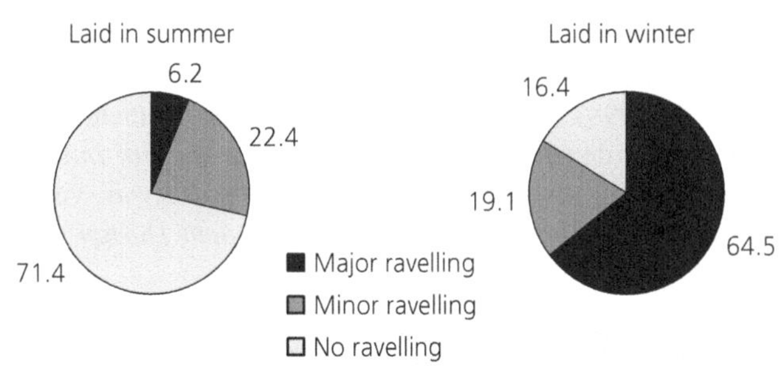

10.8. Summary

Design is a multi-faceted thing and often it's a compromise. In terms of material selection, it is sometimes a trade-off between deformation resistance, crack resistance, ease of construction (and therefore good reliability) and cost. An optimised design is a balanced design. That is to say, there is no point giving the pavement excellent rut resistance if it cracks early or if it violates the drainage principles to be outlined in Chapter 24. Each possible mode of failure has to be given appropriate weight, including those that are difficult to quantify, such as top-down cracking.

REFERENCES

AASHTO (American Association of State Highway and Transportation Officials) (1993) *Guide for Design of Pavement Structures*. AASHTO, Washington, DC, USA.

AASHTO (2007) *Guide for Mechanistic-Empirical Design of New and Rehabilitated Pavement Structures*. AASHTO, Washington, DC, USA.

Abed A, Rahman M, Thom N, *et al.* (2023) Predicting pavement performance using distress deterioration curves. *Road Materials and Pavement Design*. See https://doi.org/10.1080/14680629.2023.2238094 (accessed 30/03/2024).

Asphalt Institute (2008) *Thickness Design: Asphalt Pavements for Highways and Streets. Manual Series No. 1 (MS-1)*, 9th edn. Asphalt Institute, Lexington, KY, USA.

ASTM (2016) D4887: Standard practice for preparation of viscosity blends for hot recycled asphalt materials. ASTM International, West Conshohocken, PA, USA.

BBA (British Board of Agrément) (2004) Torque bond test. In *Guidelines Document for the Assessment and Certification of Thin Surfacing Systems for Highways*. BBA, Watford, UK, Report SG3/05/234.

Boussinesq J (1883) *Application des Potentials à l'Étude de l'Équilibre et du Mouvement des Solides Élastiques*. Gautier-Villars, Paris, France (in French).

BSI (1998) BS 598-104:1989: Sampling and examination of bituminous mixtures for roads and other paved areas. Methods of test for the determination of density and compaction. BSI, London, UK.

Casey DB, Airey GD and Grenfell JR (2016) A comparison of uniform and 3-D tyre contact pressure representations using a finite element method. *Transportation Research Procedia* 14: 2402–2410.

De Beer M (2006) Reconsideration of tyre-pavement input parameters for the structural design of flexible pavements. *Proceedings of the 10th International Conference on Asphalt Pavements*. International Society of Asphalt Pavements, Lino Lakes, MN, USA.

De Beer M, Fisher C and Jooste FJ (1997) Determination of pneumatic tyre/pavement interface contact stresses under moving wheel loads and some effects on pavements with thin asphalt surfacings. *Proceedings of the 8th International Conference on Structural Design of Asphalt Pavements*, Seattle, WA, USA, vol. 2, pp. 1643–1658.

Defence Estates (2006) *A Guide to Airfield Pavement Design and Evaluation*, 2nd edn. Defence Estates, London, UK, Design and Maintenance Guide 27.

Elsing A and Sobolewski J (1998) Asphalt-layer polymer reinforcement: long-term experience, new design method, recent developments. *Proceedings of the 5th International Conference on the Bearing Capacity of Roads and Airfields, Trondheim, Norway*, vol. 3, pp. 1523–1532.

FAA (Federal Aviation Administration) (2021) *Airport Pavement Design and Evaluation*. FAA, Washington, DC, USA, AC150/5370-6G.

Huhnholz M (1996) Asphalt reinforcement in practice. *Proceedings of the 3rd International RILEM Conference on Reflective Cracking in Pavements, Maastricht, the Netherlands*, pp. 456–463.

Liang C, Zhu D, Wang G and Shan M (2021) Experimental study on the tire-road dynamic contact pressure distribution using FTIR imaging. *International Journal of Automotive Technology* 22(5): 1305–1317.

Khojinian A (2019) *Strategy Management for Thin Surface Course Systems*. PhD thesis, University of Nottingham, Nottingham, UK.

Knapton J (2007) *The Structural Design of Heavy Duty Pavements for Ports and Other Industries*, 4th edn. Interpave, Leicester, UK.

National Highways (2020a) *Manual of Contract Documents for Highway Works. Vol. 1: Specification for Highway Works*. Stationery Office, London, UK.

National Highways (2020b) CD 224: Traffic assessment. *Design Manual for Roads and Bridges. Pavement Design*. Stationery Office, London, UK.

National Highways (2020c) CD 225: Design for new pavement foundations. *Design Manual for Roads and Bridges. Pavement Design*. Stationery Office, London, UK.

National Highways (2021a) CD 226: Design for new pavement construction. *Design Manual for Roads and Bridges. Pavement Design*. Stationery Office, London, UK.

National Highways (2021b) *Manual of Contract Documents for Highway Works. Vol. 1: Specification for Highway Works. Series 900: Road Pavements – Bituminous Bound Materials*. Stationery Office, London, UK.

Odemark N (1949) *Undersökning av elasticitetsegenskaperna hos olika jordarter samt teori för beräkning av beläggningar enligt elasticitetsteorin*. Statens Väginstitut, Stockholm, Sweden (in Swedish).

Paris PC and Erdogan FA (1963) A critical analysis of crack propagation laws. *Journal of Basic Engineering, ASME* **85**: 528–534.

Powell WD, Potter JF, Mayhew HC and Nunn ME (1984) *The Structural Design of Bituminous Roads*. Transport and Road Research Laboratory, Crowthorne, UK, Report LR1132.

Shell (1978) *Shell Pavement Design Manual – Asphalt Pavements and Overlays for Road Traffic*. Shell International Petroleum Company Ltd, London, UK.

Stöckert U (2001) Schichtenverbund – Prüfung und Bewertungshintergrund. *Straße und Autobahn* **11**: 624–631 (in German).

Thom NH (2000) A simplified computer model for grid reinforced asphalt overlays. *Proceedings of the 4th International RILEM Conference on Reflective Cracking in Pavements, Ottawa, Canada*, pp. 37–46.

Thom NH (2003) Grid reinforced overlays: predicting the unpredictable. *Proceedings of MAIREPAV'03, Guimaraes, Portugal*, pp. 317–326.

Nick Thom
ISBN 978-1-83549-711-1
https://doi.org/10.1108/978-1-83549-710-420243012

Chapter 11
Asphalt surface design

Previous chapters have concentrated on the structural strength of a pavement, ensuring that neither deformation nor cracking is excessive and that material properties do not deteriorate with time. However, the most sensitive element in a pavement from the point of view of a user is obviously the surface.

11.1. Material options

Section 9.7 introduced the different generic types of asphalt and Figure 10.5 illustrated some of the different forms of surface. It is now time to get down to the detail. In essence, the options are

- asphalt concrete – a dense mix, minimised binder content, typically 5–8% voids
- stone mastic asphalt (SMA) – also dense, two-part gradation, higher binder content, should be <6% voids
- porous asphalt – open-graded, relatively low binder content, around 18–20% voids
- hot-rolled asphalt (HRA) – very dense, two-part gradation, high binder content, around 3% voids
- mastic asphalt (e.g. Gussasphalt) – only fine aggregate, very high binder content, extremely low voids
- surface dressing – single or double layer of coarse stones set in bitumen
- grouted macadam – a porous asphalt with voids filled with a cementitious grout.

Each of these comes in different varieties depending on the stone size. The usual maximum stone size varies from about 14 mm down to 6 mm, less in the case of mastic asphalt. Generally, the smaller the stone size the higher the required binder content (and therefore the higher the cost per tonne), but a small stone size also gives greater workability to the mix, allows a thinner layer and is likely to give a reduced voids content.

- Asphalt concrete is a safe option in that contractors can easily produce and lay it. It represents a minimum-cost surface.
- SMA relies on an interlocking skeleton of larger stones, giving good rut resistance, and the gaps are filled in by a finer bituminous mastic, giving good crack resistance. If done correctly, this is an excellent material – more expensive than asphalt concrete but longer lasting.
- Porous asphalt is a special case in that it deliberately violates the golden rule of drainage that permeability should increase with depth. It means that rainwater will collect within the surface course (good from the points of view of spray and noise) and so there has to be drainage provision at the side of the road to allow it to escape. The material itself is a little expensive and is not easy to repair well.
- HRA is really an extreme type of SMA with a relatively low stone content and more mastic – and so more cost. This means that the stones no longer form a skeleton; that relies on the quality of the sand-sized particles within the mastic. It is a very easy material to pave and compact and has excellent crack resistance, but there is a risk of rutting if the sand fraction is

unable to do its job properly. It also has no texture and so it is necessary to spread bitumen-coated stone chippings over the surface before compaction.

- An even more extreme mix type is represented by mastic asphalt, basically HRA without any large stones at all and so a very high bitumen content and a very high cost – but it can be placed in very thin layers. It is most commonly used on structures (bridge decks, multi-storey car parks) and if texture is needed, additional stone chippings are required.
- Surface dressing was introduced in Section 7.6.1 as a topping for an unbound pavement. It would be unusual to apply it to a new asphalt pavement since it represents an additional construction process, but it is commonly used as a maintenance treatment.
- Grouted macadam is very much a premium material, typically used on pavements subject to slow-moving heavy vehicle loads such as bus stations or aircraft stands. It has little texture and so is not suited to high-speed locations.

11.2. Ride quality

One must never forget that excellent ride quality is not always necessary – but on high-speed roads it most definitely is. Typical tolerance limits specified at the surface of such a road would be ±3 mm over a 3 m horizontal distance; in order to achieve this an appropriate 'system design' is needed.

The uppermost layer of an asphalt road has to be of relatively low thickness in order to achieve tolerances suited to high-speed usage; 50 mm is just about the upper limit and 30–40 mm is more common. Even with a suitably thin surfacing, it will only be possible to achieve the specified tolerances (for high speed) if the surface of the underlying layer is also reasonably even – and this principle applies right the way through the pavement. The evenness of the surface of each layer can be constructed slightly better than that of the one below, but only slightly; the logic, therefore, is that thick and not particularly accurately laid materials can be permitted low down in the structure, with ever thinner and more accurately finished layers as the final running surface is approached. This is the reason why it is common practice to use a so-called 'binder course' between the base and the surface course. The current UK National Highways tolerances at each level are given in Table 11.1.

However, surface evenness is not the only factor contributing to ride quality; the material itself also matters. Table 11.2 summarises the impact of each surface type on likely ride quality, omitting grouted macadam since it is not used where vehicles travel at speed. The table is only intended as a crude summary. However, the key points are that a negative texture minimises disturbance to the tyre tread, and therefore gives low vibration, while low material stiffness leads to high energy absorption, and therefore vibration damping. 'Negative texture' refers to the 'skew' or 'asymmetry' of the surface profile around a mean level. Negatively textured surfaces have much of their profile a little above the mean, with deeply penetrating cavities at intervals; in positively textured pavements, much of the surface is below the mean level, with protruding asperities at intervals. From the point of view of ride quality, porous asphalt is the perfect surface.

Table 11.1 UK tolerances (National Highways, 2020)

Level	Tolerance: mm
Pavement surface	±6
Binder course	±6
Base	±15
Sub-base	+10–30

Table 11.2 Asphalt surface ride quality

Surface type	Vibration generation	Energy absorption	Ride-quality ranking
Asphalt concrete	Medium	Medium	3
Stone mastic asphalt	Low	High	2
Porous asphalt	Low	Very high	1
Rolled asphalt + chippings	Medium	Low	5
Mastic asphalt	Low	Medium	4
Surface dressing	High	Low	6

11.3. Strength and durability

The word 'strength' is used here in a very general sense and refers to the structural properties of the surface material and the surface's contribution to the load-bearing capacity of the pavement as a whole.

11.3.1 Asphalt surfaces

Base-quality asphalt is designed to give an optimum combination of stiffness, deformation resistance and fatigue strength in the most cost-effective way possible, and this commonly means a low-to-medium bitumen content and relatively large stone size (compared to the surface). The surface course, on the other hand, has to have relatively small stone size due to its low layer thickness (which is required in order to achieve level tolerances) and, consequently, a relatively high bitumen content. This leads to a less stiff material, although deformation resistance and fatigue strength can be as high or higher than those of the base. Table 11.3 lists the likely surface course properties relative to those of an asphalt concrete base. Surface dressing is omitted since it does not really comprise an actual asphalt course.

Table 11.3 is a rough guide to a very large range of possible materials. However, the fact is that the stiffness of the surface course is often no more than about 60% that of the base (assuming the same binder grade), while the binder course is often considered to be of the same stiffness as the base. This reduced stiffness can be taken into account directly in multi-layer linear elastic pavement analysis or, if an estimate is needed, it can be assumed that a proportional difference in stiffness is

Table 11.3 Asphalt surface course structural properties

Surface type	Stiffness	Deformation resistance	Fatigue strength
Asphalt concrete	Medium	High	Medium
Rolled asphalt + chippings	Medium	Low	High
Stone mastic asphalt	Medium–low	High	Medium–high
Porous asphalt	Low	Medium–high	Medium–low
Mastic asphalt	Very low	Low	Very high
Grouted macadam	Very high	Very high	Medium
Base	High	High	Medium

approximately equivalent to the inverse square of a corresponding difference in thickness. Thus, a factor of 2 difference in stiffness is equivalent to a factor of around 1.4 difference in thickness. In simple terms, a 40 mm surfacing is structurally equivalent to about 30 mm of conventional asphalt base – nearer 25 mm if porous asphalt is used.

Turning to the other measures of quality, surface course materials should generally have at least as good fatigue properties as an asphalt base. However, it is not uncommon for both the surface course and binder course to have relatively poorer deformation resistance because of their smaller stone size. Although deformation problems often stem from the binder course, as was illustrated in the example in Section 10.4.1, the designer should be aware that there is a higher risk of deformation problems with some surface types, particularly HRA.

11.3.2 Performance under high tyre contact stress

All surfaces are sensitive to applied pressure, even though whole-pavement analysis may not be. Calculation of the stress or strain at the bottom of a structural layer is primarily a function of total wheel load, with little effect from contact stress, whereas the stress level near the surface is almost independent of total load and entirely dependent on contact stress. If a sophisticated analysis of crack propagation is carried out, including top-down crack propagation, then these facts will properly affect the result; in most design analyses, however, they will not.

The usual approach, therefore, is to rely on experience and empirical design. The chief danger is deformation, to combat which asphalt relies heavily on aggregate interlock, a difficult property to assess. A common approach for those using the Marshall method of mixture design (see Section 9.7.2.1) is to insist on a very high Marshall stability value for the surface course in high-stress applications, sometimes as high as 20 kN. For those relying on the repeated load axial test, a strain of under 1% is required after 3600 load applications at 100 kPa if pressures of 1.5 MPa or more are to be accommodated. An extreme solution to the problem is to use grouted macadam which is, in effect, a fully deformation-resistant asphalt.

11.4. Skid resistance

Chapter 3 outlined the influence that skidding resistance requirements have on the choice of pavement type. This section explores the technical details of skidding resistance further.

11.4.1 Micro-texture

Micro-texture is the term usually applied to describe the intrinsic frictional properties of the surface. In an asphalt, the micro-texture in question relates to the aggregate particles at the surface. The term micro-texture is used because the surface features that give rise to friction are invisibly small. A stone may appear rough to the eye, but that is little indication of the level of friction achievable between tyre rubber and the stone surface.

Micro-texture can be measured in the form of a coefficient of friction, or a number related to it. A common technique is the 'pendulum test' (BSI, 2011), illustrated in Figure 11.1 which is generally carried out with the surface in a wetted state. A typical coefficient of wet friction would be between 0.5 and 0.7. It is the micro-texture that represents the ultimate skid resistance potential of a surface – the level applying without any intervening dirt, bitumen or ice lens. In theory, a friction coefficient of 0.5 would allow a vehicle travelling at 25 m/s (around 60 mph) to come to a halt in 64 m; at a friction coefficient of 0.7, this distance becomes just 46 m. It is logical, therefore, to insist on improved micro-texture at sensitive locations such as approaches to pedestrian crossings

Figure 11.1 Micro-texture evaluation

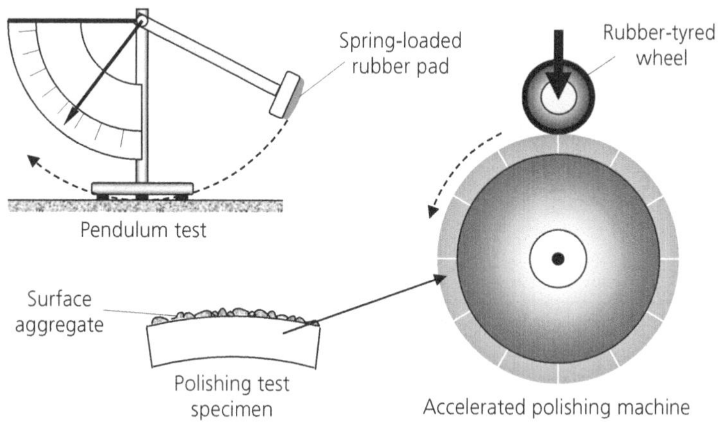

and roundabouts, and this approach is adopted by highway authorities all over the world. Certain aggregate types such as 'gritstones' therefore take on a premium value because of their excellent micro-texture.

11.4.2 Polishing

It is an observed fact that the frictional properties of a surface tend to change under the action of traffic. In dry weather, they are 'polished' by the relative motion of tyre and surface, activated partly by tyre vibration. The effect is a reduction in micro-texture and a reduction in the coefficient of friction. It is standard practice in many countries, including the UK, to assess the so-called 'polished stone value' (BSI, 2020a) of an aggregate by first subjecting it to an accelerated polishing regimen (see Figure 11.1) before measuring the frictional properties using the pendulum test. The result of such tests is that many stone types are found to be inadequate for applications where high skid resistance is required.

This approach is realistic and acknowledges the reduced friction generated by tyre action. However, there is some evidence that it may be unfairly penalising certain surface types. The point seems to be that polishing is strongly linked to the energy dissipated by the tyre during contact with the pavement surface, and that this is inversely related to the amount of energy absorbed by the pavement. Thus, an energy-absorbing surface such as porous asphalt or stone mastic asphalt, which induces less tyre vibration, also suffers less polishing. Initially, this may not be a good thing, as bitumen remaining on the surface of the stone may take a while to rub off, causing an initial period of low friction. However, once this has occurred, the lack of polishing of the aggregate particles themselves is likely to preserve a relatively high skid resistance. Anecdotal evidence is that this deduction is supported by accident statistics, although further study is required before the case can be considered proven.

Micro-texture is seasonal – see the example in Figure 11.2. Polishing occurs mainly in dry weather, whereas wet weather actually restores frictional properties to some extent due to the abrading effect of small particles of grit that are present in surface water. For this reason, skid resistance should preferably be assessed in summer or during the dry season.

Figure 11.2 Example skid resistance data (measured by a 'Griptester')

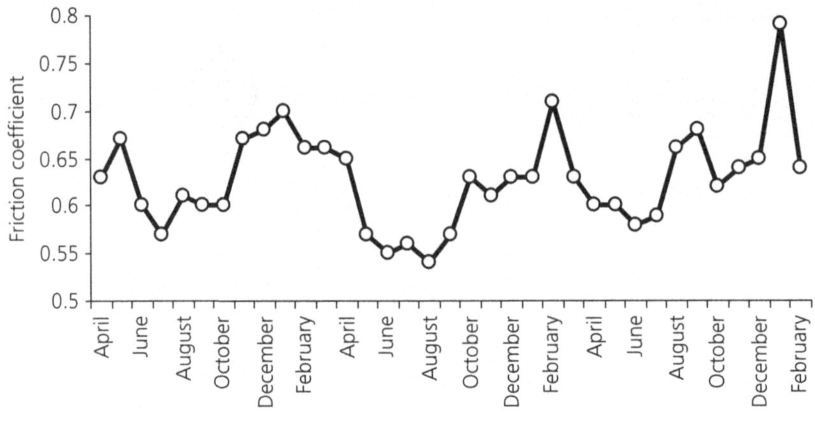

11.4.3 Macro-texture

If it never rained it would not be necessary to introduce any visible texture to a pavement surface or for tyres to include tread depth. Neither tread nor visible surface texture make the smallest contribution to basic skid resistance – they are only present to ensure that surface water has somewhere to go. The point is that direct contact is needed between the tyre and the surface in order for friction to be activated; if a water film remains in between, the vehicle will 'aquaplane' as soon as brakes are applied. It is not a precise science, but the principle is clear enough (Figure 11.3). The faster the vehicle, the less time is available for the water to be forced away from the contact point. An optimised macro-texture therefore ensures that there is only a short distance between individual contact points and regions where water can be accommodated without danger. It also ensures that there is enough storage volume in these regions to cope with the envisaged quantity of surface water. It is the macro-texture requirement that gives rise to the wide variety of surface finishes already introduced, all of which, in one way or another, have to ensure that a pavement surface includes

Figure 11.3 The importance of macro-texture

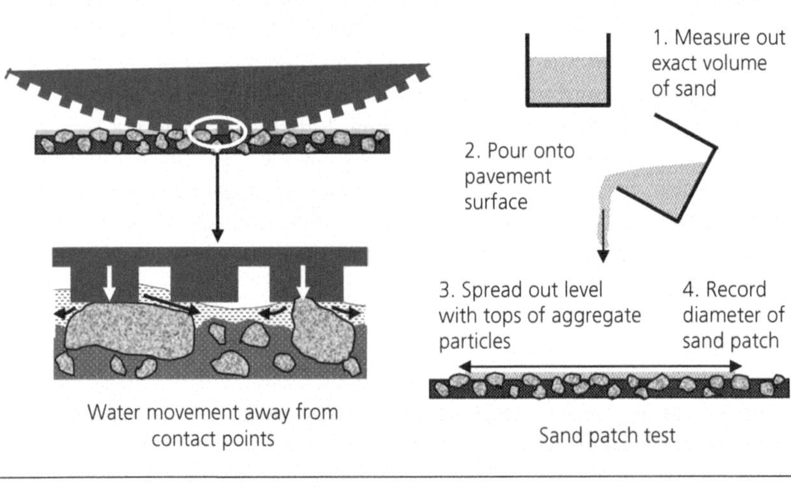

high points, where tyre contact takes place, and low points, where water can lie. Experience suggests that water should not be asked to travel more than a very short distance (i.e. millimetres only) during the few milliseconds of contact time, thus giving a further reason to restrict the aggregate size in surface course mixtures.

Macro-texture is generally expressed as a 'texture depth' in millimetres. The basic measure comes from a procedure known as the 'sand patch test' (see Figure 11.3), although there are numerous laser-based pieces of equipment on the market for rapid, sometimes traffic-speed, measurement. The latest version of the sand patch test (BSI, 2010) allows the use of glass beads instead of sand.

Unfortunately, macro-texture can deteriorate under traffic loading. The useful wet-season abrasion that restores micro-texture also reduces the height by which individual aggregate particles protrude, eventually reducing the texture depth excessively. For this reason, it is usual to specify abrasion resistance by one of several available techniques, such as the Los Angeles abrasion value (AASHTO, 2019; ASTM, 2016, 2020; BSI, 2020b).

11.4.4 Analysis of tyre–road contact in wet conditions

This section is included to throw light on the interaction between micro-texture and macro-texture, bringing in the effect of tyre tread and vehicle speed.

The most fundamental process of skid resistance loss is understood here to be as shown in Figure 11.4. A simplifying assumption is that the load is spread evenly across the flattened lower part of the tyre whether water is present or not, including the leading zone which will become separated from the surface when excess water is present. The implication is that the average water pressure where such separation has occurred balances the average pressure applied by the tyre – which is taken to be approximately equal to the internal tyre pressure.

The next assumption concerns water movement. Self-evidently, every element of water that is pushed away from the contact will eventually have to travel laterally since there can be no cumulative build-up of water in front of the tyre beyond a quasi-steady-state bow wave, which is unlikely to be significant except in deep water or at low vehicle speed. A reasonable simplifying assumption is therefore that water movement is entirely lateral, as shown in Figure 11.5.

A third assumption is that water flow is laminar. Considering both shear resistance and inertia, an expression can be developed for average water pressure, which is then assumed to be equal to the internal tyre pressure. The end result is an expression for the gradient of the contact (dh/dx):

Figure 11.4 Longitudinal section through tyre–water–surface contact region

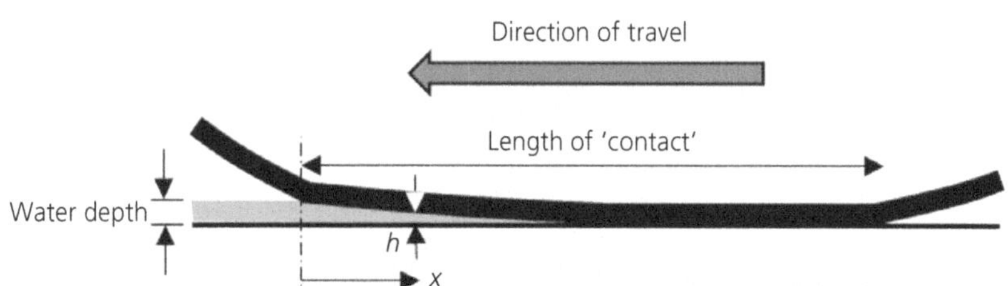

Figure 11.5 Transverse section through tyre–water contact region

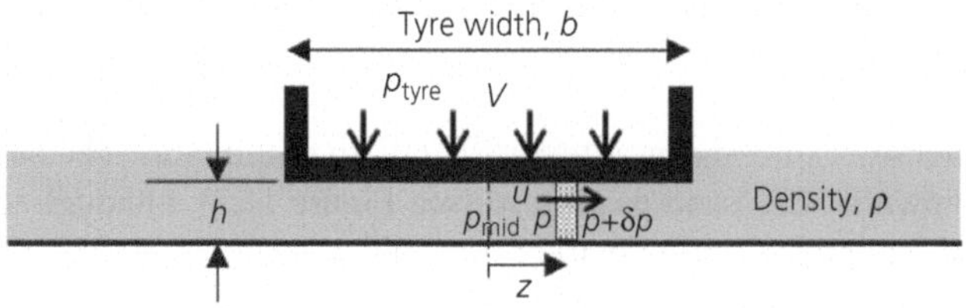

$$dh / dx = (3h / bV\rho) \times [-2b\eta / h^2 + (4b^2\eta^2 / h^4 + 4p_{tyre}\rho / 3)^{0.5}] \tag{11.1}$$

where V is vehicle velocity, ρ is the density of water, η is the viscosity of water and p_{tyre} is the tyre pressure.

Equation 11.1 can now be solved numerically. However, with no tread and no road texture, zero contact (i.e. aquaplaning) will be predicted. Thus, the next essential step is to include road texture, which relates to the volume available for water storage beneath the surface per unit area, typically in the 0.5–1.5 mm range. Furthermore, since lateral flow within the texture will depend on the type of material present, a 'connectivity' factor can be applied to describe this, quantifying a reduction in effective texture depth. For example, a 75% connectivity factor would reduce a 1 mm texture depth to 0.75 mm in flow computations, although the 1 mm would still apply to water storage capacity. Figure 11.6 shows example predictions made on this basis.

In Figure 11.6, reducing either the texture depth or connectivity increases the loss of contact, which in these examples ranges from 31% to 68%, effectively reducing the available skidding resistance by similar percentages. However, at this stage, the tyre is still bald. The majority of modern tyres include three or four circumferential grooves, representing the primary water pressure-relieving mechanism, together with complex patterns of much smaller secondary grooves and sipes. This simplified analysis will take account of the primary grooves only.

Figure 11.6 Example predictions – texture effects

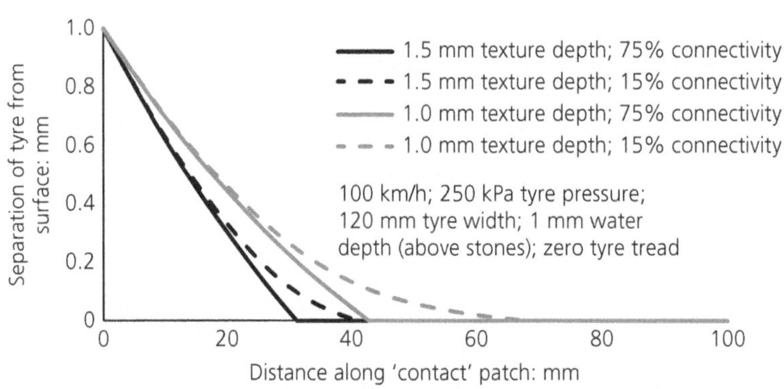

Figure 11.7 Water movement to and along tread grooves in tyre–surface contact

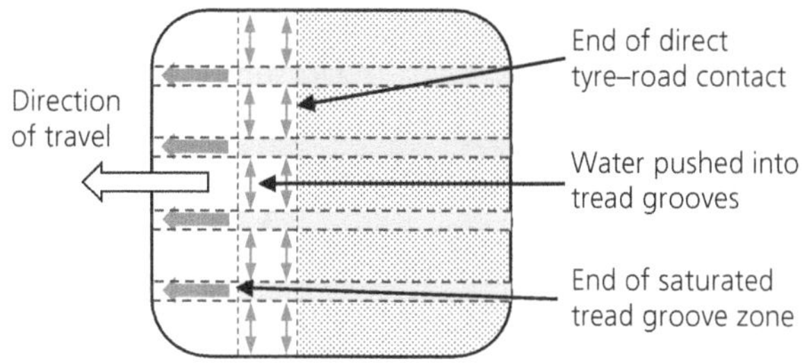

If the primary grooves are wide enough and deep enough to accommodate the necessary volume of water, then the problem analysed previously is simply reduced to one with several narrow strips of tyre (between grooves) rather than a single width. If the primary grooves are inadequate then, at some point along the contact patch, the grooves will be full and water will be ejected under pressure either at the front or rear of the tyre. Any water that the grooves are not able to accommodate and eject will pass laterally to the next groove and eventually to the edge of the tyre. Assuming that water movement is still entirely lateral until a tread groove is encountered, a simple volume balance allows computation of the point where grooves become full. It is then necessary to calculate how much water can be dissipated by the grooves. Figure 11.7 illustrates the situation in plan view.

At the front of the tyre–water contact region, where the grooves are not full, the tyre is reduced to a number (five in Figure 11.7) of narrow strips with zero water pressure either side of each strip. Beyond this point the water within the grooves takes pressure and is accelerated longitudinally in order to escape. The calculation required is therefore to find the pressure gradient in each groove such that all water flows balance. Figure 11.8 shows results from such a calculation.

One further refinement could be made. Figure 11.8 describes the shape of the tyre–water interface in front of the point where direct contact is made with the road satisfactorily, but not the situation behind this point. Behind this point, the tyre will continue to pressurise the water within the road

Figure 11.8 The predicted effect of different treads

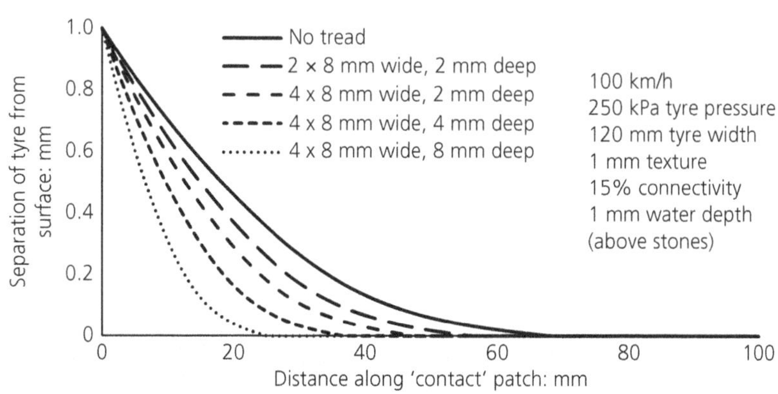

Figure 11.9 Skid resistance loss against tyre tread depth and road surface texture (100 km/h, 250 kPa tyre pressure; 120 mm width; 1 mm water depth (above stones); 75% texture connectivity)

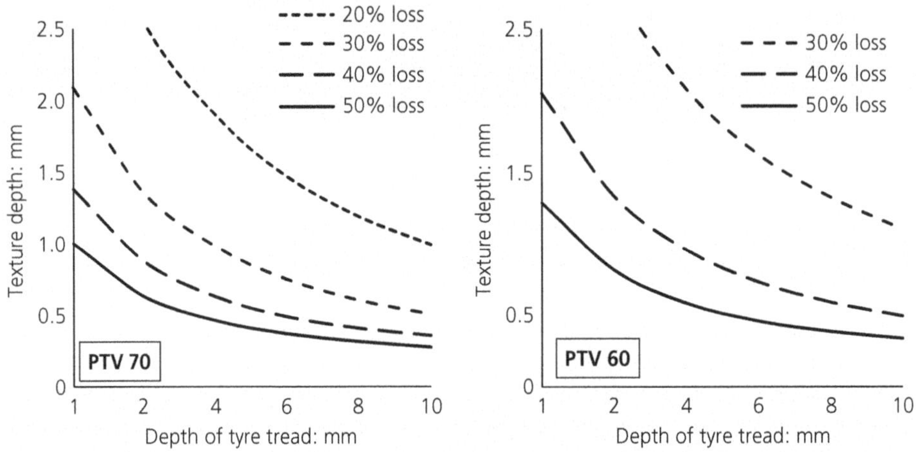

surface texture as the rubber 'drapes' around the discrete points of contact. This additional water pressure will further reduce the skid resistance available. Of course, assumptions and simplifications have to be made; the predictions in Figure 11.8 have assumed a tyre rubber modulus of 10 MPa and a Poisson's ratio of 0.5.

With this sort of analytical approach, different combinations of micro-texture (measured by the pendulum test value (PTV)), tread, tyre pressure, road surface texture, water depth, vehicle speed and so on can be evaluated for their effect on skid resistance loss. Figure 11.9 shows comparative predictions.

Figure 11.9 illustrates the conversation that should be taking place among tyre manufacturers and road producers, and among those specifying legal tread depth limits and minimum road texture depth. For example, if a 40% loss of skid resistance is considered acceptable for a 100 km/h vehicle travelling through 1 mm of standing water, this could be achieved by 1.6 mm of tread (the UK legal limit), 0.9 mm texture depth and PTV = 70 stone; or by 1.4 mm texture depth and PTV = 60 stone. Highway specifiers can therefore be aware of the trade-off between stone quality and texture depth.

Broadening the conversation, however, if the minimum legal tread depth was increased to 3 mm, the required surface texture depths would reduce to 0.75 mm with PTV = 70 stone or 1.1 mm with PTV = 60 stone, with associated road surfacing cost benefits.

11.5. Spray

Spray from surface water is a serious safety hazard. The analysis presented above made the point that once a tyre tread reaches its water-carrying capacity then excess water has to be ejected, and much of that will be in a rearward direction towards any following vehicle. There will also be side spray. Hot-rolled asphalt (HRA), with rolled-in chippings, has developed a particularly bad reputation for spray, and in the UK this was one of the driving forces behind the use of stone mastic asphalt (SMA) type surfacings. Basically, if water cannot flow easily across the surface of a pavement – low connectivity in the foregoing analysis – it will be available to form spray. The problem with HRA plus chippings is that each individual chipping sits in its own small indentation in the asphalt surface, allowing a small 'pond' of water to remain around it until it either evaporates or is dispersed in the form of spray. Most other surfaces have greater connectivity, such that water can flow around the protrusions and make its way tortuously sideways. Asphalt concrete, SMA and surface dressing therefore generate much less spray than HRA. However, for real excellence, porous asphalt is undoubtedly the premier material. Porous asphalt allows water to drain straight into the pavement itself and then to pass laterally through it, below the level of the tyre–surface contact. The result – virtually no spray at all. But it's a matter of value, a cost that's quantifiable against a benefit that is much less so (i.e. it's down to policy).

11.6. Noise

Another powerful political driver in some countries is the reduction of tyre noise, either to ease the discomfort of motorists or to protect the interests of nearby residents. This is a highly complex field and it is not necessary for pavement engineers to appreciate the exact acoustic mechanisms involved – which is just as well because these are not wholly understood. What is clear, however, is the experience of users. Figure 11.10 presents a series of measurements taken in the immediate proximity of a tyre. More information on low-noise surfacings can be found elsewhere (FEHRL, 2006).

As expected, noise level generally depends on the texture depth (i.e. roughness). However, the picture is clearly more complicated than this, with a 10 dB(A) difference – a factor of 3 in actual sound-pressure magnitude – between block paving and porous asphalt for the same texture depth. Furthermore, surface dressing seems to buck the trend, giving a relatively steady noise level whatever the texture depth. Of the more common asphalt surfaces, SMA is evidently the quietest at normal texture depths (around 1 mm).

Figure 11.10 Noise measurement data (derived from Sandberg and Ejsmont (2002))

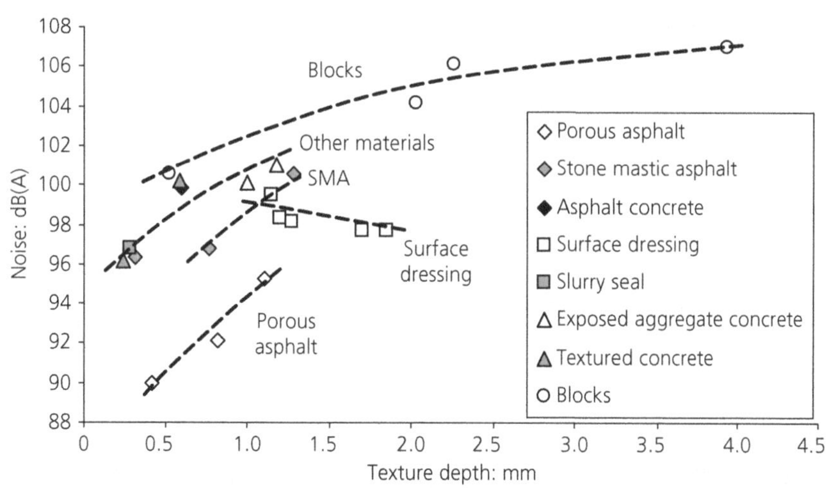

Conceptually, it is not hard to understand why this ranking order should be. Noise is caused by vibration, principally of the tyre-tread elements. The pavement surface elements will vibrate as well, but mainly at frequencies well outside the audible range. However, the surface type affects both the amplitude and, to a much lesser extent, the frequency of tyre-tread vibration. Basically, a rough surface will induce a high amplitude of tyre vibration, and therefore high noise. Some of this noise will then be absorbed by the surface, and this will depend principally on the hardness of the surface material. Porous asphalt, with a low stiffness, therefore causes little excitation to the tyre tread elements; it also has excellent noise-absorption properties.

11.7. Summary

The surface layer of a pavement is almost always the most expensive. This is no surprise considering the variety of roles that it has to fulfil, but it does mean that the choice and design of this layer carries a high economic weighting. This is amplified further in that it commonly lasts for a shorter time than lower layers, being either replaced or overlaid within about 10–15 years on most major highways.

Efficient pavement engineering will, therefore, demand that the requirements in a given case are clearly understood. Low noise is unlikely to be an issue for an industrial pavement; likewise skid resistance. However, high strength probably is. Airfield runways demand good skid resistance – but not taxiways or aprons. On high-speed roads, spray may be a serious issue, leading to the consideration of porous asphalt, but this is much less important on low-speed roads. Ride quality is always a compromise between what is desirable and what is achievable, but there is no point over-specifying in situations for which excellent ride quality is not really required.

All of this means that it is economically worthwhile devoting a little thought to surface layer policy.

REFERENCES

AASHTO (American Association of State Highway and Transportation Officials) (2019) T96-02: Resistance to degradation of small-size coarse aggregate by abrasion and impact in the Los Angeles machine. AASHTO, Washington, DC, USA.

ASTM (2016) C535-16: Standard test method for resistance to degradation of large-size coarse aggregate by abrasion and impact in the Los Angeles machine. ASTM International, West Conshohocken, PA, USA.

ASTM (2020) C131/C131M-14: Standard test method for resistance to degradation of small-size coarse aggregate by abrasion and impact in the Los Angeles machine. ASTM International, West Conshohocken, PA, USA.

BSI (2010) BS EN 13036-1:2010: Road and airfield surface characteristics. Test methods. Measurement of pavement surface macro-texture depth using a volumetric patch technique. BSI, London, UK.

BSI (2011) BS EN 13036-4:2011: Road and airfield surface characteristics. Test methods. Method for measurement of slip/skid resistance of a surface: The pendulum test. BSI, London, UK.

BSI (2020a) BS EN 1097-8:2020: Tests for mechanical and physical properties of aggregates - determination of the polished stone value. BSI, London, UK.

BSI (2020b) BS EN 1097-2:2020: Tests for mechanical and physical properties of aggregates. Methods for the determination of resistance to fragmentation. BSI, London, UK.

FEHRL (Forum of European National Highway Research Laboratories) (2006) *Guidance Manual for the Implementation of Low Noise Road Surfaces*. FEHRL, Brussels, Belgium, Report 2006/02.

National Highways (2020) *Manual of Contract Documents for Highway Works. Vol. 1: Specification for Highway Works*. Stationery Office, London, UK.

Sandberg U and Ejsmont JA (2002) *Tyre/Road Noise Reference Book*. Informex, Kisa, Sweden.

Nick Thom
ISBN 978-1-83549-711-1
https://doi.org/10.1108/978-1-83549-710-420243014

Chapter 12
Maintenance of flexible pavements

Pavements are designed to achieve a certain life, 40 years for example. But (*a*) that won't happen unless a certain amount of maintenance is carried out and (*b*) the road authority would probably like the road to last longer than 40 years if possible. So this chapter will take a look at the various things that are likely to go wrong and what can be done to put them right.

12.1. Deterioration mechanisms
12.1.1 Surface texture deterioration
More than one effect can lead to deterioration of both micro-texture and macro-texture, as follows.

- No stone is infinitely strong. Over the course of time it will abrade. Typically, as stated in Chapter 11, the polishing action of tyres reduces the micro-texture while the abrasion caused by dirt particles restores it, although this is also a function of stone mineralogy. Abrasion also causes a steady reduction in macro-texture.
- Dirt can accumulate within the macro-texture, reducing the effective texture depth and therefore the wet skid resistance.
- Bleeding is the phenomenon of bitumen oozing up from the asphalt under repeated tyre loading. It happens when densification of the asphalt reaches very low voids contents or when there is simply too much bitumen in the surface course mix.

In general, experience is that the texture depth tends to reduce steadily over time, as illustrated in Figure 12.1 by the average loss of texture from several thousand major road sites in the UK (Abed *et al.*, 2023).

Figure 12.1 Average trend for loss of texture depth (Abed *et al.*, 2023)

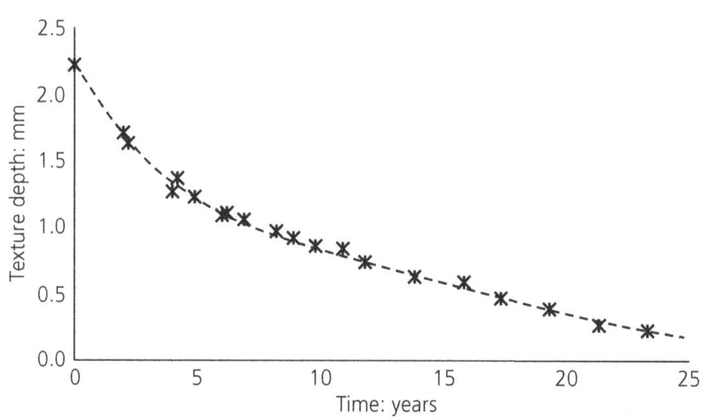

12.1.2 Ravelling

Abed *et al.* (2023) found that a minority of sites, about 5% of those sampled, were different. They showed an increasing texture and the mechanism that can cause this is ravelling – the loss of stones from a road surface. This could be an indication of an aged binder at the surface, poor bitumen–stone adhesion for some reason or simply a material with more voids that it should have. Left unchecked, ravelling is eventually likely to lead to the development of potholes.

12.1.3 Cracking

Cracks can be longitudinal or transverse, close or widely spaced, wide or narrow, short or long – and every crack has a meaning. There will often be a combination of transverse and longitudinal cracks present because, unsurprisingly, more than one mode of damage may be taking place at the same time. Deterioration of the base may be masked by top-down cracking, for example. In old pavements that have already been subject to previous strengthening treatments, these issues become ever more complicated. Cracks may simply be reflective cracking from an older pavement construction. Crack patterns can tell an engineer a lot – but not everything. Here are some observations on different crack types.

- If wheel-path cracking predominates, then it is the traffic that is responsible, whatever the cracking may look like. Similarly, if cracking extends more or less equally over all areas of the pavement, then the traffic is irrelevant and the cause is environmental or due to a general material defect.
- Transverse cracks extending right across the pavement are thermally generated. Either the climate is severe enough to cause low-temperature cracks in asphalt or the asphalt surface overlies a cracked (or jointed) hydraulically bound base and reflective cracking has occurred. If there are irregular connecting longitudinal cracks, then low-temperature cracking is most likely. If the cracks are regular, then an underlying hydraulically bound layer, possibly a pre-existing jointed pavement-quality concrete pavement, is present.
- Short, frequent transverse cracks restricted to the area immediately around the wheel-path may represent defects built in during construction (roller cracks, poor-quality asphalt) that have been developed by subsequent trafficking.
- If there is a single well-developed longitudinal crack in the wheel-path, this indicates traffic-induced fatigue of a thick asphalt pavement with no significant layer debonding. The crack direction is almost certainly top-down and the crack depth is often no more than 50 mm or so.
- Multiple cracking (crazing) in the wheel-path implies a shallow failure. Either the asphalt is thin or the upper layer, or two layers, have become debonded and are cracking independently of the rest, possibly due to deterioration at one of the layer interfaces.
- If slurry is pumping up to the surface through the cracks, this means that water has become trapped within the pavement, either within the bound materials, for example, at an interface, or else in the upper part of an unbound foundation. Damage is likely to accelerate rapidly.
- All crack types become wider under freeze–thaw action as ice occupies more volume than water and a crack that widens during freezing will not recover after the thaw.
- Crack growth is also driven by the presence of detritus. Day–night temperature change will result in alternate narrowing and widening of the crack, allowing the detritus to penetrate ever deeper, reducing the daytime narrowing.

12.1.4 Rutting

As with cracks, not all ruts are the same. A rut is a function of deformation taking place in all the various layers of a pavement and, depending on which layer or layers are dominant, the shape will be different. For example, if it is the surface course that is primarily to blame, the rut will typically

Figure 12.2 Different types of rutting

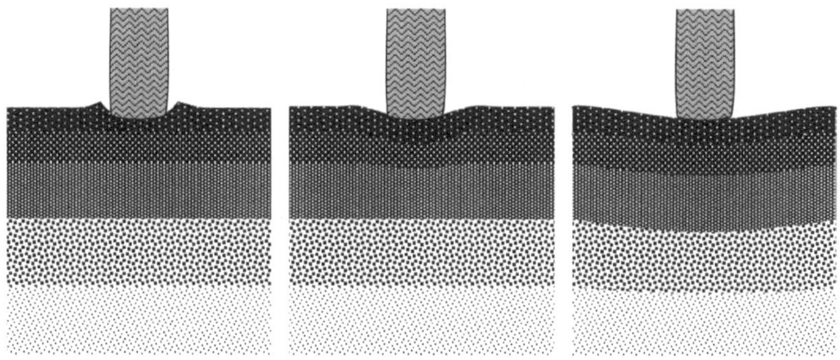

Surface course problem | Binder course/base problem | Foundation problem

be narrow with marked shoulders either side. It is possible to judge from the width of a rut, taking into account the likely lateral distribution of wheel locations, as well as the overall shape, approximately where the problem lies (Figure 12.2).

The distinction between the first two cases shown in Figure 12.2 is particularly important but is sometimes not easy to discern visually. It is common for the assumption to be made that any narrow rut must be due to problems in the surface course, whereas this is frequently not the case, and merely replacing the surface course will not solve the long-term problem if it is a function of the binder course or the base; ruts will reappear.

12.1.5 Potholes

In climates such as that in the UK, it is commonly observed that cold wet weather results in a rapid proliferation of potholes in pavements with an asphalt surface, resulting in a large annual repair bill and the question could legitimately be asked as to whether this problem could be minimised by sensible design or construction practice. This, in turn, means that pothole formation needs to be properly understood. With the current state of knowledge, it is not practical to try to predict pothole occurrence or compute any specific pavement life before they occur, but it most certainly is practical, indeed essential, to understand the mechanisms at work.

12.1.5.1 Factors leading to pothole formation

A pothole is a localised area in which the pavement has disintegrated and the asphalt materials have then been removed by repeated tyre loading. But why should such concentrated damage occur when the surrounding pavement is often substantially intact? Can there really be such massive differences in the quality of the materials from place to place? Well, there can, especially adjacent to construction joints where poor compaction is common. But actually potholes can occur without significant differences in material quality because a pavement in a wet climate is in 'unstable equilibrium'. Once local damage starts, a downward spiral of deterioration is activated. Typically, the sequence of events is as follows.

- Slight localised damage occurs, for example, loss of aggregate from the surface course or the beginnings of top-down cracking.
- Rainwater ponds in the slight depressions left by lost aggregate and/or penetrates into top-down cracks.

- Vehicle tyres pressurise the water during each pass. Local pressures could theoretically be much higher than the internal tyre pressure.
- Water at high pressure breaks the bonds between individual aggregate particles and, where a top-down crack has penetrated significantly, also between pavement layers. This results in a gradual fragmentation of the asphalt.
- Freeze–thaw cycles result in expansion of water- or ice-filled voids, exacerbating such fragmentation.
- Dynamic traffic load scatters some of the fragments, resulting in a deepening and widening depression in the road.
- As the depression becomes deeper, so it becomes increasingly able to retain water following rain. It also acts as a drain, helping the surrounding pavement to avoid the same fate while accelerating the damage within the depression itself.
- At some stage the water will begin to penetrate to the pavement foundation at a sufficient rate to reduce the support stiffness being offered, increasing asphalt flexure and further accelerating deterioration.

A relatively minor defect can thus end up becoming a serious local pavement failure and, although it may be local to begin with, it will expand as pothole edges break off and as the effect of a low-stiffness foundation becomes more widely felt. Water/ice is obviously the key driver, so pothole formation tends to be seasonal. Also self-evidently, the porosity and permeability of the asphalt will have a significant effect. The worst-case scenario is for there to be sufficient permeability to allow water to penetrate significantly but for those pores to be narrow, leading to high water pressure under traffic load. Also, the situation in which a permeable layer overlies a much less permeable material is highly undesirable.

12.1.5.2 Possible measures to avoid potholes

Designers, specifiers, supervising engineers, contractors and equipment producers all have a role to play in minimising the potential for pothole formation. It is hoped that the following points contain useful advice.

- Either incorporate a near-impermeable surface course, such as a well-designed stone mastic asphalt (SMA) or hot-rolled asphalt (HRA), or else use a high-porosity surface course (porous asphalt) over a near-impermeable binder course.
- Bond the layers together well.
- Avoid joints where possible, for example, by paving in echelon formation (Section 10.1.4). If joints like this cannot be avoided entirely, they should not be located in a wheel-path.
- Faces of joints should be cut back to dense material. There is some evidence that these cuts are best inclined to the vertical, although this is still subject to ongoing investigation.
- Control compaction properly. Make sure that rolling is started at the right temperature, which depends on the grade of the binder being used, and that it is not continued once the material has cooled excessively.
- Also ensure that rolling leaves sufficient crossfall over the entire pavement surface to avoid water ponding. This point and the previous one require a skilled roller operator as well as good supervision.
- Avoid 'cold spots'. These form where lumps of asphalt either cool excessively during transportation or else become lodged in the paver for some time before reaching the paved layer. The result is a small zone that remains poorly compacted. Cold spots are probably responsible for many of the potholes that appear for no apparent reason on otherwise intact pavements.
- Finally, make sure the foundation is well drained.

12.2. Retexturing

12.2.1 Micro-texture

As noted previously, the micro-texture of surface aggregate, the property that gives basic frictional grip, tends to reduce during the life of a surface due to the polishing action of tyres, and the natural tendency for micro-texture to recover during winter months due to the action of grit within surface water may not be sufficient to counteract this. 'Bush-hammering' represents a way of abrading the surface aggregate sufficiently to restore its original micro-texture. It is done using an automated machine and consists of controlled impacts delivered to the surface, impacts of relatively high velocity but low inertial mass, inducing local fracturing of the surface of the aggregate rather than any deeper breakage. Alternatives are shot-blasting (firing steel balls at the surface, which are recovered by vacuuming and reused) or impacting by means of rotating discs.

There is no doubt that these techniques work and that they can sometimes extend the safe working life of a surface by several years. There is also no doubt that they will reduce the texture depth present, which may impact on the wet-weather skid resistance because of lack of water drainage. It is, therefore, a technique for particular circumstances rather than being suited to all situations. The engineer has to weigh up the advantages and disadvantages carefully.

12.2.2 Macro-texture

A number of techniques exist to restore macro-texture, depending on circumstances.

12.2.2.1 Jet-blasting

This is the counterpart to bush-hammering. Whereas bush-hammering is designed to chip small fragments off the aggregate, jet-blasting, using a high-pressure water jet, removes some of the bituminous mortar material from between the aggregate pieces, thereby increasing the texture depth. As with bush-hammering, there is no doubt that the technique works; as with bush-hammering, there are accompanying dangers. Depending on the type of surface, it may be that some of the aggregate pieces become dislodged as their surrounding mortar is partially removed. Thus, here too, the engineer has to weigh up the pros and cons carefully before deciding to go ahead.

12.2.2.2 Grooving

Grooving is achieved by saw-cutting hardened material, and so is not cheap. However, it is an effective way to improve the macro-texture. It is regularly used on airfield runways where maintenance of suitable macro-texture is absolutely essential. It is less likely to be cost-effective on highways, but is sometimes used.

12.2.2.3 Gritting

There are two circumstances where gritting (i.e. application of a spread of approximately 3 mm size particles) might be required. The first is during the early life of certain surface courses (typically stone mastic asphalt or porous asphalt) in which the individual surface particles have been well coated with bitumen (sometimes tough, modified bitumen), and this bitumen takes some time to wear away and expose the micro-texture of the aggregate itself. It is, therefore, a practical short-term safety measure. The second circumstance is in response to bleeding, usually during hot weather, in which the surface becomes a smooth textureless plane, ideal for skidding in the wet. Gritting represents a short-term emergency response to this dangerous situation.

12.3. Surface treatments

12.3.1 Crack sealing

Sealant technology is a complex field, and is one in which significant advances have been made over the years. For the pavement engineer, sealing is an important, if not particularly exciting, activity. Water is the enemy of the pavement engineer; it will degrade asphalt, weaken hydraulically bound materials and reduce the strength and stiffness of unbound layers. It expands on freezing, breaking materials apart, destroying soil structure and even causing the whole pavement to 'heave' upwards. There is no doubt about it – water genuinely is the enemy and every effort should be made to limit how much of it is present in a pavement; which means that sealing is important.

In simple terms, you get what you pay for. The quickest and cheapest solution is to paint or pour on a thin film of bitumen or bituminous mastic (bitumen plus filler) without preparing the crack in any way other than general cleaning of the surface. The problem is that there will be very little penetration of bitumen into the crack, which means that it will be relatively easy for the film of sealant to re-fracture under combinations of thermal and traffic-induced stress and strain. This solution can only be considered as short term therefore, with no lasting reduction in the ability of the crack to transmit water.

A more time-consuming treatment consists of first scabbling the surface around each crack, thereby presenting a suitably rough material onto which the sealant mortar can bond. It also allows a slightly greater thickness of sealant, presenting a much increased resistance to re-fracturing. There will still be little penetration of bitumen into the crack, however.

More time-consuming still are treatments that consist of routing out a channel along the crack and then filling it with a sealant material, generally a high-durability bitumen–fines mixture. These treatments give real depth to the seal, making it much harder for the crack to re-propagate through it. The weak point becomes the bond between the sealant and the routed faces.

It is impossible to give general rules here about the most appropriate technique to use. It depends on the type of crack (transverse thermally induced or wheel-path traffic induced), whether it is likely to have penetrated full depth, how water-sensitive the foundations are, what the long-term strategy for a particular pavement may be and so on. Two key points must be borne in mind. First, active cracks (i.e. cracks with large relative movement between the faces) are difficult to seal effectively; second, crack sealing is most important where underlying layers are water-susceptible. The rate of deterioration of a pavement can often be slowed significantly by judicious application of crack sealing and, when whole-life costing is considered, the price is probably worth paying.

12.3.2 Slurry sealing

Individual crack sealing is a sensible option when crack numbers are few. It also has the advantage that the overall surface properties (particularly skid resistance) are only affected very locally. However, in cases where there are large numbers of cracks or where skid resistance is not particularly important, a more cost-effective solution may be to 'slurry seal'; that is, to apply a coat of bituminous slurry over the whole pavement surface. This is well suited to footways, cycleways, car parks, low-speed city roads and airfield taxiways.

A slurry seal generally consists of a mixture of bitumen emulsion (see Chapter 13) with filler and fine aggregate and it is commonly spread to a thickness of 3–8 mm. Specialist spreading machines can be used, with paving screeds as in conventional asphalt paving. As well as speed of operation and certainty of coverage, this technique has the advantage that a degree of surface regulating can be achieved (i.e. minor improvement to the profile of the surface). The natural surface texture will be relatively smooth, but it can be roughened by means of brushing.

As the slurry seal material will have a high binder content, as a consequence of the lack of coarse aggregate, it will form a relatively crack-resistant layer, and may therefore provide several years of additional life to the pavement before the cracks start to re-occur.

12.3.3 Preservative treatments

Many highway authorities around the world have come to the conclusion that regular application (e.g. every 5 years) of a so-called *preservative* treatment is a highly cost-effective means of extending the life of a heavily trafficked pavement. The principle is shown in Figure 12.3. The treatment is a spray-applied liquid that contains elements that are sufficiently fluid to carry the liquid into the surface voids and any lower voids that are in any way connected to the surface. It also contains an ultra-hard bitumen, usually a naturally occurring asphaltic material known as gilsonite, which forms a hard skin across the surface, filling minor cracks and gripping individual surface stones. The result is that the surface is given a significant additional toughness without changing the nature of the surface texture – although a very small loss of texture depth is to be expected. It also greatly inhibits further ageing of the existing surface course.

Because the nature of the surface remains unchanged, a preservative treatment will do nothing to restore skid resistance. It is therefore only suited to surfaces where skid resistance is not a problem. It is also vital that it is carried out *before* significant deterioration has occurred. Occasional stone loss is acceptable. Minor cracks can be filled; deep cracks will not be; potholes certainly will not be!

12.3.4 Rejuvenation treatments

These are also spray-applied, but they are entirely different from preservative treatments – almost the opposite in fact. While preservatives aim to form a really tough skin, rejuvenators aim to soften the bitumen. The point is that as the binder in the surface course ages it becomes ever more brittle and therefore susceptible to top-down cracking and stone loss (i.e. ravelling). A rejuvenator includes a low-viscosity component and is designed to diffuse into the bitumen and restore its crack resistance; it may also achieve a degree of crack healing – but only of minor cracking. Dosage has to be carefully applied as excess softening would obviously present a danger of rutting and/or ravelling. As with a preservative treatment, no change is made to the nature of the surface and no improvement is made to skid resistance.

This book can give no 'rules' as to when to use what but, logically, a surface should be almost intact if benefit is to be derived from a rejuvenator.

Figure 12.3 Spray-on surface treatments

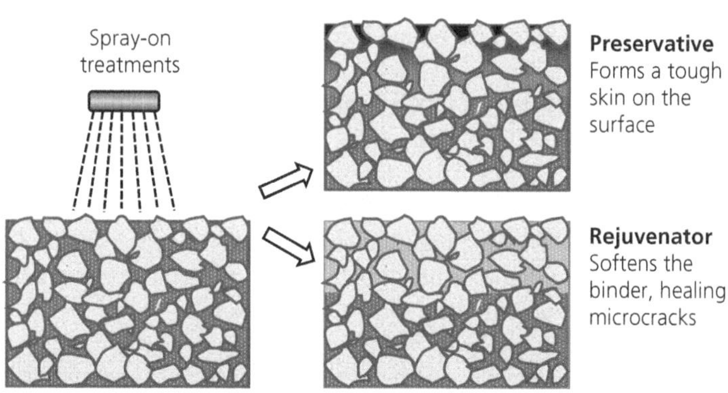

12.3.5 Surface dressing

A full description of the process of surface dressing is provided in Section 7.6.1. As a maintenance treatment, surface dressing both seals the surface and restores both micro-texture and macro-texture. For many pavement types it is therefore a powerful and cost-effective means of prolonging a pavement's life. Indeed, on minor roads, repeated surface dressings are often the only treatments carried out, giving almost indefinite life if carried out at optimal times.

The negative aspect to surface dressing – besides being more expensive than a preservative or rejuvenator treatment – is ride quality, making it an unpopular choice for road users on high-speed roads.

12.4. Local repairs

None of the surface treatments presented in Section 12.3 can do anything significant for the structural condition of a pavement; they merely serve to reduce the deterioration rate. However, the nature of pavement construction is that there will always be certain areas that, due to material defects, lack of thickness, concentrated load (e.g. due to cornering), drainage problems or some other reason, begin to fail much earlier than the rest of the pavement. These local failures may start as individual cracks that can be sealed, but they will progress to multiple cracking (crazing), surface spalling, potholes, edge defects and so on. As local phenomena, it clearly makes sense to carry out local repairs.

12.4.1 Emergency pothole repairs

Potholes on relatively high-speed roads present a serious danger to the travelling public. Vehicle suspensions can be damaged; accidents can even be caused. They therefore need to be repaired rapidly. This is unplanned maintenance, probably required to be carried out within a fairly tight time window and the appropriate material will be a cold-mix asphalt (see Chapter 13). The repair has to be seen as temporary, restoring the serviceability of the road until such time as a fuller programme of works can be planned. Nevertheless, despite being classified as 'temporary', the repair may need to last several months under heavy traffic load. The key points that need to be attended to if this is to be achieved are

- the hole must be cleaned of loose material and standing water – although there are systems on the market that claim to be usable even under water
- the base and faces of the hole should be sprayed with bitumen emulsion; the repair material should be well compacted – although some systems claim to be self-compacting
- a seal should ideally be applied around the edge of the repair (although this is no substitute for covering the faces of the hole with bitumen emulsion).

The next subsection describes the type of repair required to give a more permanent solution.

12.4.2 Asphalt patching

With proper processes and materials, it is possible to achieve relatively long-lasting repairs to potholes and other local defects. The first requirement is to determine the depth of patch required. It is, of course, easiest simply to plane or break out an area of surface course and replace it but, if problems extend deeper, this will only be a short-term measure.

Traditionally, the procedure is to cut or break the repair area back to sound material, usually forming a rectangular area, apply a bitumen coat to the base and sides and then to compact in the repair material. Patching materials can be either hot or cold. Hot-mix asphalt can be stored in a 'hot box' housed on the back of a vehicle; cold-mix asphalt is more user-friendly in that it needs no special storage and has a much longer 'shelf life'. But studies consistently show that cold-mix repairs last less long on average than hot-mix or warm-mix materials.

Compaction of a patch is always difficult, relying on small manually controlled rollers or vibrating plates, and a thin layer of hot-mix cools rapidly and so does not give much time for compaction to be carried out. The result is that low density is almost inevitable, particularly around the edges of the repair, although this problem can be reduced by preheating the sides before filling. Cold-mix imposes no such time constraints, allowing compaction to be continued for longer if necessary, particularly in awkward corners. And even awkward corners can be avoided – machines have been developed that deliver a curved cut, allowing much more even compaction.

Figure 12.4 illustrates the patch repair process. In addition to compaction, three further features contribute significantly to the life of a patch. The first is the 'stepping' shown between layers, making it more difficult for water to penetrate; the second is the use of a bond coat (i.e. bitumen film) on the bottom and sides; the third is the surface seal at the joint between old and new material.

12.4.3 Asphalt edge repairs

An edge repair is simply a patch at the side of the pavement. The edge is always a vulnerable location, particularly on narrow roads subject to heavy goods vehicle traffic. The result is that edge deterioration is a common feature of such roads. The repair is known as a 'haunch'.

The key to successful haunching is the understanding that it is possible to improve the situation by local thickening (or some other form of strengthening). Section 10.6.7 suggested that the tensile stress induced at the bottom of a bound layer in an edge situation will be around 80% higher than the equivalent stress remote from an edge. A useful rule of thumb here, too approximate for normal pavement design but ideal for the rather indeterminate case of pavement edge strengthening, is that tensile stress or strain is approximately related to the square of layer thickness. This means that

Figure 12.4 Asphalt patching method

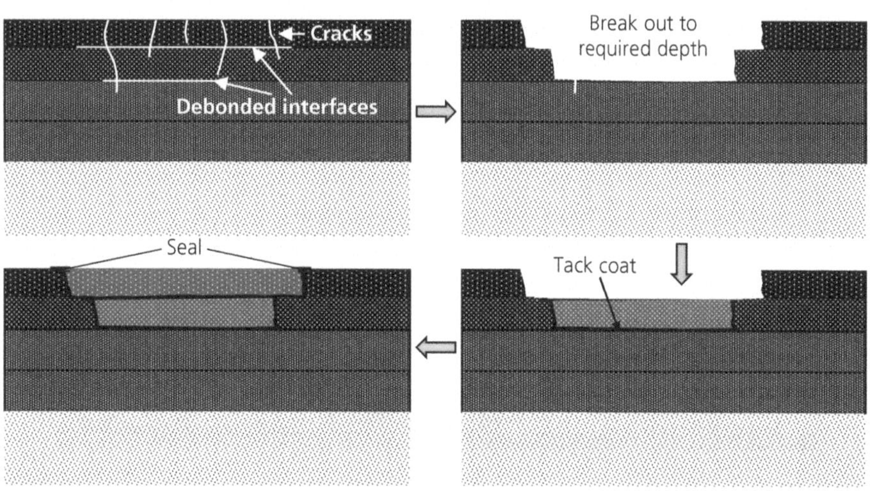

the bound layer thickness at a trafficked edge should be about 35% greater than elsewhere in the pavement in order to give comparable life. This is an appallingly crude analysis of course, but the problem is a highly complex one and also subject to plenty of unknowns (wheel location, variation in foundation quality etc.). It is also, frankly, not worth spending a long time on the design of individual edge failures; thus a general recommendation for a 35% bound layer thickness increase compared with the rest of the pavement is not unreasonable. One could argue for a greater thickening still if local softening of the foundation has occurred and drainage is either non-existent or malfunctioning. An example of a haunching thickness design guide is provided by the County Surveyors Society (CSS, 1994).

12.5. Hot/warm in situ recycling

Hot/warm in situ recycling represents a significant challenge for equipment manufacturers; however, the pavement engineering is much less challenging. Basically, the aim is to transform an aged and partially failed asphalt surface layer into a good-as-new product. The process requires that the surface of the pavement is heated to a temperature at which it can be reworked, healing all cracks and reshaping any deformed profile. In the so-called 'repave' process, the uppermost materials are scarified (i.e. loosened) while hot (above about 90°C) and additional new hot-mix or warm-mix asphalt material is added, typically increasing the level by around 20 mm. The hope is that the heated old material blends with the new to form a single contiguous surface course with fully restored properties.

'Remix' is the name given to a slightly more sophisticated process in which the heated existing surface material is physically scraped up and mixed together with fresh hot or warm material before being re-laid as a new surface course. This ensures a fully integrated and blended layer. The bitumen grade and mixture composition of the added material are selected to balance out any defects in the composition of the existing material. Figure 12.5 illustrates the concept of the remix process.

Both these processes are excellent in that they actually do transform a failing surface into an intact new layer. However, they have their limitations.

- The existing surface material must be suitably uniform in composition (i.e. not too many patches or other material changes).
- The transformation is only effective to the depth to which the asphalt can be heated. This means that the treatment can potentially cure rutting, top-down cracking or ravelling in the surface course, but does nothing to overcome deeper problems.

The depth to which the treatments can extend depends on the efficiency of the heating that can be achieved directly (by naked flame or infrared) or indirectly (by feeding superheated gases over the

Figure 12.5 Schematic diagram of hot in situ recycling of asphalt

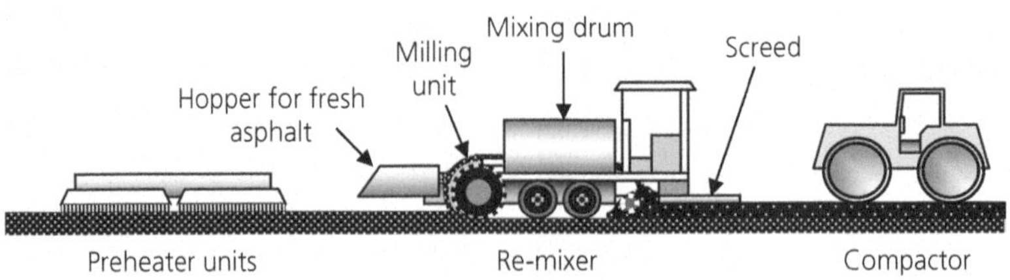

pavement surface). Direct heating has traditionally only managed a workable depth of 20–30 mm; indirect heating is claimed to be able to heat to at least 75 mm. Clearly, if 75 mm is 'restored' and a further 20 mm is added, this provides a new 95 mm-thick layer, which would make a serious contribution to the strength of the pavement as a whole; that is, the process is becoming more than a mere surface course treatment.

Naturally the process is not cheap. The equipment forms a long train and is not suited to tight corners. This is therefore a process that only becomes economic where there is a long stretch of major highway to be treated.

12.6. Summary

As with any engineering undertaking, the true cost of a highway or other pavement infrastructure is much more than its initial construction cost. It has to be carefully managed in order to achieve its intended life, and that means timely maintenance. The treatments described in this chapter are all intended to forestall the progression of damage rather than add greatly to the overall strength. But, if not carried out when needed, the intended design life will not be achieved.

REFERENCES

Abed A, Rahman M, Thom N *et al.* (2023) Predicting pavement performance using distress deterioration curves. *Road Materials and Pavement Design.* See https://doi.org/10.1080/14680629.2023.2238094 (accessed 30/03/2024).

CSS (County Surveyors Society) (1994) *Road Haunches – A Guide to Maintenance Practice.* Transport and Road Research Laboratory, Crowthorne, UK, Report PA/SCR243.

Part 4

Cold-mix asphalt pavements

Nick Thom
ISBN 978-1-83549-711-1
https://doi.org/10.1108/978-1-83549-710-420243015

Chapter 13
Cold-mix asphalt

Excellent though bitumen is as a binder of aggregate particles, one of its drawbacks is that it will only adhere to aggregate properly if that aggregate is heated sufficiently to drive off all moisture, and this is a costly and energy-intensive procedure. Thus, there are strong economic and environmental drivers towards so-called *cold-mix* technology, utilising a form of bitumen that is workable at ambient temperatures and which can mix with cold, wet aggregate. Cold-mix technology opens the door to a brave new world of possibilities.

In this book, cold-mix asphalt has been separated out from hot and warm mixes because it genuinely is different, although numerous references will be made back to the chapters on hot/warm mixes. Furthermore, it is a moving target, with development continuing apace. It is also a vital component of recycling and in the use of aggregates that would not be suitable in hot or warm mixes. The topic is one that is therefore likely to grow in importance – which means that engineers really should not ignore it.

13.1. Cold-mix binders

13.1.1 Bitumen emulsion

An emulsion is a suspension of one liquid in another, in this case of bitumen in water. It is manufactured by putting hot bitumen through a high-shear colloid mill, which reduces it to tiny droplets, no more than a few tens of micrometres across, at the same time as feeding in an emulsifying agent and water (Figure 13.1). The emulsifying agent comprises molecules with charge at one end, either positive or negative, and a long polymer tail with strong affinity to bitumen. As these emulsifier

Figure 13.1 Bitumen emulsion

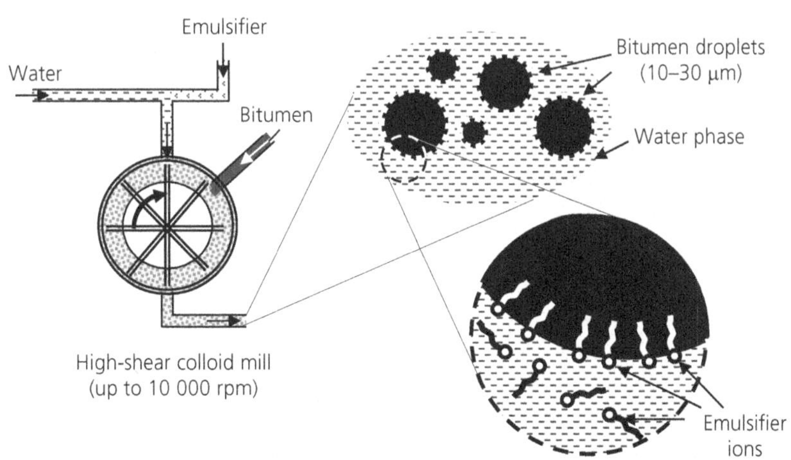

ions attach themselves to the bitumen droplets, these are converted into charged particles, each repelling its neighbour; the droplets therefore become suspended in the water and are prevented from coalescing. The specific gravities of bitumen and water are very similar (1.02–1.03 for bitumen), so there is little tendency for the droplets to sink under gravity, and the result is a very stable product. Occasional light stirring is sufficient to ensure a long 'shelf life'.

As the water phase is continuous, the fluidity of bitumen emulsion is not much reduced from that of water, despite the bitumen comprising 40–70% of the total volume. This means it can readily be piped, sprayed and mixed with aggregate. The difficulty, of course, is to achieve the transformation from an unbound material with water containing individual bitumen droplets to a genuine bitumen-bound mixture. This is discussed further in Section 13.2.

13.1.2 Foamed bitumen

An alternative form of cold-mix asphalt uses bitumen in the form of a foam. Figure 13.2 illustrates the method of production. The foam is produced by the very rapid transformation of water into steam as it comes into contact with hot (typically 160–180°C) bitumen, aided by the addition of air. The water content within the foam is generally 2–5% of the bitumen volume. However, once transformed into steam, the foam expands to many times the volume of the bitumen itself and, in this state, it can readily be mixed into aggregate. The ratio of peak foam volume to original bitumen volume is known as the expansion ratio (ER), while the time until this peak volume is halved is known as the half-life. The ER is typically between 5 and 20; the half-life is typically 5–20 s. A high foaming water content gives a high ER but a short half-life and vice versa. It is generally recommended (e.g. Sabita, 2020) that the foaming water content should be chosen to give an ER of at least 10. Recommended half-lives are more varied but most authorities go for at least 8 s, allowing time for adequate mixing, although this in turn depends on the efficiency of the mixer. Foamed bitumen therefore contrasts with emulsion. Emulsions can be stored for months before use, whereas foamed bitumen has to be used within a few tens of seconds of production.

13.1.3 Bitumen grades

Bitumen behaviour was covered in Section 9.1, where it was noted that bitumen grade is most commonly expressed in terms of its 'penetration', with a high penetration meaning a softer grade. It was also observed that modifiers, typically polymers or recycled tyre rubber, are sometimes added to improve performance against rutting or cracking or both.

Figure 13.2 Foamed bitumen

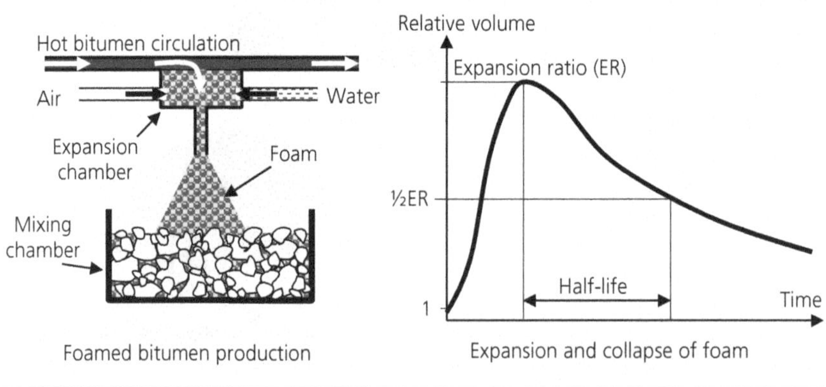

A bitumen emulsion can be made with bitumen of any grade and including any modifier. If the emulsion is to be used as a tack coat or bond coat, or in surface dressing, the hardness or softness of the bitumen presents no problem at all. However, when used in a cold-mix asphalt it is generally found that harder grades hinder the development of a properly bound material. Good performance can be achieved typically down to about 50 pen, with or without modifier; harder grades should be avoided.

Foamed bitumen is a little more choosy about which grades give good performance. Anything harder than about 70 pen is difficult to foam well, which means it is difficult to mix well and is not advised.

13.2. Mixing and laying

13.2.1 The role of water

A key difference in mixture designs for cold mixes is that an additional component has to be allowed for, namely water. Cold aggregate inevitably contains water, even if it has been protected from the rain for some time; furthermore, cold mixes need water in order for bitumen droplets to disperse within the mixture during mixing. In an emulsion, the water phase already carries the bitumen within it, but it is the presence of water in the aggregate that allows the bitumen droplets to move freely during mixing. In the case of a foamed bitumen mixture, the water in the aggregate is equally necessary to allow the foam to disperse properly. Furthermore, an optimum quantity of low-viscosity liquid is needed to allow good compaction. In an unbound material, this liquid is entirely water; in a hot- or warm-mix asphalt it is bitumen; in a cold-mix asphalt it is a blend of water and bitumen. The upshot is that *water* is an *absolutely essential* ingredient in the mixture.

But, if there is water in the mixture then, compared with a hot mix, there has to be less of something else – and that something usually includes bitumen. Take, for example, an idealised dense hot-mix asphalt, with 85% of the final volume taken up by aggregate, 10% by bitumen (about 4–5% by mass) and 5% by air voids. If the same aggregate gradation is used in a cold mix, then, in order to achieve good compaction, there still needs to be about 10% fluid content by volume. Perhaps this can be pushed up to around 13%, but it is hard to squeeze the air-voids content down to less than about 2% during compaction. However, even 13% fluid by volume is only about 5–6% by mass, about half of which comes from the water present in the cold aggregate (and in emulsion where it is used) in order to facilitate binder movement. This leaves no more than 3% actual bitumen content (by mass). Frankly, this is less than ideally required to bind the particles together properly – and it leaves cold-mix designers with an insurmountable problem: it is *physically impossible* to replicate exactly a well-compacted dense asphalt using cold-mix technology; there just isn't room.

These are the alternative approaches that may be considered (see Figure 13.3).

- Accept a low bitumen content. Make sure that the filler content is also correspondingly low, or else the contact between the bitumen and large aggregates will be poor, resulting in poor mixture properties. The result will be a stiff and deformation-resistant mixture, but one with relatively low durability and fatigue resistance.
- Increase the bitumen content to a more suitable 4% by mass. This means that the final voids in mixed aggregate (VMA), bitumen + water + air, will have to increase significantly, reducing stiffness and deformation resistance.
- Change to a more open gradation, with a naturally higher voids content at full compaction. This allows a higher bitumen percentage to be used, but should still allow good aggregate interlock and so reasonable deformation resistance; however, it will have a reduced stiffness in comparison with a dense gradation.

Figure 13.3 Alternative cold-mix asphalt volumetrics

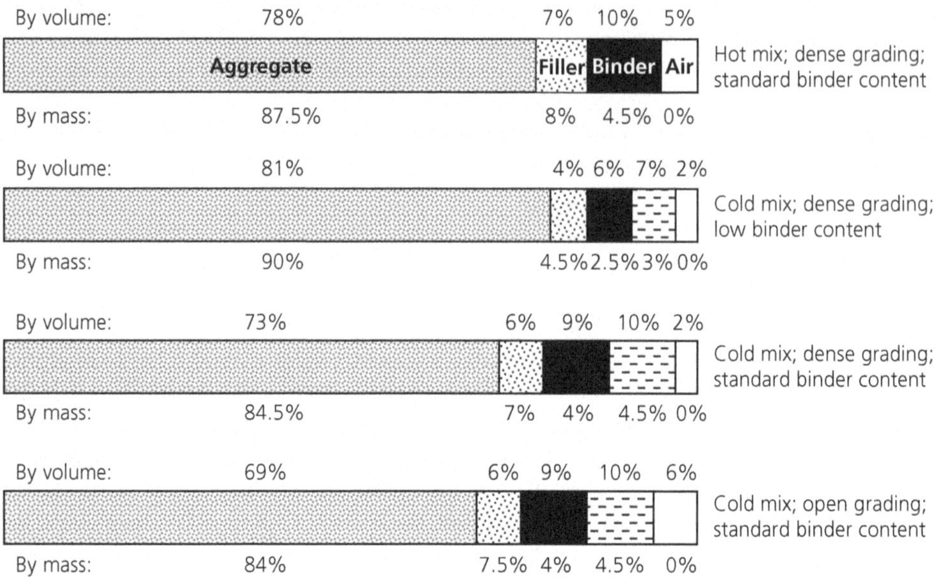

- Incorporate recycled asphalt planings (RAP) as part of the aggregate. These already contain bitumen, and so a higher overall binder content can be achieved with a moderate quantity of added emulsion or foamed bitumen. The two bitumens (one in the RAP and the other added) will not blend, but both will contribute to asphalt performance. Oke (2010) carried out an extensive laboratory investigation on the inclusion of RAP in emulsion-bound mixtures, finding that durability and fatigue strength were enhanced compared with mixtures containing only virgin aggregate, as long as the recovered penetration of the RAP was above about 10 dmm.
- Include a small percentage of cement or other hydraulic binder (see Section 13.2.2).

Unfortunately, the presence of water during compaction means there is no way to avoid a relatively high air-voids content once the water has evaporated. It is very tempting for cold-mix practitioners to pretend that they can produce a direct equivalent to a hot mix, using identical gradation and bitumen content, but this is a very dangerous claim to make; the aggregate skeleton will be less efficient than that in a hot mix because of the lower aggregate density achieved due to the presence of water. Mix design is explored in more detail in Section 13.3.

13.2.2 Curing

Another inherent reality of cold-mix asphalts is that they take time to achieve their full potential. Whether emulsified or foamed, the binder is temporarily rendered fluid (by the action of water or a foaming process), in which state it is mixed with an aggregate. The result is that the bitumen is dispersed throughout the mixture in the form of very small droplets or flecks of foam within the water phase. In this form there is no continuous binder phase and adhesion to the aggregate is initially negligible; the mixture is still workable and can be stored loose for days or weeks before use – an attractive property.

Once compacted into the pavement, however, the process of strength gain known as 'curing' commences; the action of compaction itself pressurises some of the binder between aggregate particles, promoting binder adhesion. Over a period of time, anywhere between 2 weeks and 6 months depending on temperature and degree of exposure to the atmosphere, much of the water evaporates and the water that remains is unable to prevent bitumen droplets coalescing onto the aggregate; the result is a steady increase in strength and stiffness. Figure 13.4 presents examples of early curing data. Note, however, that curing is more than simply evaporation of water; the bitumen droplets or flakelets need to be able to flow and coalesce and this is of course a more rapid process at higher temperature. As demonstrated in Figure 13.4, curing at 5°C is likely to result in a poorer material than curing at higher temperature, even once the water evaporates. Tests on fully wrapped specimens, where evaporation was prevented, have demonstrated that, even at 40°C, curing can be very slow indeed. Clearly, this delayed strength gain – together with the uncertainty involved – demands some serious risk management. The options would appear to be

- avoid overlaying a cold-mix course during early life, allowing maximum evaporation
- restrict usage to locations where little early-life traffic is anticipated
- accelerate the curing process in some way.

So how to accelerate curing and mitigate risk? In the current state of knowledge there is only one method that has been proven to work, and that is the addition of a hydraulic binder. Hydraulic binders will be discussed in more detail in Chapter 16, but their key quality in the context of cold-mix asphalt is that they react with water. Portland cement is the best known such binder; lime is another, so are industrial by-products like fly ash and ground granulated blast-furnace slag, especially when activated by a little lime. The inclusion of as little as 1% hydraulic binder dramatically alters the properties of the finished mixture as well as reducing the time required to reach those properties, as illustrated in Figure 13.5 for emulsion mixes. Figure 13.6 shows the effect of higher dosages in relation to curing time, showing in effect the transition from a bitumen-bound material to a hydraulically bound material.

Figure 13.4 The influence of curing temperature and water content on foamed bitumen mixtures (Kuna, 2014)

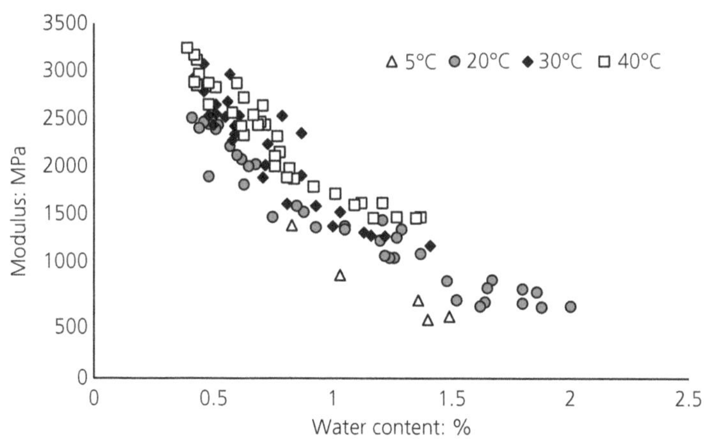

Figure 13.5 Effect of 1% hydraulic binder on strength at 28 days (Ojum, 2015)

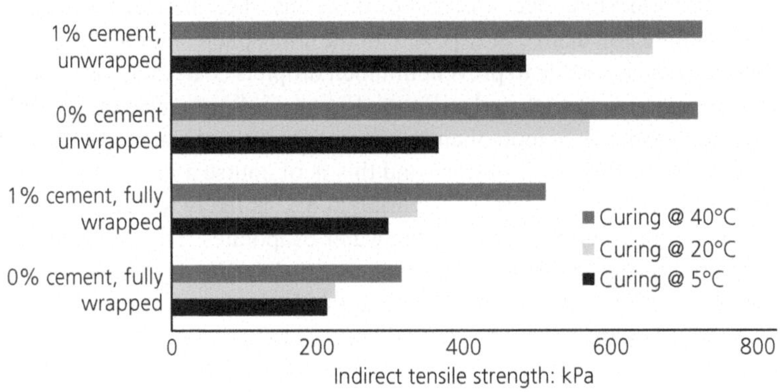

However, one property that does not improve with cement addition is toughness (the energy needed to cause failure) and this signals the fact that hydraulic binders, forming cementitious bonds, lead to an embrittlement of the mixture; in effect, it is on the way to becoming a concrete rather than an asphalt, albeit not very far down that way. Note also that the formation of hydration products is relatively rapid so, as with concrete, a mixture with hydraulic binder cannot be stored; if the binder is Portland cement it has to be used within about 2 hours, rather longer for alternative slower-setting binders.

13.3. Mix design

Essentially, cold-mix design is very similar to that of hot mix. First, the aggregate gradation has to be chosen, followed by selection of the most appropriate binder content. The difference is that the water cannot be ignored. The following sections outline the process.

Figure 13.6 The effect of different dosages of Portland cement on a cold emulsion mixture (Nassar, 2016)

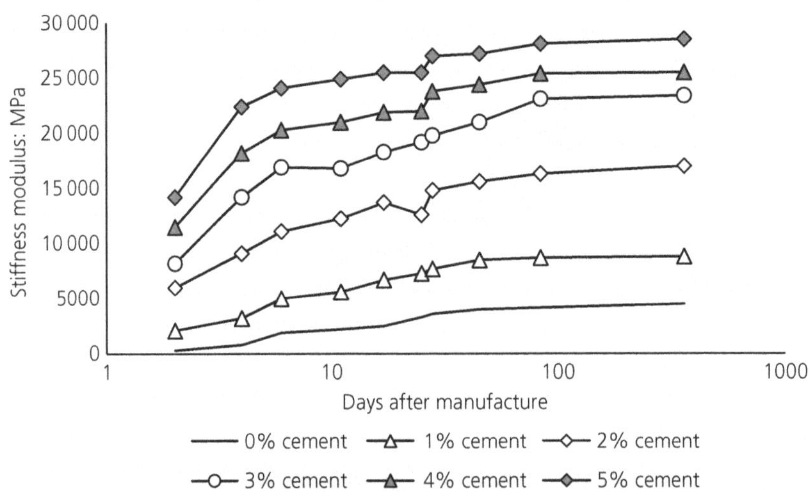

13.3.1 Aggregate gradation

The options and the restrictions imposed by the presence of water were discussed in Section 13.2.1. Basically, an asphalt concrete type mix is possible but it will always have a higher voids content than the equivalent hot mix. The alternative is a slightly more open-graded mix (i.e. one with a relatively lower fine aggregate percentage), potentially leading to more rapid curing due to increased water evaporation.

13.3.1.1 Use of recycled asphalt planings (RAP)

The need for water places a limit on bitumen content – that is a fact of life. So, why not cheat and bring in bitumen with the aggregate? If, for example, 60% of the aggregate comprised RAP with a 5% binder content, this would form 3% of the finished mixture. To achieve, say, a 5.5% final binder content would require another 2.5%, which is entirely achievable either with emulsion or foamed bitumen.

As a bonus, it would appear that the relative deformability of RAP aggregate compared with solid rock allows a higher density to be achieved – depending on how hardened the RAP binder is and the ambient temperature during mixing and compaction. Table 13.1 shows air-voids data from compaction of emulsion mixtures containing RAP of different hardnesses, mixed at two different temperatures. Unfortunately, the inclusion of RAP also means that deformation resistance in service is likely to be slightly poorer.

Admittedly, hardened RAP binder, which will not in any way blend with the new emulsion or foamed bitumen binder, may have poorer properties than new binder, but it can still be 'active' – again, depending on how hardened it is. Oke (2010) also found that the softer RAP binder, 20 pen, which is typical of RAP from temperate climates, gave a marginally enhanced dry stiffness modulus compared with the use of virgin aggregate, while harder RAP did not. The reason may be improved adhesion between the new cold binder and RAP compared with rock.

More dramatically, softer RAP was also found to give much enhanced resistance to water attack and an improved fatigue resistance, although to be honest it will still be poorer than that of a hot-mix asphalt, at least in the laboratory.

13.3.1.2 Hydraulic binder

If hydraulic binder is to be added then it effectively forms part of the 'filler' in the mix, and an equivalent volume should be subtracted from the amount of conventional inert filler used. Note the importance of volume here rather than mass and the likelihood that hydraulic binder has a different specific gravity from rock fines.

Table 13.1 Effect of RAP quality and temperature on achieved air voids (Oke, 2010)

Compaction temperature: °C	Air-voids content: %				
	Virgin aggregate	RAP aggregate			
		5 pen	10 pen	15 pen	20 pen
20	14.6	14.5	15.7	13.7	11.9
32	13.1	12.3	11.5	9.1	8.3

Figure 13.7 Defining optimum fluid content (Kuna, 2014)

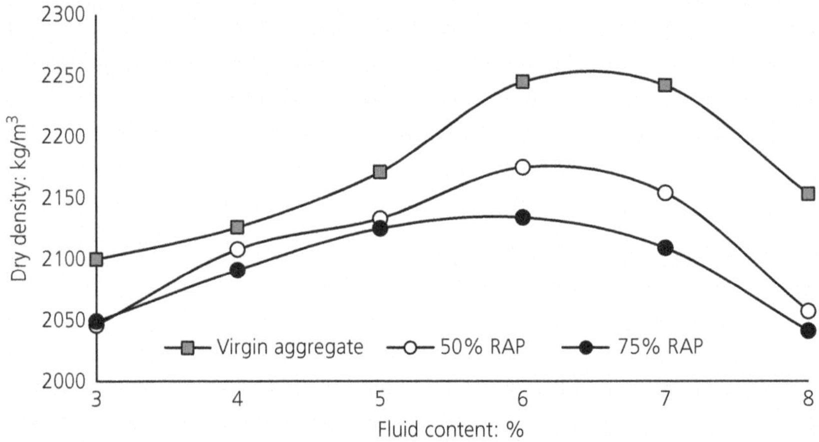

13.3.2 Optimum fluid content

When a cold-mix asphalt is compacted it behaves essentially like an unbound material, for which Chapter 6 already introduced the concept of an optimum water content for compaction. In the case of a cold-mix asphalt, water is replaced by 'fluid'; that is, the combination of water and the droplets or flakelets of bitumen carried by that water. In fact, it is common to carry out compaction tests using only water and to assume that the water content that gives the maximum dry density (the densest aggregate packing) is approximately equivalent to the optimum fluid content. Figure 13.7 shows examples with different RAP contents; the approximately 6–7% optimum fluid content shown is typical for relatively dense mixes.

13.3.3 Selection of binder content: emulsion mixes

How much of the fluid should be bitumen? How much physically can it be? There are no absolute rules here. However, the Marshall mix design approach (see Section 9.7.2.1) is a sensible one, in which a number of laboratory specimens are made at different emulsion contents, making up the remaining fluid (to reach optimum fluids content for compaction) with so-called pre-wetting water (water within the aggregate). In the case of emulsion there is also additional water within the emulsion – typically a 60:40 bitumen to water ratio used in cold-mix asphalt.

Specimens then need to be cured in a reasonable time frame, and the most common procedure is to carry out accelerated curing at 40°C for 3 days (e.g. Wirtgen, 2004). The resulting material properties probably represent a best possible case, but for mix design this is not a problem. To achieve realistic long-term properties, Merrill *et al.* (2004) recommend a 28-day curing period, either at 20°C for rapid-setting mixtures (i.e. including hydraulic binder) or 40°C for slow-setting mixtures.

The specimens are then tested in some way. Most commonly, cylindrical specimens are tested for indirect tensile strength and/or stiffness, both dry and soaked (again 40°C for 3 days is common practice). The decision on design bitumen content is always a compromise. Figure 13.8 shows typical test data for a soaked emulsion mixture, showing an optimum of just over 4%.

Figure 13.8 Properties after soaking: emulsion mixture (Ojum, 2015)

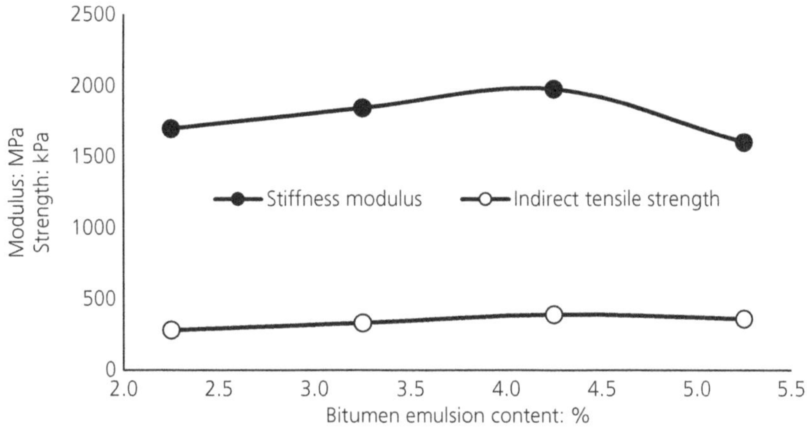

13.3.4 Selection of binder content: foamed bitumen mixes

The approach in the previous section undoubtedly works well for emulsion, but less well for foamed bitumen. It seems that, while the droplets of bitumen within emulsion are lost within the water phase and so contribute fully to total fluid content, this is only partially true for foamed bitumen flakelets. Based on work by Kuna (2014), the amount of pre-wetting water should be no more than 1% less than the optimum water content for compaction no matter how much foamed bitumen is added. Any further reduction will give poorer compaction. But the problem is that the foamed bitumen still occupies volume and so the dry density of the aggregate will be much less than that given at optimum water content and zero bitumen.

This is a bit of a nuisance. It means that a foamed bitumen mix with a high bitumen content will probably end up with a low dry density and so a high voids content – which means reduced strength, stiffness and durability. A low bitumen content will allow good compaction but there will be a price to pay in terms of fatigue resistance and durability. As stated previously, the decision will be a compromise, but the most useful foamed asphalts tend to be those with relatively low foamed bitumen contents, say 2–3%, but supplementing this with bitumen from RAP, sometimes at up to 90% of the total aggregate.

13.4. Summary

I hope the reader now has a realistic expectation of what can be achieved with cold-mix technology. The main issue is the water – needed for compaction but then leaving a high air-voids content when (and if) it evaporates. An important secondary issue is the time taken to cure to full strength. To some extent these drawbacks can be combatted by the use of a small amount of hydraulic binder and RAP aggregate. The result is that it is indeed possible to produce a mix with equivalent stiffness and deformation resistance to a conventional hot mix, but probably not an equivalent fatigue resistance or durability.

Of course, these negatives have to be set against some very considerable positives.

- Energy consumption and therefore carbon dioxide footprint are reduced.
- There is no differential cooling problem and so much thicker layers are possible.
- Cold-mix technology opens the door to deep in situ recycling, something that is entirely impossible using hot mixtures.
- Mixes without hydraulic binder, although they may be weaker, have the great advantage that they can be stored for weeks before use.

Furthermore, Chapter 14 will also make the point that cold mixes are genuinely different materials from hot mixes and they degrade in a different, and potentially less damaging, way.

REFERENCES

Kuna KK (2014) *Mix Design Considerations and Performance Characteristics of Foamed Bitumen Mixtures (FBMs)*. PhD thesis, University of Nottingham, Nottingham, UK.

Merrill D, Nunn ME and Carswell I (2004) *A Guide to the Use and Specification of Cold Recycled Materials for the Maintenance of Road Pavements*. Transport Research Laboratory, Crowthorne, UK Report, TRL611.

Nassar AIM (2016) *Enhancing the Performance of Cold Bitumen Emulsion Mixture using Supplementary Cementitious Materials*. PhD thesis, University of Nottingham, Nottingham, UK.

Ojum CK (2015) *The design and optimization of cold asphalt emulsion mixtures*. PhD thesis, University of Nottingham, Nottingham, UK.

Oke OL (2010) *A study on the development of guidelines for the production of bitumen emulsion stabilised RAPs for roads in the tropics*. PhD thesis, University of Nottingham, Nottingham, UK.

Sabita (Southern African Bitumen Association) (2020) *Technical Guideline: Bitumen Stabilised Materials*, 3rd edn. Sabita, Cape Town, South Africa, TG2.

Wirtgen (2004) *Cold Recycling Manual*, 3rd edn. Wirtgen, Windhagen, Germany.

Nick Thom
ISBN 978-1-83549-711-1
https://doi.org/10.1108/978-1-83549-710-420243016

Chapter 14
Cold-mix asphalt pavement – structural design

14.1. Construction practicalities
In general, cold-mix asphalt is much easier to work with than hot-mix asphalt. The relevant practicalities are almost all to ensure the most rapid curing possible.

14.1.1 Layer thickness
Regarding compaction, the limit here is the same as for an unbound material; that is, dependent on the compactor being used. With modern plant this means that 200 mm is no problem and 250 mm may be possible.

But compaction isn't the only thing to worry about. Figure 13.4 made the point that when water is prevented from escaping, curing is much slower and probably never reaches its true potential. Water can evaporate and escape from the upper surface readily enough, but finds it much harder to work its way up through narrow void spaces as water vapour. It's impossible to predict this effect with any confidence, but an approximate evaluation suggests that if it takes, say, 1 day to 'dry' to 50 mm then it takes 4 days to dry to 100 mm, 11 days to reach 150 mm, 24 days for 200 mm and well over a month to get to 250 mm – and that's without any additional rain! Note that 'dry' does not mean totally dry but rather that enough evaporation has taken place to encourage curing.

Thus, it may be preferable to subdivide a thick layer into two or three separate courses, allowing each to dry before placing the next and so shortening the overall curing period. But this will be more expensive and, in the case of in situ recycling, logistically very difficult.

14.1.2 Working time
This is a comparatively simple topic. If there is no hydraulic binder then you have as long as you like! If Portland cement is included then work should be finished within about 2 hours; for slower-setting binders such as lime/fly ash combinations this can be extended to several hours.

14.1.3 Surfacing
This section is included simply to reiterate the point that underlying layers should be allowed to dry out significantly before surfacing is added, whether that surfacing is another cold-mix layer or a hot/warm-mix layer.

14.2. Current practice
In the UK, the document normally referred to for cold-mix asphalt pavement design is Transport Research Laboratory report TRL611 (Merrill *et al.*, 2004). It is assumed that the cold-mix asphalt forms the base and a hot/warm-mix surfacing will be added. The basis is the same conventional

analytical pavement design described in Section 10.6.2, assigning stiffness modulus values for the cold-mix asphalt of between 2000 MPa and 3000 MPa. For the user, convenient charts are provided. Another excellent and very scientific design guide, both for the materials and the pavement, is given by Sabita (2020), based on a concept similar to the 'structural number' in the AASHTO (1993) method, but bringing in an analytical approach.

14.3. Predicting pavement life

14.3.1 Early-life damage

This is a very tricky issue, and it is impossible to generalise, but there is an obvious danger that the long-term in-service condition will have reduced strength and stiffness due to enlarged microcracks, possibly even fully developed cracking, if significant early traffic is allowed. In the early stages of curing of cold-mix asphalt, the material is undeniably very weak indeed. The first issue is that if significant load is to be taken during the first few hours, then it is essential that the aggregate itself has sufficient shear strength to avoid rutting. Once this challenge has been met, the next issue is the effect of early trafficking on long-term properties. With regard to the long-term stiffness modulus, experience from the author's evaluations of cold-mix asphalt pavements is that values can range from about 4000 MPa (at 20 °C) down to about 500 MPa, although how much stiffness loss is due to early-life damage and how much to unfavourable curing conditions is hard to determine. Under normal curing conditions it is unlikely that the long-term modulus would be less than 1500 MPa. Nevertheless, the conclusion has to be that this type of pavement is really best suited to cases where early-life trafficking is light.

14.3.2 In-service deterioration

There are basically two approaches to pavement performance prediction where cold-mix asphalt is involved: either the material can be treated in the same way as a hot/warm mix, in which case the design principles already presented with respect to rutting, cracking and durability all apply; or it can be considered as an ultra-high-quality unbound material. If considered in the same way as a hot/warm mix, then design against cracking is required and, as the stiffness will be lower than that of a hot/warm mix and the fatigue resistance may also be slightly poorer, this means that the design thickness will be greater. The advantage of treating cold mix as a high-quality unbound material is that no design against cold-mix material cracking is required. It is assumed, in effect, that cracking has taken place already. Figure 14.1 illustrates the two alternatives.

Figure 14.1 Design with cold-mix asphalt base

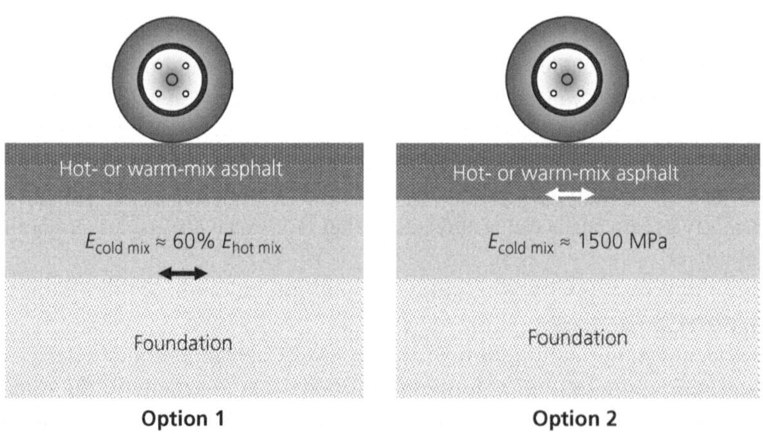

The actual numbers in Figure 14.1 are only indicative, and the designer has to take account of the properties anticipated in each case. The estimate in option 1 of 60% of the stiffness of an equivalent hot/warm mix (i.e. one with the same grading, binder content and grade) is based on the assumption that the voids content of the cold mix will be some 5% higher. With only 60% of the stiffness of hot/warm mix, the cold-mix design thickness in option 1 will typically be some 20–25% greater than it would have been if hot/warm mix had been used.

But which alternative is right? There is a real difference between hot/warm-mix and cold-mix asphalt, and it can be seen in fatigue data, of which Figure 14.2 is an example – from indirect tensile testing in this case. The point is that, while both materials lose stiffness during fatigue due to the development of microcracking, the transition to macrocracks occurs earlier in hot-mix asphalt. The reason is almost certainly the fact that a cold-mix asphalt is, in reality, only partially bound to begin with (i.e. a proportion of particle-to-particle contacts have no binder); to put it another way, a cold mix is full of microcracks from day one. The effect is that the material really does have a long phase of its life during which it loses stiffness modulus but remains more or less a continuous and uniform layer, and when macrocracks form they tend to be closer together, as shown in the insert in Figure 14.2.

So what does all this mean? In a fully cold-mix pavement – that is, one with a cold-mix surface as well as the base – it means that the failure condition should probably be taken as the point where about 70% of the original stiffness has been lost, rather than the 50% usually taken for hot/warm mix; or, to put it another way, the fatigue life should be assumed to be at least 50% longer than a traditional fatigue test would suggest.

In a pavement with a hot/warm-mix asphalt surface, as shown in Figure 14.1, the lack of major macrocracks in the base implies it is reasonable to treat it as a uniform low-modulus layer. The 1500 MPa shown in design option 2 in Figure 14.1 is intended to be a reasonable long-term value, accepting that trafficking during early life will limit the final stiffness achieved and that a significant number of microcracks or missing interparticle bonds will be present. The figure is a little

Figure 14.2 Fatigue test data (Jiménez *et al.*, 2016)

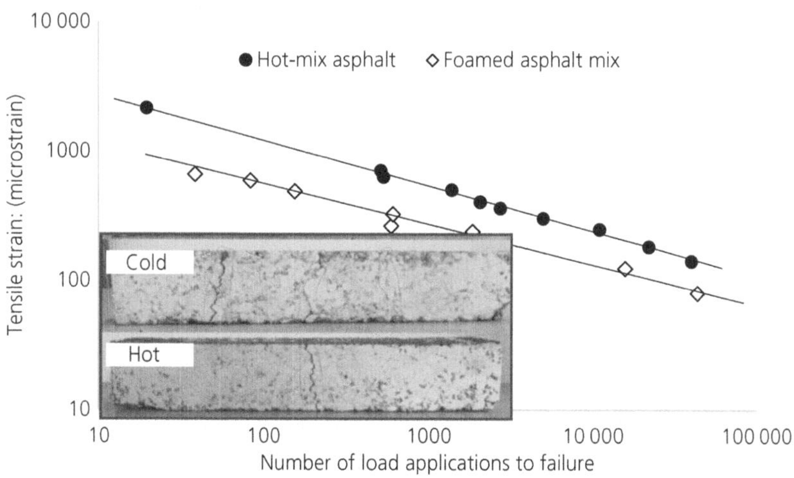

lower than would probably be found as an average for the layer because it is intended to represent the least intact locations. The hot/warm-mix surfacing can now be treated independently using one of the flexible pavement design approaches introduced in Chapter 10.

Despite the understandable keenness of the cold-mix industry to promote full hot/warm-mix equivalence, it is advised here that option 2 is the safer and the more realistic. By nature, a cold mix has discontinuities in it that are not present in a hot/warm mix, and it seems sensible to acknowledge this in design. As it happens, the resulting design thickness is often no greater for option 2 than for option 1.

With respect to design against rutting within the cold-mix material itself, however, it is advised that hot/warm-mix principles are followed. Either a wheel-tracking test or a repeated load axial test will give a comparative measure of deformation resistance. It will usually be lower than for a hot/warm mix because of the higher voids content, which may mean that a hot/warm-mix surfacing is required; this will depend on traffic levels.

14.4. In situ or ex situ recycling

This is one of the areas where cold-mix technology really scores well because of the self-evident difficulties in heating any significant depth of existing pavement. In its simplest format, in situ recycling consists of milling the existing pavement to a selected depth – machines are capable of reaching to over 300 mm – and simultaneously mixing in the chosen binder type, emulsion or foamed bitumen, together with a small amount of hydraulic binder. The concept is simple and is shown in Figure 14.3; in reality, the machine is far from simple of course.

Alternatively, the process can be broken down into an initial pass of a milling machine and a second pass by a re-mixer/paver unit that picks up the milled particles, mixes with the selected binder and paves. The advantage is that mixing is likely to be more uniform; the finished layer is therefore more consistent and the risk is reduced. Furthermore, it opens the door to the addition of fresh aggregate as well as binder. The disadvantage is that, as a two-stage process, costs will almost certainly be higher.

Taking things one step further, it is of course possible to mill the existing pavement and transport the arisings a short distance to an adjacent mixer unit (i.e. a mini asphalt plant). Mixed materials can then be paved back into the road. This is the 'ex situ' process. The additional work phases mean added costs but control should be excellent, and it gives the opportunity to pave back in multiple thinner layers rather than a single one. This gives better level control and allows more rapid curing.

Figure 14.3 Cold in situ recycling

Pavement design is the same as for any other cold-mix asphalt pavement, as discussed in Sections 14.2 and 14.3, with the same decision to be made as to whether the material should be treated as a sightly inferior asphalt or a very superior unbound material. The stiffness assumed for an in situ recycled layer should normally be assumed to be lower than that for an equivalent plant-mixed material, perhaps by around 10%, because of the greater variability of in situ treated mixtures, as well as the likelihood of slower and possibly incomplete curing.

14.5. Maximising reliability

The potential risks with cold-mix asphalt are intrinsically greater than with hot/warm mix, which means this section is particularly important.

14.5.1 Specification

A new hot/warm-mix asphalt can be specified in terms of 'recipe' (mixture proportions, binder grade) and 'method' (compaction equipment, number of passes) and the specifier can reasonably expect a product of a certain standard. There will be variations due to the differences between different aggregates and the tolerances permitted within the specification, but overall the level of assurance is relatively high. It can be raised higher still by specifying and checking compaction and/or by insisting on cores being taken and mechanical testing being carried out, a so-called 'end-product' specification.

But with a cold mix, the difficulties really mount up. For a start, so many cold-mix asphalts make use of recycled asphalt planings (RAP) in varying proportions. Then, there is the issue of the use of hydraulic binder. An individual producer may determine their own specification with their own supplies of materials and their own experience, but it is hardly practical to roll out the same specification everywhere. It is a fast-moving field and rigid specifications stifle innovation. So perhaps it's best to avoid the prescriptive recipe and method approach and to go for end-product specification. But there we have an even greater problem. While hot/warm-mix asphalts stiffen up nicely after a matter of hours, cold mixes take weeks, meaning that (*a*) coring is not practical in the early days and (*b*) in situ stiffness testing using a falling weight deflectometer (Section 25.6.3) or lightweight deflectometer (Section 6.3.10) will give a misleadingly low result.

Unfortunately, there is no magic answer here. In the interests of promoting reliability, testing of some sort is surely desirable. If early testing is either impossible or meaningless, then all that remains is to carry out tests long after construction, say after a month. In most practical uses of cold-mix asphalt, roads have to be open almost immediately, so this means a further visit to the site, requiring traffic control and – if coring is carried out – the acceptance by the road authority of a slightly blemished surface. And then of course, the contractual implications of non-compliance have to be carefully considered, probably a scaled financial penalty, possibly a financial reward for exceeding expectations.

But what sort of tests should be carried out anyway? Deflectometer testing is just too temperature dependent, as well as being affected by the stiffness of the underlying subgrade, so difficult to use contractually despite providing potentially useful information. Coring and laboratory measurements of stiffness and fracture strength (with and without soaking) are probably the most practical, useful and relevant. If not done, there is little in the short term to incentivise the contractor – although of course success or failure in the longer term would work through and affect reputation.

Nevertheless, it is far from straightforward to know what limits to set. If a high strength or stiffness is specified, this can readily be achieved by increasing the hydraulic binder content, in effect resulting in a weak concrete. But this would be too brittle. A better specification strategy would be to require both a strength and a measure of toughness, most conveniently a minimum strain at failure. This is a topic on which no agreement has yet been reached.

14.5.2 Curing

A good specification will encourage timely curing. But how should the contractor set about meeting such a specification? The key accelerators to curing are

- allowing free air access to each layer for as long as possible
- minimising the thickness of individual layers
- using a gradation that allows inter-connected internal air voids
- using a small amount of hydraulic binder.

All of these techniques come at a cost – either direct or through extending the construction programme – but if reliability is to be maximised then, for most pavements at least, it is crucial to expedite the curing process somehow. A key issue is when to allow an overlying hot/warm-mix asphalt or surface dressing layer to be applied. As soon as this occurs, the primary water evaporation route is closed off and, while curing will continue, it will proceed more slowly and, according to laboratory tests at least, may never to reach its full potential.

14.5.3 Water and drainage

As noted already, cold-mix asphalts are usually more susceptible to water damage than hot or warm mixtures, which means it is more important than ever to make sure there is no accumulation of water within the material. The principles to be set out in Chapter 24, namely that the surface should be as well sealed as possible and that permeability should increase with depth to a drainage layer of some sort, definitely apply to cold-mix pavements.

A related point worth emphasising is that, when determining the bitumen content of a cold-mix asphalt, particular attention should be given to strength and/or stiffness tests after soaking rather than dry.

14.5.4 Construction conditions

The criticality of water evaporation to the success of cold-mix asphalts means there is a realistic lower limit for ambient temperature, typically 10 °C (e.g. Asphalt Institute, 1997). In temperate climates, therefore, there is a definite season for cold-mix construction.

The other potential problem is rainfall. A small amount is not an issue for a material that already contains plenty of water, but if it rains enough to materially change the water content, say by more than 0.5% by mass (about 1% by volume), then that will begin to slow the development of curing. For example, in a 100 mm layer that means more than about 1 mm of rain could potentially cause problems. Specifications should be worded to ensure that such problems are avoided. In particular, if unexpected rainfall leads to a significantly raised water content, then application of overlying material should be prohibited until it has had the opportunity to start drying out.

14.6. Summary

There are so many complexities to cold-mix asphalt that it's hard to summarise concisely. However, the key points that the author believes practitioners need to concentrate on are (*a*) the best way to specify the material to ensure proper performance and (*b*) the correct realistic assumptions to make regarding long-term properties. The specification needs to be performance-related – that is, real tests on real materials from site – and appropriate – which probably means more than just stiffness or strength. As for appropriate long-term assumptions, it is impossible to be specific, but the general rule is that a significantly reduced modulus must be expected compared with that from accelerated-cured laboratory testing.

REFERENCES

AASHTO (American Association of State Highway and Transportation Officials) (1993) *Guide for Design of Pavement Structures*. AASHTO, Washington, DC, USA.

Asphalt Institute (1997) *Asphalt Cold Mix Manual. Manual Series No. 14 (MS-14)*, 3rd edn. Asphalt Institute, Lexington, KY, USA.

Jiménez HIL (2016) *Airfield Pavement Design with Cold Recycled Materials*. PhD thesis, University of Nottingham, Nottingham, UK.

Merrill D, Nunn ME and Carswell I (2004) *A Guide to the Use and Specification of Cold Recycled Materials for the Maintenance of Road Pavements*. Transport Research Laboratory, Crowthorne, UK Report, TRL611.

Sabita (Southern African Bitumen Association) (2020) *Technical Guideline: Bitumen Stabilised Materials*, 3rd edn. Sabita, Cape Town, South Africa, TG2.

Nick Thom
ISBN 978-1-83549-711-1
https://doi.org/10.1108/978-1-83549-710-420243017

Chapter 15
Maintenance of cold-mix asphalt pavements

If the surface course is a hot or warm mix, then exactly the same maintenance options as detailed in Chapter 12 also apply to a cold-mix base pavement. This chapter will therefore concentrate on cases where either a cold-mix surface course or a surface dressing is used.

15.1. Deterioration mechanism

The defining feature of most cold-mix asphalts is that they have less intrinsic strength than a hot mix. Whereas a very high Marshall stability asphalt concrete is suited to aircraft tyre pressures of around 1.5 MPa, and a hot/warm-mix stone mastic asphalt is suited to heavy goods vehicles with tyre pressures of around 0.7 MPa, most cold-mix surfaces are really only suited to vehicles with tyre pressures of less than 0.5 MPa (e.g. cars and light goods vehicles). The problem is that, even on minor roads, there will be the occasional heavier vehicle – so damage is expected.

Assuming that the structural strength of the pavement is adequate, the damage expected is to the surface in the form of *ravelling*. Stones will be dislodged more easily from a cold-mix surface than from a hot/warm mix, although the presence of a surface dressing, which is in reality also a type of cold-mix layer, provides a much tougher obstacle to stone loss due its high binder content.

15.2. Treatment options

Ravelling will inexorably lead to potholes unless treated in a timely manner. The following subsections spell out possible courses of action.

15.2.1 Preservative treatments

Chapter 12 presented two spray-on treatments to extend the life of a hot/warm-mix asphalt, namely *preservative* and *rejuvenator* treatments, having almost opposite effects. Rejuvenators soften an aged bitumen and no reason is seen to apply this to a cold mix, in which the bitumen will not suffer the short-term ageing effects of heating, mixing, transporting and paving while still hot. Preservatives, on the other hand, are a logical choice. The author has to admit having no direct experience of this but, since toughness is what cold-mix asphalts lack and a preservative treatment forms an ultra-tough skin on the surface, it would seem to be a logical choice for pavements with a cold-mix surface, including those with a worn-out surface dressing (see Section 12.3.3 for further remarks.)

15.2.2 Surface dressing

Whether originally surface dressed or not, a further surface dressing application will also help to create the tough skin needed to hold stones in place as well as sealing the surface and so protecting the underlying materials from deterioration. It also restores skid resistance, something that a preservative cannot do (see Section 7.6.1 for a full description of surface dressing).

15.2.3 Local and edge repairs

In the end, despite the life extension that can be achieved through preservative treatments or surface dressing, it is expected that the surface will begin to ravel, break up around the edges and finally form increasingly deep potholes, as is also the case for hot/warm-mix asphalt. Indeed, the general advice in Section 12.4 also applies to cold-mix surfaces. The difference is that, while Section 12.4.2 advised that cold-mix repairs are likely to last less long than hot or warm, this becomes more questionable when the pavement itself comprises cold-mix materials. Joints between very different materials with different stiffness moduli often form starting points for cracks, cracks that can then become rapidly wider and deeper, so it is probably best to use cold-mix repairs in a cold-mix pavement, albeit with a hydraulic binder content to give relatively rapid curing.

15.3. Summary

A cold-mix asphalt is still an asphalt, which means that similar treatments apply to those used for hot/warm-mix materials. However, the importance of timely toughening and sealing the surface layer must be emphasised particularly strongly in the case of a cold-mix surfaced pavement, especially if it is to take significant heavy goods vehicles.

Part 5

Hydraulically bound material and composite pavements

Nick Thom
ISBN 978-1-83549-711-1
https://doi.org/10.1108/978-1-83549-710-420243018

Chapter 16
Hydraulically bound material

Hydraulic binders such as Portland cement rely on a chemical reaction with water, resulting in the generation of a hard, strongly adhesive mortar. A hydraulically bound material (HBM) is therefore entirely different from an asphalt, since it is a genuine solid. Asphalt may be strong and rut-resistant, but in the end it can deform since the binder (bitumen) remains a very viscous liquid at in-service temperatures. As a solid, an HBM cannot deform plastically unless it first cracks, which means that it is *strength* that is the key property, determining whether cracking occurs or not.

16.1. Strength
16.1.1 Test mode
The question here is: what strength? Cracking is caused by *tensile stress*, so a tensile strength would be a logical measure. On the other hand, tensile stress in a pavement layer occurs principally due to bending, in which case a so-called *flexural strength* (AASHTO, 2018; ASTM, 2022; BSI, 2019a) might be more appropriate. But the problem with both measures is that the test is a little complicated to carry out, especially the tensile test. So what about the easy option – *compressive strength* (AASHTO, 2022; ASTM, 2023; BSI, 2019b)? The test is straightforward to set up and requires no complex kit; the problem is that HBMs don't really fail in compression.

A full comparison of the different methods will be presented in Chapter 19 in relation to pavement-quality concrete, where it will be suggested that the flexural strength should be used in specification, while the compressive strength is suited to quality control. But HBM is a lower grade material and in many cases is designed with the understanding that it will develop a certain amount of cracking, so it is certainly less critical to use a premium test method, and virtually all pavement specifications understandably go for compressive strength.

So what should a designer assume? Well, the first advice is to use a large safety factor in any design where cracking needs to be prevented. It is, however, possible to make some generalisation when translating from compressive to tensile or flexural strength, depending on the aggregate type used.

- Limestone aggregate results in a low compressive/tensile strength ratio (e.g. 7).
- Granite aggregate results in moderate compressive/tensile strength ratio (e.g. 10).
- Gravel aggregate results in high compressive/tensile strength ratio (e.g. 12).

16.1.2 Strength gain with time
A useful property of all HBMs is that they become stronger with time. The initial chemical reactions may take place over a period of a few hours, but other slower reactions are set in motion that take much longer – many years in some cases. The ratio of rapid reaction to long-term strength gain depends on the chemical details of the binder, and there are many different hydraulic binders or combinations of binders in common use. Although ordinary Portland cement (OPC) is still the standard all over the world, industrial by-products such as fly ash and ground granulated

blast-furnace slag are sought after as key ingredients in slower-setting materials, usually activated by hydrated lime. While OPC comes to near full strength within a month, ash- or slag-bound mixtures may take 6 months or a year to achieve their design strengths.

For practical reasons, it is common to design OPC mixes based on a 28-day strength, as it is usual to have at least 28 days between pavement construction and opening to traffic, and it is administratively convenient to work in whole numbers of weeks. However, for quality-control purposes it is common to test at 7 days or sometimes 3 days in order to give early warning of potential problems. The appropriate ratio of 7-day to 28-day strength depends on the binder, but 0.7 is commonly assumed for OPC concrete.

In the case of slow-setting materials, the design properties are typically those at 360 or 365 days (e.g. Nunn, 2004), in which case 'early' warning of problems would commonly come from the ratio of 90-day to 365-day strength; it is impractical to specify tests more than 90 days after construction. As with cold-mix asphalt, use of such materials will result in less certainty of long-term performance during the construction phase – which should logically be compensated for by use of a greater safety margin.

16.2. Fatigue

For detail on fatigue of HBMs, the reader is referred forward to Section 19.2 on the subject of pavement-quality concrete.

The point is that, as with asphalt, cracking in HBMs develops over a period of time under repeated traffic (or thermally induced) loads, and this means that stresses within the material, specifically tensile stresses, need to be kept well below the expected tensile strength – if the material is not to crack in service. And since the tensile strength already carries plenty of uncertainty, a significant safety margin is advised. Furthermore, the way fatigue works in HBMs is different from that in asphalt. An asphalt fatigue characteristic, as introduced in Section 9.4.4, can be expressed as a power law function of tensile strain, with a power of around −4 (or −0.25 if the equation is expressed the other way round), whereas the equivalent power for an HBM, in terms of stress rather than strain, is typically around −12 in the applicable stress range. This means that HBMs are very sensitive to overloading, further increasing the required safety margin in design.

The upshot of all this is that the calculated tensile stress in an HBM should be kept at no more than about 30% of the supposed tensile strength – if cracking is to be prevented.

16.3. Durability

One of the advantages of a well-designed HBM in a sensibly designed and constructed pavement is that it should have a long life. Although the strength of such materials should continue to increase throughout their lives, as chemical bonds continue to develop, this will only be true if appropriate precautions are taken.

The key to HBM durability is *low permeability* – which means small non-interconnecting voids. The principal agent of environmental damage to the structure of HBM is water; so, if water is unable to penetrate, then such damage will not occur. The permeability of HBMs can be measured directly using a high-pressure permeameter, although this is rarely carried out in practice. Of more likely importance, however, is the permeability local to an interface between pavement layers, in particular at the bottom of a layer where compaction may not have been as effective as higher up.

16.3.1 Water damage

Pavements are rarely in a fully dry condition. Even dense asphalt materials are not 100% impermeable and there is always a tendency for voids within all types of pavement material to become partially filled with water. At the base of an imperfectly compacted layer, water is therefore likely to be able to penetrate relatively easily. The danger arises when voids become filled with water and there is no clear escape path. Figure 16.1 illustrates the potential problem conceptually.

Testing for susceptibility to water damage is not a usual part of HBM design because it is normally assumed that water will never be allowed to accumulate within the material to the extent that damage is caused. However, where it is carried out, the approach is similar to that for asphalt (i.e. to measure a so-called retained strength ratio).

$$\text{Retained strength ratio} = \frac{\text{Strength after soaking in water}}{\text{Unsoaked strength}} \tag{16.1}$$

Various soaking procedures with varying temperatures and durations have been tried and specified, a 7-day soaking period being a typical compromise between realism and practical duration. It is usual to specify that the retained strength should be no less than 80% of the unsoaked strength. However, it is important not to be lulled into a false sense of security, as the test specimen is likely still to be undergoing 'curing' (i.e. it would naturally gain in strength during the soaking period). Thus, when evaluating a result from such a test, it should be appreciated that a zero change in strength may, in reality, imply that sufficient 'damage' has occurred to approximately offset the strength gain due to curing. The use of retained strength was investigated by WRAP (2007), following the requirements by National Highways (2020), on different HBMs. The test was seen as useful in that it induced both volumetric expansion and also certain chemical effects, but the authors warned that soaking times of up to 90 days are ideally required in order that all potential problems are identified.

Figure 16.1 A mechanism of water damage in a hydraulically bound layer

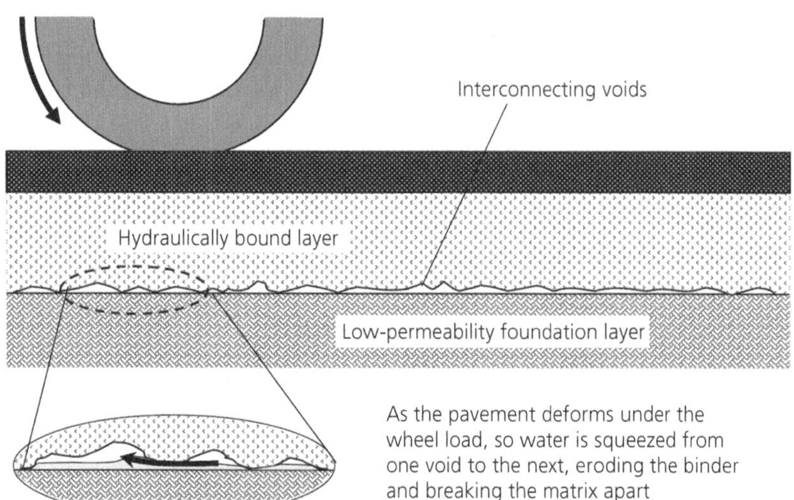

It should also be noted that a high retained strength does not directly guard against the sort of damage under traffic action illustrated in Figure 16.1 because the test does not include any simulation of dynamic loading. To protect a material against such damage, there is no substitute for proper pavement design (see Chapter 17).

16.3.2 Frost damage

Frost damage is a potentially serious problem in many climates. Water expands when it turns into ice, which means that, if the water is inside a nearly closed void at the time when it freezes, then the expansive pressure is likely to fracture the surrounding material. In a hydraulically bound base or sub-base, the material can fragment to the extent that it is reduced to nothing more than an unbound layer. The key of course is to make sure that materials within the frost zone rarely reach high levels of saturation – which comes back to the all-important topic of pavement drainage.

16.4. Thermal properties

A problem with all HBMs is that they are rigid solids, which means they are susceptible to thermal expansion and contraction. In many forms of construction this can be ignored – but not in most pavements (other than indoor pavements such as those for warehouse floors). The problem is that when a 1 km of continuous road cools by 10°C it wants to contract by about 10 cm and, as the ends will be firmly restrained by the continuous nature of road construction, the result will be a tensile failure. In a hydraulically bound base this means that discrete cracks will exist, potentially causing problems to overlying layers as the cracks open and close on a daily basis.

Thermally induced stresses and strains depend on the property known as the *coefficient of thermal expansion* (α). The value of α depends on the ingredients comprising the material, of which the large aggregate fraction is usually dominant. Thus, α depends on the rock (or other aggregate) used. Typical values are listed in Table 16.1.

16.5. Stiffness

HBMs can reasonably be considered as *linear elastic*. They therefore have a stiffness modulus, which can be used in pavement analysis.

16.5.1 Stiffness determination

It is not usual to measure the stiffness of an HBM directly. There is a European standard for obtaining stiffness from uniaxial (compression/tension) tests on hydraulically bound base layers (BSI, 2017), but this is not commonly carried out. It is more usual simply to make an assumption based on the mixture strength.

Numerous claimed relationships between strength and stiffness have been published. However, it should always be appreciated that such relationships can only be approximate for the excellent

Table 16.1 Typical values of coefficient of thermal expansion

Aggregate	Coefficient of thermal expansion, α
River gravel	13×10^{-6}/°C
Igneous rock	10×10^{-6}/°C
Limestone	7×10^{-6}/°C

Table 16.2 Stiffness predictions based on material compressive strength (CS)

Predictive equation (CEB, 1993): E (MPa) $= 21\,500C(\sigma_{c,failure}/10^{-7})^{0.333}$	Predicted elastic modulus, E: MPa			
	CS 5 MPa	CS 10 MPa	CS 20 MPa	CS 40 MPa
$C = 1.2$ (basalt, dense limestone)	20 500	25 800	32 500	40 100
$C = 1.0$ (quartzite)	17 100	21 500	27 100	34 100
$C = 0.9$ (most limestones)	15 400	19 400	24 400	30 700
$C = 0.7$ (sandstone)	11 900	15 100	19 000	23 900

reason that strength and stiffness are not directly related. Strength depends on the bond strength between particles, which is a function of the binder quality and quantity; stiffness also depends on aggregate gradation and the stiffness of individual particles. Of course, it is usual for strong mixes to make use of high-strength aggregate and such aggregate will usually also have high stiffness – but the relationship can never be other than an indirect one. Table 16.2 presents predictions based on one of the more commonly quoted relationships (CEB, 1993).

16.5.2 Influences on material stiffness

Although it is quite acceptable to consider an HBM as being homogeneous for most purposes, this is not really the case and it does not promote real understanding of the way such a mixture works. The reality is that an HBM consists of an aggregate, the particles of which are bonded together. In a sense, it is an unbound material without the possibility of slip between particles (so long as the bonds do not break). If interparticle slip is denied, the only deformation mechanism that remains is compression at particle contacts – and this is, therefore, the fundamental mode of strain within an HBM. Calculation Sheet 16.1 presents a methodology for modelling mixture stiffness on this basis.

Calculation Sheet 16.1. Stiffness modelling approach

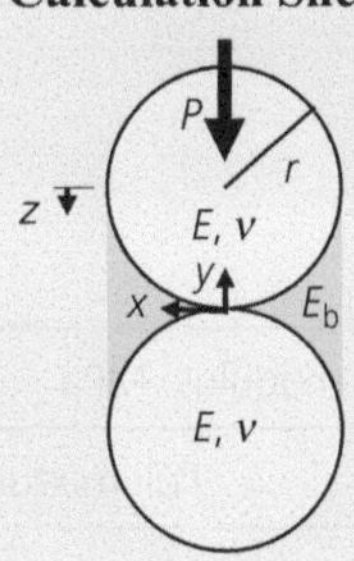

Granular matrix

Bulk stress $= \sigma$

Coordination number (average number of contacts per particle) $= n$

$\therefore$ Total average force/particle $= \sigma \times 4\pi r^2$ (assume spherical)

$\rightarrow$ Force per particle contact $P = 4\pi\sigma r^2/n$

Hertz contact equation: $z = [9P^2(1-\nu^2)^2/(2rE^2)]^{0.333}$

$$\therefore \text{Matrix bulk modulus } K_{mat} \approx \sigma/(z/2r) = \sigma/\{[144\pi^2\sigma^2 r^4(1-\nu^2)^2/(16r^4E^2n^2)]^{0.333}\}$$
$$= [(\sigma E^2 n^2)/9\pi^2(1-\nu^2)^2]^{0.333}$$

Elastic modulus $E_{mat} = 3K_{mat}(1-2\nu_{mat})$ where $\nu_{mat} = \nu$ of aggregate matrix ≈ 0.3

Including binder

(*a*) Work out approximate radius of contact area, $r_c = \sqrt{2r \times \frac{1}{2}z}$ (intersecting chords)

(*b*) With binder of modulus E_b, integrate binder compression force P_b between r_c and r, replacing the circular particle shape by a parabola of equation $y = x^2/2r$

$$\rightarrow P_b = \int (z/2y)E_b 2\pi x \, dx = \int (zr/x)2\pi E_b \, dx = 2\pi zrE_b[\ln(r) - \ln(r_c)]$$

(*c*) Calculate bound material stiffness as $E_{mat}(P + P_b)/P$

Notice that the matrix stiffness E_{mat} in Calculation Sheet 16.1 is independent of particle size. It does, however, depend on the coordination number n (the average number of contacts that each particle makes with its neighbours), which will vary according to the particle size distribution and, of course, compaction. In a single-sized aggregate, n might be around 7–9; in a graded aggregate it will be significantly more, typically around 12.

The key additional parameter affecting the stiffness of a bound material is the stiffness modulus of the mortar – another highly variable quantity. In the case of dense continuous cement paste, it is suggested that the value should be around 1000–2000 MPa, but the figures for other binders or if significant air voids are present may be considerably different. Table 16.3 gives illustrative predictions, all for an aggregate particle modulus of 200 GPa (typical of hard rock) and Poisson's ratio (of particles, not of mixture) of 0.45.

In no way should this predictive approach be taken as accurate, particularly in special cases such as porous concrete, but it is presented in order to make the reader aware of the factors that determine the stiffness of a particular HBM. It should also be appreciated that these values represent the maximum stiffness possible; the value will reduce as soon as interparticle bonds begin to break.

Table 16.3 Stiffness predictions based on aggregate gradation

Coordination number, *n*	Stress: kPa	Modulus of mortar: MPa	Predicted modulus: MPa	
			Matrix alone	Full mixture
8 (single-sized)	200	1000	2500	19000
	200	2000	2500	35000
	600	1000	3600	18000
12 (graded)	200	1000	3300	27000
	200	2000	3300	52000
	600	1000	4700	27000

Table 16.4 Typical stiffnesses of HBMs

Mixture type	Typical stiffness: MPa
Strong cement-bound base	15 000–25 000
Weak cement-bound base	5000–15 000
Slag- (etc.) bound base	3000–10 000
Hydraulically bound sub-base	2000–5000
Stabilised soil	100–300

16.5.3 Typical material stiffness values

If the foregoing section represents a reasonable picture of what is actually occurring within an HBM, then the clear conclusion is that stiffness is sensitive to mixture details, aggregate type and so on. Nevertheless, it is possible to draw out typical values for different types of material, as each type tends to make use of a restricted range of ingredients. Table 16.4 represents the author's accumulated experience, based largely on field testing. For the weaker materials, the values quoted will certainly be affected by the presence of microcracking.

A key point to appreciate is that a material of lower stiffness than those listed in Table 16.4 is not necessarily inferior. Indeed, high stiffness in an HBM attracts high stress – inviting failure unless the strength is correspondingly high, which is not always the case.

16.5.4 Effective stiffness of a discontinuous layer

A knowledge of material modulus can be useful, but in many cases an HBM is designed to end up in a cracked state – which means that its apparent stiffness in situ will inevitably be less than that of the intact material. Experience suggests that an HBM layer with an initial stiffness of 10 000–20 000 MPa can easily end up with an apparent in situ stiffness of no more than 5000 MPa. And it is this apparent stiffness that affects the way the layer spreads load to underlying materials and supports overlying layers.

Figure 16.2 illustrates a rational approach to estimation of that stiffness. Basically, the approach is to acknowledge that there are three elements to the 'compliance' (the inverse of stiffness) of a cracked layer: (*a*) bending of the intact material, (*b*) angular rotation at crack locations and (*c*) relative vertical movement across cracks. In each case, a relationship between bending moment and curvature can be estimated.

Full calculation details have been omitted from Figure 16.2, partly due to lack of space, but also because such details are not in any way fundamental; it would be possible to derive alternative, slightly different, but no less valid approximations for E_{eff}.

However, the formula in Figure 16.2 at least allows an estimate to be made. Take the common case of a cement-bound base layer of material modulus $E = 15\,000$ MPa. Such a layer commonly breaks naturally into blocks of 5 m length or less, or, nowadays, is deliberately pre-cracked into 3 m-long blocks (see Chapter 17). This pre-cracking minimises crack width, maximises interlock and therefore also maximises 'slip stiffness' (*g*). Admittedly, slip stiffness is not a commonly used quantity, but experience (e.g. from falling weight deflectometer testing – see Section 25.6.3) suggests that it may have a value of around 5000 MN/m^3 for a new pavement layer, reducing with time and traffic.

Figure 16.2 Bending of a cracked layer (developed from Thom and Cheung, 1999)

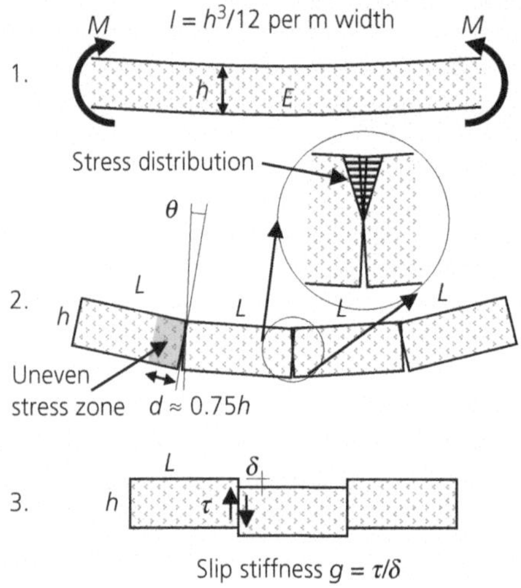

1. Curvature = $M/EI = 12M/Eh^3$

2. Strain at top = $(\theta h/4)/d \approx \theta/3$
   Stress at top = $\theta E/3$ (→ 0 at mid depth)
   Moment = $(3/256)\,\sigma h^2 = \theta h^2 E/256$
   Curvature = $\theta/L = 256M/(h^2EL)$

3. Moment = $\tau hL/2 = g\delta hL/2$
   Radius of curvature = $(L^2/4)/2\delta = L^2/8\delta$
   Curvature = $8\delta/L^2 = 16M/ghL^3$

Combine:
Curvature = $M[12/Eh^3 + 256/(h^2EL) + 16/ghL^3]$
Curvature also = $M/E_{eff}I = M[12/E_{eff}h^3]$

$\therefore E_{eff} = 1/[1/E + 64h/3EL + 4h^2/(3gL^3)]$

If the layer is 150 mm thick, then Figure 16.2 suggests an effective layer modulus (E_{eff}), when new, of 7250 MPa. In fact, the third element of the equation, namely that due to vertical slip at cracks, is almost negligible with 3 m crack spacing, so the effective stiffness is only likely to decrease significantly if additional cracks appear. Many cement-bound base design methods assume around 5000 MPa as a long-term effective stiffness (e.g. BAA, 1993).

Use of this formula is believed to be appropriate for most HBM layers, but *not* in the case of stabilised soils. In general, it is advised that a stabilised soil should be assigned a long-term stiffness no more than about three times that of the non-stabilised material beneath (following the methodology given by Powell *et al.* (1984)). In effect, it should be assumed to act as an unbound material, whatever laboratory data may suggest.

16.6. Mixture design

A fuller explanation of the mixture design of HBM is given in Section 19.6 in the context of pavement-quality concrete, including the important concept of the water/cement ratio, which is required to lie between certain limits in order to get full hydration of the cementing products. Basically, if high strength is needed and therefore high binder content, then a high water content is also required.

But, of course, the water content also affects compactibility, and compaction is critical to achieving the design strength. HBMs are almost always compacted by roller like an unbound material and, as for an unbound material (Section 7.1.3), this means there is an optimum water content (OWC) to achieve maximum density. And that OWC places an approximate limit on the amount of cement or other hydraulic binder that can be activated, and therefore on the strength that can be achieved. It's a bit complicated, and the OWC certainly depends to some extent on the aggregate gradation, but the end result is that roller compaction is ideal for HBMs with compressive strengths in the range 5–20 MPa.

Roller compaction is also an attractive option for HBMs because it removes the need for a specialist concrete paving machine, thus saving on costs and, probably, time. Placement can be either by means of a standard asphalt-type paver or, where level tolerances are less critical, by tipping and grading as for an unbound pavement layer.

16.6.1 Typical HBM mixtures

Because of the enormous range of possible hydraulic binders and combinations of binders, this section is confined to those using Portland cement. Table 16.5 gives typical proportions for different mixture types. The proportions given should, of course, be seen as indicative only, as they are highly dependent on the aggregate and gradation used. The descriptions given are generic terms and may not relate to the names used in particular specifications.

16.6.2 Slow-curing materials

Mixture designs for Portland cement concrete are not too difficult to validate as a good idea of long-term properties can be obtained at 7 days. However, for materials that take many months to reach their design strength it is unrealistic to wait the many weeks necessary before carrying out representative tests. To overcome this difficulty it is common practice to use an *accelerated curing* technique in a similar way to cold-mix asphalt. Strength gain is highly temperature dependent and a typical accelerated curing regimen therefore consists of placing specimens in an oven at 40°C (for 28 days in current UK standards; see National Highways (2020)). Although much criticised by purists, this practice is the only sensible way of obtaining early data on which to base mixture design. It does, of course, make the assumption that the actual material in situ will be sufficiently protected from the weather (notably rain and frost) and traffic (construction traffic in particular can cause irreparable damage) to achieve a similar level of performance to that of accelerated-cured specimens.

It should be noted that, while accelerated curing is undeniably useful in mixture design, it is not advised as a specification requirement. That should preferably be based on materials cured under site conditions and tested at an appropriate age, typically 90 days.

16.6.3 In situ stabilised materials

In situ mixing presents its own unique problems.

- The 'aggregate' is usually a natural soil, which inevitably varies in properties from place to place.
- The difference between laboratory and site is likely to be much more pronounced for in situ stabilised material than for plant-mixed material, as the mixing process is clearly less well controlled on site.

Table 16.5 Typical mixture proportions

	Design strength at 28 days: MPa		Cement content: kg/m³	Water content: l/m³
	Flexural[a]	Compressive		
Strong hydraulically bound mixture	(3)	20	320	150
Medium hydraulically bound mixture	—	10	220	150
Weak hydraulically bound mixture	—	5	150	150

[a]Flexural strength is particularly appropriate to strong layers that fail primarily in bending and would not normally be used in material specification

- Most in situ stabilised materials are slow-curing, compounding the uncertainty.
- Many soils include chemicals that are potentially detrimental to the long-term health of some binders.

However, despite these very real difficulties, in situ stabilisation is often the obvious and most economical way to improve a subgrade to a state suitable for pavement construction – so it has to be taken seriously. That being the case, it is important to understand the processes involved. These are modification and binding action (i.e. stabilisation proper).

Modification simply means changing a soil's plasticity characteristics. It is achieved by the action of lime (or other products that include free lime) and the result is a modification to the particulate structure of the material. Its effect is that the plastic limit value (see Section 6.1.7) increases and, if the water content remains unchanged, the behaviour at that water content will become more 'plastic' and less 'liquid'. In engineering terms, the suction will increase, and with it the strength, stiffness and resistance to deformation. As the hydration of the lime also results in a (probably temporary) reduction in water content, the short-term effect is made even more impressive. In the field, modification is optimised by allowing the mixed material to lie in a loose state for about 24 h, a phase known as 'mellowing'. Once the material properties have been modified, compaction can take place or, in some cases, stabilisation by the addition of further binder. For example, it is not uncommon to apply 2–3% lime to modify a soil, followed a day later by Portland cement as a stabiliser.

Binding action (i.e. *stabilisation*) occurs following modification if there is surplus binder, which means if more than about 3% lime has been added or if a second binder has been used following modification. Here, the chemical reaction that takes place produces similar cementitious products to those produced in conventional HBMs, and the result is that soil particles are actually bonded together. The effect is chemically irreversible – although this does not of course prevent bond breakage under physical stress.

So, in situ stabilisation is conceptually not difficult to grasp. Unfortunately, confident design is much less easy, which means that a reasonably generous safety margin should be maintained. The following steps should ensure a satisfactory product.

- Check the chemical composition of the soil, particularly for sulfates and sulfides. These cause expansion and consequent disruption to the structure of the material (see National Highways, 2020). All stabilisation specifications include limits on sulfur in one form or another.
- Mix and compact test specimens in the laboratory (usually cylinders or cubes).
- Apply an accelerated curing regime (e.g. 28 days at 40°C), except where Portland cement is the main binder.
- Carry out mechanical tests (compressive strength for stronger materials; California bearing ratio (CBR) for weaker materials).
- Soak specimens to assess durability (many different procedures exist). The same soaking regimen may also be used to evaluate whether excessive volumetric expansion is likely (see below).
- Retest to evaluate potential strength loss under water action.

With the exception of the question of volumetric expansion, the procedure is relatively straightforward. Specifications vary regarding mixing procedure, compaction technique, specimen shape and size, soaking procedure and testing regimen, but conceptually they are all aimed at proving the physical properties of the material under controlled conditions, including an evaluation of the effect of water. However, the combination of unknowns present on site (soil properties, water content, mixing efficiency, curing conditions, stressing during early life) means that the properties measured in the laboratory certainly *cannot* be assumed on site. A strength (or CBR) of 50–75% of that measured in the laboratory is advised as realistic unless site testing confirms otherwise.

This leaves the tricky problem of *volumetric expansion*. It certainly happens – there have been several high-profile failures during construction due to this mechanism. Behaviour appears to be sensitive to the exact combination of chemistry, water content and temperature, not to mention the type of test regimen applied. The well-recognised development of ettringite crystals, for instance, occurs when the temperature drops to 15°C or below, immediately invalidating those many soaking regimens where 20°C is specified. Expansion is also highly sensitive to water availability. In tests where the specimen is encased in a steel mould and water is only given access through a porous plate at the base, there may be insufficient supply to generate significant expansion in the laboratory, whereas there may be plenty on site. The steel walls of a mould also have the effect of resisting expansion by friction. A further important question in laboratory evaluation is just how soon after specimen preparation the water should be applied. In short, this is an area where research is still very much in progress, and engineers should certainly be ready to question test data rather than automatically accept results. As in many fields, the fact that a test is carried out to some national or international standard unfortunately does not prove material suitability. There are two current EU norms that can be applied to soaking of hydraulically stabilised soils. In BS EN 13286:47 (BSI, 2021), a specimen enclosed in a CBR mould is wetted from below for 4 days, commencing 3 days after manufacture. In BS EN 13286:49 (BSI, 2004), the unprotected specimen is submerged in water after a period of just 1.5 to 2 times the workability period, and the soaking period is 7 days (at 40°C). Unsurprisingly, the effect can be quite different in the two cases.

16.7. Summary

In a sense, HBMs are not as complex as unbound materials or asphalts. However, what they lack in complexity in material characterisation they more than make up for in terms of the sheer range of different materials possible, from weak soil–cement to strong base layers, which means that it is not easy to generalise. Nevertheless, the following are some of the key points pertaining.

- There is always a strength gain process, which may be one of weeks in the case of OPC-bound materials, or months in the case of slow-setting HBMs. The environmental and loading conditions during this 'curing' process are therefore critical to the eventual performance of the material.
- Appropriate water content is essential to achieving design strength, and is also constrained by the needs of material compaction.
- The general mode of failure for all HBMs is tensile, occurring when the tensile stress (or strain) limit is exceeded. However, failure can occur under multi-cyclic loading at stresses much lower than the tensile limit, due to the phenomenon known as fatigue.
- Slow-setting mixtures, particularly those mixed in situ, present additional problems in ensuring a product of suitable quality.

REFERENCES

AASHTO (American Association of State Highway and Transportation Officials) (2018) T97: Standard method of test for flexural strength of concrete (using simple beam with third point loading). AASHTO, Washington, DC, USA.

AASHTO (2022) T22M/T22: Standard method of test for compressive strength of cylindrical concrete specimens. AASHTO, Washington, DC, USA.

ASTM (2022) C78/C78M-22: Standard test method for flexural strength of concrete (using simple beam with third point loading). ASTM, West Conshohocken, PA, USA.

ASTM (2023) C39/C39M: Standard test method for compressive strength of cylindrical concrete specimens. ASTM, West Conshohocken, PA, USA.

BAA (British Airports Authority) (1993) *Pavement Design Guide for Heavy Aircraft Loading*. BAA, Group Technical Services Aircraft Pavements, London, UK.

BSI (2004) BS EN 13286-49:2004: Unbound and hydraulically bound mixtures. Accelerated swelling test for soil treated by lime and/or hydraulic binder. BSI, London, UK.

BSI (2017) BS EN 13286-43:2003: Unbound and hydraulically bound mixtures. Test method for the determination of the modulus of elasticity of hydraulically bound mixtures. BSI, London, UK.

BSI (2019a) BS EN 12390-5:2019: Testing hardened concrete. Flexural strength of test specimens. BSI, London, UK.

BSI (2019b) BS EN 12390-3:2019: Testing hardened concrete. Compressive strength of test specimens. BSI, London, UK.

BSI (2021) BS EN 13286:47:2021: Unbound and hydraulically bound mixtures. Test method for the determination of California bearing ratio, immediate bearing index and linear swelling. BSI, London, UK.

CEB (Comité Euro-International du Béton) (1993) *CEB-FIP Model Code 1990: Design Code*. Thomas Telford, London, UK.

National Highways (2020) *Manual of Contract Documents for Highway Works. Vol. 1: Specification for Highway Works*. Stationery Office, London, UK.

Nunn ME (2004) *Development of a More Versatile Approach to Flexible and Flexible Composite Pavement Design*. Transport Research Laboratory, Crowthorne, UK, Report TRL615.

Powell WD, Potter JF, Mayhew HC and Nunn ME (1984) *The Structural Design of Bituminous Roads*. Transport and Road Research Laboratory, Crowthorne, UK, Report LR1132.

Thom NH and Cheung LW (1999) Relating in situ properties of cement bound bases to their performance. *Transportation Research Record* **1673**: 3–8.

WRAP (Waste and Resources Action Programme) (2007) *Development of Durability Tests for HBMs*. WRAP, Banbury, UK, Report A66-79-005.

Nick Thom
ISBN 978-1-83549-711-1
https://doi.org/10.1108/978-1-83549-710-420243019

Chapter 17
Composite pavement – structural design

Composite pavements comprise a hydraulically bound base and an asphalt surface. They therefore have the potential to deteriorate in various different ways, and what happens in the base layer will obviously affect what happens in the overlying asphalt. This will therefore be a chapter of many parts.

17.1. Construction practicalities

17.1.1 Particle shape and size distribution

Hydraulically bound materials (HBMs) do not, in the long term, rely significantly on particle–particle interaction for their engineering properties, and so the shape and size distribution are less critical than is the case for unbound material. For example, it is common to blend large aggregates with sand in an HBM. The resulting rather discontinuous gradation would not be suitable for an unbound layer but produces excellent concrete. Similarly, rounded gravel, although highly undesirable in unbound materials, is perfectly suitable in most HBMs. The principal exception is the case of slow-setting materials, which rely on particle–particle contacts for stability during the early days and weeks of life, particularly during the construction phase of a project.

In general, it is undesirable for there to be a high voids content, as water can attack the hardened material if it can gain access through voids. There should, therefore, always be enough fine material in the mixture to ensure that large voids do not occur. Broadly speaking, the denser the mix the higher its resulting strength will be – and strength is the property that really matters. However, dense mixtures can be achieved using a wide range of particle sizes; the maximum size is typically 40 mm or less and should be no more than around 30% of the layer thickness to ensure mixture workability.

17.1.2 Batching and mixing

HBMs require careful proportioning and mixing. The usual technique is to mix in batches, with a certain measured volume of water and a certain weight of hydraulic binder such as Portland cement. Mixing takes about 30 s and, on completion of the mixing, the material is dropped into the back of a waiting wagon and transported to site. It is also common to use ready-mix suppliers, in which case the concrete is delivered in mixing trucks.

An alternative to batch mixing is to mix continuously in a drum mixer. A controlled feed of aggregate and binder is fed into one end of the drum, together with a measured flow of water; the rotation of the drum then mixes the materials together while baffles on the inside of the drum draw the mixture to the far end and out, whence it can be fed directly into a waiting wagon. The advantage is that higher productivity can be achieved. The technique is well suited to large production runs of relatively dry mixes.

17.1.3 Placement and compaction

Most HBMs are designed to be roller-compacted, and placement and compaction techniques are exactly the same as for unbound materials. The same restrictions on layer thickness apply because of the need to achieve density, an even more critical matter for HBMs than for unbound materials. This means that layers up to about 200 mm thick can comfortably be placed in a single lift; thicker layers need multiple lifts in which case the bond between lifts becomes critical. Cameron and Matthias (2001) found that an overlying layer should ideally be placed on a brushed surface on the same day as the layer below; any delay leads to an increasingly weak bond.

Wet-forming of hydraulically bound bases is uncommon these days but can be done using a slip-former, in which case the above thickness restriction no longer applies.

17.1.4 In situ stabilisation

All hydraulic binders lend themselves to in situ stabilisation work. The machinery needed is basically an adaptation of the agricultural rotavator – plant that is capable of mixing to the required depth, sometimes up to 350 mm. However, it is an inescapable fact that the process cannot be controlled as well as plant mixing. The distribution of binder and water is unlikely to be uniform and the result is a layer that, although it may be fit for purpose, is certainly more varied than an equivalent plant-mixed material. This must be borne in mind in the pavement design.

17.1.5 Curing

A particularly important issue for all HBMs is 'curing' – the process of strength gain. The chemical reactions that lead to cementing action take time, generate heat and cause shrinkage of the cementitious mortar, and these facts pose the danger of *shrinkage cracking*. The problems are as follows.

- The rate of set of the mortar is dependent on both temperature and water supply.
- Unless special measures are taken, the upper surface will dry quickest and potentially form shrinkage cracks, especially if exposed to sunshine or wind.

The most common curing aid used for HBMs is to spray the finished surface with bitumen emulsion. This keeps the water in and so avoids premature set at the surface.

It should be appreciated that curing problems really only become an issue when using relatively rapid-setting binders such as Portland cement. Slow-setting materials present minimal danger of differential set; also, if any internal damage takes place, there will still be plenty of unused binder available to 'heal' any minor cracks that occur. In base or sub-base mixes, it is therefore often acceptable or even desirable to use very slow-setting mixtures, for example, incorporating blast-furnace slag, fly ash or lime.

17.1.6 Pre-cracking

Careful curing may produce a layer with an approximately uniform material quality, but one of those qualities is that a hardened HBM layer is a solid, contrasting with asphalt in which one of the components (bitumen) is still a very viscous liquid. This means that there is no mechanism to dissipate the longitudinal stresses generated by the restraint (due to pavement continuity over considerable lengths) of daily thermal expansion and contraction. And it can readily be demonstrated that, in virtually all climates, these thermal stresses will be sufficient to cause cracking. Furthermore, if the HBM layer cracks, those cracks will almost certainly 'reflect' through the overlying asphalt surface. So what can be done?

The answer is to create joints, or to '*pre-crack*'. This is not difficult. It is simply necessary to create a plane of weakness in the layer and this is usually achieved by cutting a slot (typically to one-third depth) in the wet material and inserting either a thin polythene strip or a spray of bitumen emulsion (Figure 17.1). This is followed by compaction of the material, closing the slot but leaving the weakness. Subsequent thermal, and later traffic-induced, stresses will ensure that the weakness becomes a crack – but a crack in a predetermined location.

This technique was pioneered in France (Bonnot, 1988) during the 1980s and is now widely accepted; National Highways (2020a) includes a specification. But what spacing should be used?

The first point is that excellent interlock across the pre-crack is essential in order to transfer load from one side to the other and therefore protect the asphalt from unacceptable stress concentrations. For this, ideally, a crack opening of no more than about 0.5 mm is desirable, which demands a pre-crack spacing of no more than about 2.5 m. Against this must be set the fact that the effective stiffness of the layer as a whole will decrease at such a low crack spacing, as explained in Section 16.5.4, reducing its effectiveness as a protection to foundation layers; it will become slightly 'flexible' in nature, allowing local deformation of the foundation to take place. The accepted compromise nowadays is a spacing of 3 m, although this is sometimes reduced. The resulting effective modulus of the layer should be at least 40% of the value for intact material. Figure 17.2 shows an example prediction, computed for the case of a 150 mm-thick layer of strong HBM base, and assuming a relationship between crack width and load-transfer stiffness based on Figure 22.2.

Difficult though this aspect of design is, it is important. Load-transfer efficiency has a significant effect on the time before a crack reflects through an asphalt surface, and reflective cracking is a serious design issue. If unchecked, water will enter the crack, accelerating deterioration of the crack faces as well as potentially softening the pavement foundation. Thus, it is worth trying to ensure that material of the right strength is used and that cracks are kept narrow (generally by pre-cracking). The alternative to pre-cracking is post-cracking, carried out within a few days of

Figure 17.1 Pre-cracking of strong HBM bases

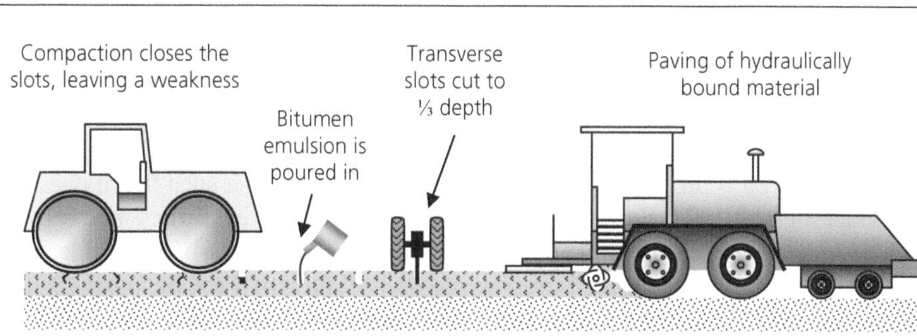

Figure 17.2 The effect of crack spacing in a strong HBM layer

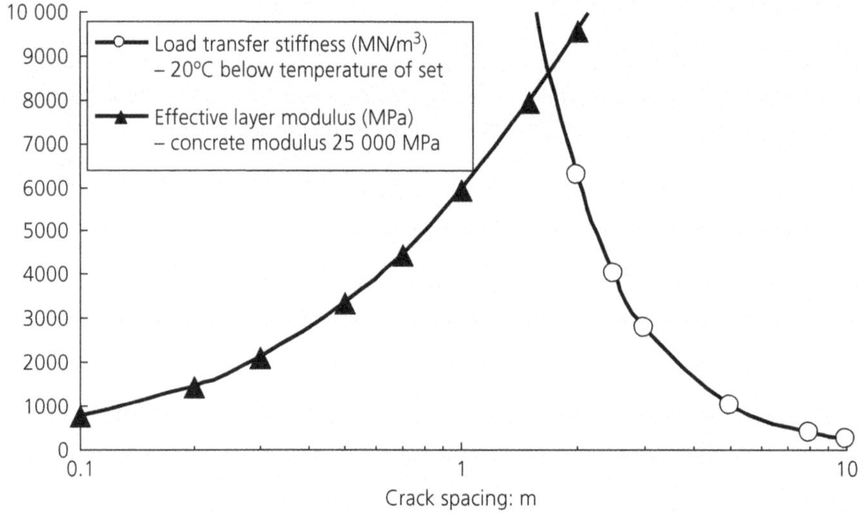

construction. This can be achieved by controlled rolling of the layer, although experience suggests such control is very difficult to achieve. Rolling is particularly applicable where close-spaced cracks are required, usually where the overlying asphalt will be thin.

17.2. Current design practice

All major design methods, including those mentioned in Section 10.2, provide composite pavement options. In the language of airfield design, the description 'flexible' also covers composite pavements; indeed HBM bases are the norm. The approach used for composite pavements is generally similar to that for flexible pavements, simply using different equations or charts, and always calibrated against experience. However, it is probably fair to say that, other than for heavy-duty designs (airfields and industrial pavements), composite pavements are a 'poor relation' in terms of the detail in which they are treated, especially bearing in mind the wide range of strengths available.

The remainder of this chapter will, however, take a more fundamental approach to understanding composite pavement performance and design.

17.3. Cracking of HBM base

17.3.1 Initial condition

As has become clear, it is difficult to design any HBM pavement layer such that it does not form transverse cracks. Every 1°C drop in temperature induces a tensile stress sufficient to resist a strain of around 10^{-5}, depending on the aggregate used. If, for example, the early-life stiffness modulus of the material was 10 GPa, the induced stress would be 100 kPa (10 GPa × 10^{-5}) per 1°C temperature drop. As the early-life tensile strength may not be more than 1 MPa, cracking would occur with a temperature drop of just 10°C. In short, it will crack; hence the need to pre-crack, as described in Section 17.1.6.

Very slow-setting materials are the exception. In such materials, early thermally-induced cracks will be partially repaired as hydraulic reactions proceed and the eventual result should be a layer with zero visible cracking but a significantly reduced stiffness. The inference is that there are numerous minute (and therefore invisible) cracks, all with excellent load transfer, giving a layer that is, in practice, flexible rather than rigid. The effective layer modulus will typically be much lower than that displayed by an intact laboratory specimen. UK National Highways design standard CD 225 (National Highways, 2020b) assigns 20% of the laboratory modulus for cement-bound materials (pre-cracked or naturally cracked) but only 10% for slower-setting HBMs.

17.3.2 Strong HBM layers – traffic loading

As noted above, HBM layers will usually include thermally-induced cracks in the transverse direction. These represent points of weakness but, as long as longitudinal wheel-path cracking is avoided, the layer still forms an excellent support to overlying asphalt materials and gives excellent protection to an underlying subgrade. That all changes as soon as a longitudinal wheel-path crack forms.

The usual way to guard against such longitudinal cracking is to carry out a multi-layer linear elastic analysis and to compute the tensile stress at the base of the layer under a design-load application. This can then be divided by the tensile strength expected from the material to give a stress ratio. The number of load applications to failure could be determined using a fatigue characteristic such as the so-called Packard line, to be introduced in Section 20.3.4. However, this depends on having confidence in a tensile strength estimate, which is not usually possible for an HBM base layer, so it is probably more sensible simply to follow the advice in Section 16.2 and keep the stress/strength ratio at no more than about 30%.

The computation should be carried out using a stiffness for the HBM layer representing the value for *fully intact material* (i.e. no reduction due to the presence of transverse cracks). Remember that the higher the layer stiffness, the greater the computed tensile stress will be – although the strain will be lower.

The computed stress must then be increased by a factor to take account of the increased layer bending that occurs near a transverse crack due to loss of continuity. This factor may be as little as 1.2 for cracks with excellent interlock but as high as 1.5 for wide cracks; unfortunately there are no absolute rules here so the designer should play it safe.

Most strong HBMs are designed not to crack longitudinally, and a composite pavement incorporating such a layer can be considered as having exhausted 'phase 1' of its life as soon as longitudinal cracks occur. However, slow-setting materials are different. They will already include minute longitudinal cracks as well as transverse ones but, because there will be numerous such cracks, lying at many different orientations, the effect on overlying asphalt or underlying subgrade will be much less significant. The material is already, in effect, acting as an outstandingly effective unbound layer and, because the long-term intrinsic strength is high, little further damage is likely. In theory, the load transfer across cracks should be highly efficient and should not deteriorate appreciably.

Figure 17.3 Design guidance – strong HBM layers

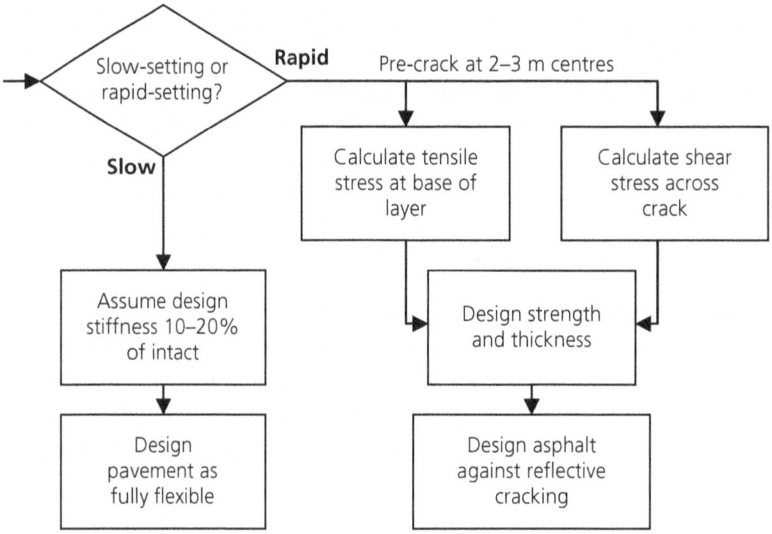

Because of the numerous pitfalls possible when designing with strong HBM layers, Figure 17.3 is included as a guide. Note that for rapid-setting mixtures, the figure also includes a check on the shear stress across transverse cracks, which can be computed, approximately, using multi-layer linear elastic analysis. Based on the laboratory-determined crack deterioration data presented fully in Section 22.1.3, this should be kept to less than 10% of the design flexural strength.

17.3.3 Weak HBM layers – traffic loading

This design case is a trivial one. The description 'weak' here means that the strength is not intended to be sufficient to prevent cracking under traffic loading. Actual compressive strengths will generally be well under 10 MPa on highways, but stronger materials could still fall under the classification 'weak' when used on airfields and industrial pavements where load levels are much higher.

As cracking is expected, no design against cracking is required. The only task is to assign a realistic long-term stiffness modulus to the layer so that overlying layers can be designed correctly. The stiffness computation for a discontinuous layer introduced in Section 16.5.4 can be used, but this depends on sensible inputs to the calculation.

- *Crack spacing*. It is reasonable to assume that a layer subject to stresses sufficient to break it will, eventually, reach a state where cracks are separated by about two to three times the layer thickness (see Calculation Sheet 17.1).
- *Slip stiffness at cracks*. A value of 5000 MN/m^3 represents a crack with good interlock, probably including angular crushed rock aggregate, whereas a value of 500 MN/m^3 represents a material with poor interlock, for example having a very small particle size or very low material strength.
- *Intrinsic material modulus*. Based on material strength (see Section 16.5.3).

Calculation Sheet 17.1. Layer fragmentation – see Section 20.3.3 for the method

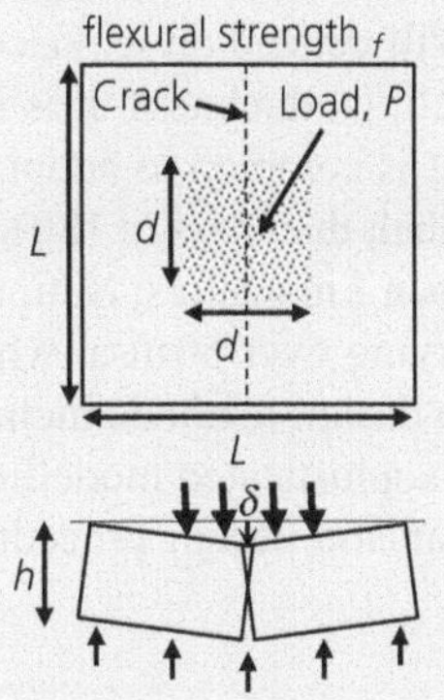

Work done by load $= P\delta(L/2 - d/4)/(L/2)$

(assuming uniform pressure)

Energy absorbed by foundation $= P\delta/2$

(assuming uniform stress)

Crack energy $= ML2\delta/(L/2)$

(M is the moment at failure)

Balance and rewrite: $P(1 - d/2L) = 4M + P/2$

$\rightarrow P(1 - d/L) = 8M$

$\rightarrow M = P(1 - d/L)/8$

But $M = \sigma h^2/6$ (see Section 19.1.2) $\rightarrow \sigma = 6M/h^2$

$\therefore \sigma = 3P(1 - d/L)/4h^2 \;\rightarrow\; L = d/(1 - 4\sigma h^2/3P)$

Examples

$d = 0.5$ m; $\sigma_f = 0.5$ MPa; $h = 0.15$ m; $P = 40$ kN $\rightarrow L = 0.8$ m (0.4–0.8 m pieces)

$d = 0.5$ m; $\sigma_f = 0.2$ MPa; $h = 0.2$ m; $P = 50$ kN $\rightarrow L = 0.64$ m (0.32–0.64 m pieces)

Applying the discontinuous layer computation from Section 16.5.4 to the two examples in Calculation Sheet 17.1, with a material modulus of 10 GPa and a slip stiffness of 1000 MN/m^3, gives effective layer moduli ranges of 730–1790 MPa and 330–1030 MPa, respectively. These are typical ranges for such materials.

17.4. Reflective cracking

The asphalt failure mode that is particular to composite construction is 'reflective cracking'. Reflective cracking can occur when any cracked or jointed pavement layer is overlaid with asphalt, and it is likely to occur when asphalt is placed over most HBM bases, as they almost always form transverse cracks, either naturally or as pre-cracks. Exceptions already noted are very slow-setting base materials, where microcracking is spread throughout the material, and very weak bases, where the layer rapidly breaks into relatively small pieces. In these cases, design of the asphalt layer can follow the principles for asphalt pavements given in Chapter 10 and the hydraulically bound base can be considered as a continuous layer with a single stiffness modulus value. However, in the majority of cases, where asphalt overlies a relatively strong HBM, reflective cracking can occur.

17.4.1 Thermally induced reflective cracking

Each intact length of HBM will undergo a daily thermal cycle, which will cause each crack to open and close. Therefore, the layer of asphalt above the crack has to accommodate this opening and closing movement, and clearly the degree to which it is able to do this will depend on its viscosity, its fracture resistance, the magnitude of crack opening and, most of all, its thickness. It is not a particularly simple situation to model accurately and, for this reason, it is common to adopt an experience-based design. In the UK, the current official view is that an asphalt thickness of 180 mm is sufficient as long as the underlying HBM layer is cracked at no more than 3 m centres; National Highways (2021) gives 180 mm of asphalt in all composite designs carrying over 50 msa, while 200 mm is a figure sometimes quoted for a more general situation. Some design methods include simple predictive equations. However, they will not usually be based on sophisticated modelling. Even the otherwise frighteningly sophisticated AASHTO (2007) mechanistic design procedure includes nothing more than the following rather crude formula.

$$\%\,\text{cracking} = 100/(1 + e^{a+bt}) \tag{17.1}$$

$$a = 3.5 + 0.75(h + f) \tag{17.2}$$

$$b = 0.688584 - 3.37302(h + f)0.915469 \tag{17.3}$$

where t is time (in years) and h is asphalt thickness (in inches); $f = 0$ for asphalt underlay, $f = 1$ for concrete underlay in good condition and $f = 3$ for concrete underlay in poor condition.

However, there is an alternative. Mechanically, the problem is not an impossible one, although a degree of approximation is required to solve the equations resulting from balancing forces and from viscous flow. Calculation Sheet 17.2 details a solution.

Calculation Sheet 17.2. Thermal stresses

One-dimensional heat flow

Thermal conductivity $= k$

Specific heat capacity $= C$

Density $= \rho$

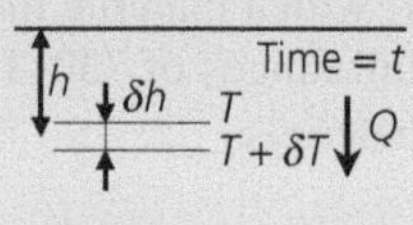

Heat flow

$Q = k\, dT/dh$

$dT/dt = Cr\, dQ/dh$

$\therefore dT/dt = (k/Cr)\, d^2T/dh^2$

$\rightarrow T = T_{av} + T_s e^{-Ah} \sin(Ah + Bt)$

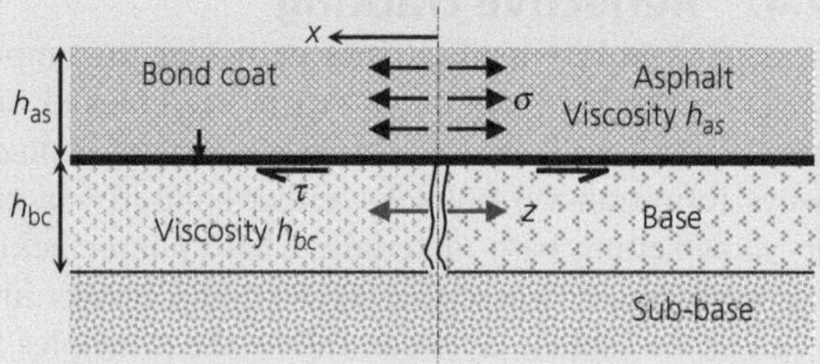

where

T_{av} = average temperature

T_s = surface temperature variation

$A = \sqrt{C\rho B/2k}$

$B = 2\pi/86400$

$\rightarrow$ expansion/contraction z

Solving the mechanism

Assume σ only varies with $x \therefore \mathrm{d}\sigma/\mathrm{d}x = \tau/h_{as}$

If $\sigma = \sigma_{max}e^{-ax}$ (a is a constant), then $\tau = (a\sigma_{max}/h_{as})e^{-ax}$

If y is the relative movement between asphalt and base then

$\mathrm{d}y/\mathrm{d}t = \tau h_{bc}/\eta_{bc} = \int (\sigma/\eta_{as})\,\mathrm{d}x\ (\infty \text{ to } x)$

Solve and equate: $a^2 = h_{as}\eta_{bc}/h_{bc}\eta_{as}$

At $x = 0$

$\mathrm{d}y/\mathrm{d}t = \mathrm{d}z/\mathrm{d}t = \sigma_{max}/a\eta_{as} \rightarrow \sigma_{max} = a\eta_{as}(\mathrm{d}z/\mathrm{d}t)$

The analysis summarised in Calculation Sheet 17.2 is τadmittedly simplified, and depends greatly on the assumptions made for the thickness and viscosity of the bond coat between the asphalt and the base. Nevertheless, it is realistic enough to allow the relative effect of different variables to be explored, and is well suited to spreadsheet work, in which the viscosities of both asphalt and bond coat can be varied as the temperature changes. Once calibrated against experience, the model should be capable of generating satisfactory predictions for a wide range of cases.

As in the case of low-temperature asphalt cracking, it should be remembered that the problem is really one of fatigue. Thus, it is reasonable to postulate a fatigue characteristic for the asphalt, assuming a slope of -4 (or -0.25) on a plot of $\log_e(\sigma)$ against $\log_e(N)$, and to consider the build-up of fatigue damage to occur during the course of a winter season. During the summer, although the opening and closing of the underlying cracks may be just as great or greater, the ability of the bitumen to self-heal will effectively repair the damage. Thus, for example, if a total of ten worst-case thermal cycles are expected during a winter, the permissible stress should be taken as around 0.56 (i.e. $10^{-0.25}$) times the ultimate tensile fracture strength.

One problem, of course, is that the worst-case daily temperature cycle needs to be estimated. Records of air temperature are readily available – but pavement temperature is less frequently recorded.

17.4.2 Traffic-induced reflective cracking

In the minds of many engineers, all reflective cracking is thermally induced. This is certainly not true. There are numerous examples of two-lane carriageways with considerably more reflective cracking in the more heavily trafficked lane. And this is entirely logical. If nothing else, a crack in an underlying HBM layer represents a local reduction in support stiffness. But the truth is that such a crack can represent much more than this. If the crack develops even the smallest 'step' (i.e. relative vertical movement) under the passage of a load, then the overlying asphalt has to accommodate a serious strain. Experience has suggested to many researchers that a practical way of predicting this strain is to carry out multi-layer linear elastic analysis with an artificially low stiffness for the underlying cracked material, 500 MPa being commonly quoted and used. This is usually much lower than the 'effective layer modulus' – the average flexural stiffness that would be measured in situ – or predicted using the approach in Section 16.5.4; it is a purely artificial value with no other justification than that of experience. However, several researchers and practitioners have come to the conclusion that the figure of 500 MPa is a sensible conservative value to use, and a standard analytical pavement design computation can then be carried out to determine life before traffic-induced reflective cracking occurs. Where a designer has confidence in the integrity of joints and cracks in the underlying materials, then a higher (but still artificial) stiffness may be justifiable, but no figure of greater than 1000 MPa can be recommended here.

For those who wish to look beyond this simple estimate, however, it is necessary to think in more detail about the actual mechanisms involved in the production of traffic-induced reflective cracking. These are illustrated in Figure 17.4.

Note the assumption that the adhesion between the cracked or jointed HBM layer and the overlying asphalt is insufficient to prevent separation. It is not too difficult to demonstrate that this assumption is justified. If full bond is assumed, then the stresses – both within the asphalt and at the interface with the underlying layer – become extremely large, demanding failure somewhere in the

Figure 17.4 Traffic-induced reflective cracking mechanisms

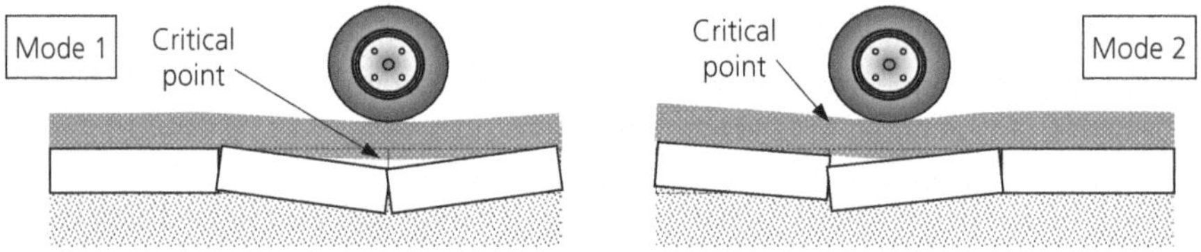

Analysis assumptions

- two-dimensional problem (i.e. 1 m width)
- underlying material broken into uniform rigid elements
- zero vertical movement at boundaries
- asphalt separates readily from underlying material
- foundation described by modulus of subgrade reaction (MN/m^3)
- cracks described by a 'load transfer stiffness' (MN/m^3)

Computation procedure

- Balance forces on each element.
- Treat asphalt as a beam in bending.
- Derive asphalt bending moment → stress → strain → crack prediction.

system; as the bond is the weak link in the chain, significantly weaker than the asphalt itself, that is where failure will take place. Besides, there is anecdotal evidence from sites where 'debonding' either side of a reflective crack has been noted.

Once debonding is allowed, the analysis illustrated in Figure 17.4 is relatively straightforward; however, implementation is much more problematic. It is necessary to make several engineering approximations in order to generate a meaningful prediction of cracking. The key points that have to be addressed are as follows.

- The analysis is two-dimensional; the calculated asphalt bending moment (and therefore strain) must be adjusted to reflect real three-dimensional life.
- If a crack propagation calculation is to be attempted, the local reduction in effective asphalt layer stiffness (due to a crack) has to be taken into account.
- Allowance should also be made for bending of the individual elements in the cracked supporting layer.
- Mode 2 in Figure 17.4 occurs twice during every passage of a wheel – both before and after the crack. The second occurrence is the mirror image of the first.
- One of the key inputs to the analysis is the load-transfer stiffness; this cannot easily be assigned, particularly because it will vary with crack width (and therefore with temperature in an HBM layer). Experience suggests that 5000 MN/m^3 represents an excellent condition whereas 500 MN/m^3 is extremely poor, but the real state can only be estimated approximately.

The truth is that this approach, like all pavement design, requires calibration. However, as it genuinely addresses the mechanisms that produce reflective cracking, it is inherently likely to give realistic predictions, even outside any directly calibrated range. On that basis, it is therefore recommended to the reader as a sensible method for reflective cracking analysis and the basis of this approach is provided by Thom (2000). It also enables the engineer to appreciate the relative importance of different variables, including asphalt stiffness and thickness, but most particularly the load-transfer efficiency across a crack. Where mode 1 (bottom-up) is dominant, the load transfer is irrelevant; if mode 2 is dominant, it is of critical importance.

17.4.3 Geogrid reinforcement

It has been proven many times (Elsing and Sobolewski, 1998; Huhnholz, 1996) that geogrid reinforcement of asphalt can give a significant life extension with respect to reflective cracking. Section 10.6.6 suggested that a geogrid is likely to be worth about 30–50 mm of asphalt in a flexible pavement. In a composite pavement, it depends on whether the primary mechanism of reflective cracking is thermal or traffic-induced. If traffic-induced, then the same 30–50 mm is likely to apply; if thermally driven, then the saving can be much greater. The optimum location for a geogrid to combat thermal action is somewhere in the middle of the layer. If traffic is the main problem, then it should theoretically be placed at or near the bottom of the asphalt layer.

17.5. Heavy-duty block pavements

Blocks form a discontinuous layer and so add little to the structural strength of a pavement. If an analytical design is to be carried out, then a layer stiffness of 500 MPa is commonly used – see Section 7.7 for a justification for this figure.

Immediately beneath, there will be a layer of either bedding sand or sand–cement mortar, and this may be considered part of the block layer for analysis purposes. However, the block layer will readily deform unless it is supported by a base of adequate strength, and heavy-duty pavements will require a hydraulically bound base. The potential failure mode will be cracking of the base, followed by disintegration and then deformation, and design against cracking was covered in Section 17.3. Note, however, that use of block paving means there is little benefit in controlling cracking in the hydraulically bound base by means of joints.

Thus, structural design presents no more challenge than any other composite pavement type; the difficulties lie in the details, as described in the next two sections.

17.5.1 Bedding and jointing

Traditionally, blocks have been placed on bedding sand, and experience has led to the use of 30–50mm. Sand is also used to fill the joints between blocks. This is a subject where engineering design achieves only so much; success depends critically on workmanship – that is, ensuring a sound, even support to each block. However, that does not excuse the pavement engineer from rational thought. One of the most common forms of deterioration of block paving is settlement of individual blocks under repeated load and an approximate analysis of the problem is certainly possible. The issue (Figure 17.5) is one of failure within the layer of sand bedding, and this can only occur if the failure strength of the material is exceeded or, at least, closely approached. Obviously the quality of the sand in terms of its angle of internal friction is critical, so too is the thickness of the blocks themselves and also the sand layer thickness – too thick and failure occurs more easily. However, one of the key influences is the effectiveness of joint filling and the block pattern used. The pattern shown in Figure 17.5, known as 'herringbone', is one of the most effective at generating the required interlock.

Failure within the sand bedding has to be recognised as a potential mechanism and this places a limit on the tyre pressure that should be permitted on this sort of structure. In heavy-duty applications, it is necessary to use something much stronger than sand if block settlement is to be avoided, namely cementitious mortar.

17.5.2 Avoiding block fracture

Blocks can fracture when a non-uniform stress distribution is applied, and this will occur where there are defects within the bedding. Naturally, this cannot be predicted with any confidence, but Calculation Sheet 17.3 gives examples of two possible defective situations, with sand and mortar bedding, respectively. Despite the uncertainties involved, one factor emerges as being of critical importance – the ratio of block length to block thickness.

Figure 17.5 Sand bedding failure

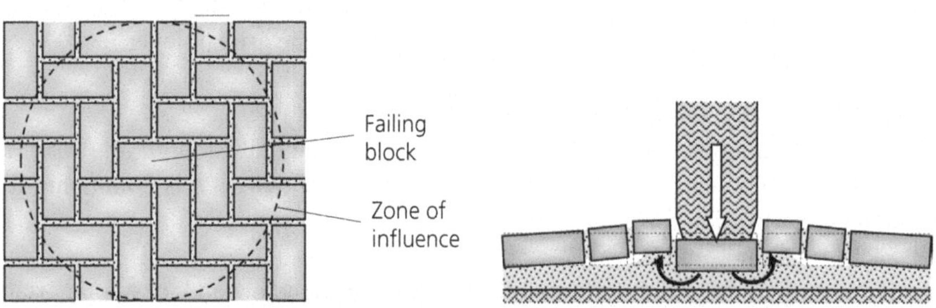

Calculation Sheet 17.3. Block fracture

Sand bedding: poor support at block ends

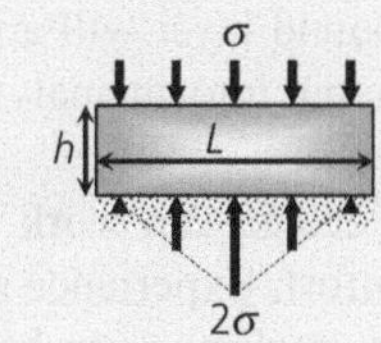

Moment/m $= \sigma(L/2)(L/4) - \sigma(L/2)(L/6) = \sigma L^2/24$

Also, moment/m $= \sigma_t h^2/6$ (σ_t is the tensile stress in block; see Figure 19.3 for equation)

Therefore $\sigma_t h^2/6 = \sigma L^2/24 \rightarrow \sigma_t = \sigma L^2/4h^2$

Example $\sigma = 600$ kPa; $L = 0.3$ m; $h = 0.06$ mm $\rightarrow \sigma_t = 3.75$ MPa (block in serious danger)

Mortar bedding: third of block unsupported

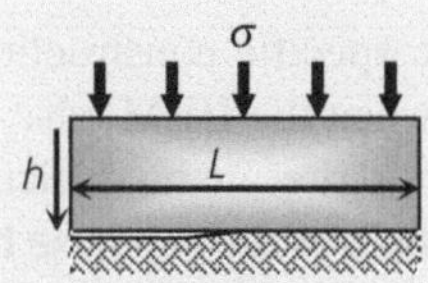

Moment/m $= \sigma(L/3)(L/6) = \sigma L^2/18$

Also, moment/m $= \sigma_t h^2/6$ (σ_t is the tensile stress in block)

Therefore $\sigma_t h^2/6 = \sigma L^2/18 \rightarrow \sigma_t = \sigma L^2/3h^2$

Example $\sigma = 600$ kPa; $L = 0.2$ m; $h = 0.08$ mm $\rightarrow \sigma_t = 1.25$ MPa (block should survive)

One further point is that uneven support to a block on cementitious mortar will also tend to stress the adhesive bond between the block and the mortar. In the second example in Calculation Sheet 17.3, if the tyre load only covered the unsupported part of the block, the resulting bending moment would cause a tensile stress of 450 kPa at the block–mortar interface. This has to be resisted by means of a specialist bonding product; otherwise, loose blocks will result, eventually leading to an uneven surface as detritus works its way in under the block.

17.6. Inverted pavements

If reflective cracking is such a problem, it might reasonably be asked whether anything can be done to protect against it directly, and the answer is yes. The term 'inverted pavement' refers to a pavement type that includes an unbound layer between two bound layers, a form of construction that has found frequent usage in South Africa (DoT, 1996) and which has now also been employed on several projects across the world. The particular design problem, for which this type of pavement may be useful, is as follows.

- High pavement temperatures mean that thick asphalt is likely to lead to rutting.
- Very heavy, slow-moving traffic also presents the danger of rutting but demands that a thick, strong pavement is used.
- Pavement-quality concrete is not considered a viable option (e.g. due to cost or equipment non-availability).

As thick asphalt is not suitable, an HBM layer is definitely required. However, if the asphalt is placed directly over a hydraulically bound base, then reflective cracking is to be expected, unless a reasonable thickness (say 180 mm) is used – but this thickness is most undesirable due to the threat of rutting. A practical solution is to separate the asphalt from the base by means of an unbound interlayer which is thick enough to prevent reflective cracking but not so thick as to lose the benefit of the high stiffness support provided by the HBM base. Because of this excellent support, the unbound layer will appear to have an unusually high stiffness, perhaps 300 MPa, which can then be input into an analytical pavement design to predict asphalt surfacing life.

No design can work miracles. There are too many conflicting requirements here for complete comfort. Experience suggests that the granular interlayer must be at least 120 mm thick if reflective cracking is to be avoided, preferably 150 mm, yet the thicker it is the poorer the support to the asphalt and the greater the danger of fatigue cracking. The obvious response, thickening up the asphalt, then increases the danger of rutting. The design therefore has to be a practical compromise.

One specific construction issue should be mentioned. It is important to achieve good interlock between the HBM base and the unbound interlayer. If the HBM is dry and roller-compacted, this should present no difficulties; if it is laid wet, then it is prudent to spread a layer of stones over the surface of the wet base before it sets.

17.7. Water and drainage

Water has the potential to cause significant damage to all types of pavement, and there is nothing particularly different about a composite pavement. As was the case with asphalt, an HBM that is allowed to reach saturation for long periods is expected to deteriorate relatively quickly, and so the principles are the same as for other pavement types: seal the surface as well as possible and maintain a drainage path down to a relatively permeable layer from which water can escape.

Inverted pavements present a particular drainage problem. It is very important that water is not allowed to accumulate within the unbound interlayer as this will rapidly destroy its properties and lead to asphalt fatigue. Therefore, as water will certainly penetrate the asphalt, it must be able to drain laterally through the unbound layer and out at the sides. It is most unlikely that any system of drainage holes through the underlying HBM layer will suffice.

17.8. Maximising reliability

Composite pavements are very commonly used all over the world for major highways and airfield pavements due to their relative cost-effectiveness. However, that cost-effectiveness depends on their achieving a high level of reliability. The following issues are those that require particular attention.

17.8.1 Specification

As intimated previously, it is usual to specify an HBM base in terms of a compressive strength, despite that being a less meaningful measure than flexural strength. This immediately introduces an element of risk, although this can be mitigated by over-specifying the compressive strength of course. This works in the case of a strong HBM base, but where the design deliberately calls for something weaker, aiming to avoid discrete cracks that could give rise to reflective cracking, it is difficult to be confident that the design assumptions will be achieved.

The other issue is aggregate type. As well as affecting the flexural to compressive strength ratio, aggregate mineralogy is also the key determinant of thermal expansion coefficient and therefore of maximum crack width. In principle, if a 3 m pre-crack spacing is correct for a limestone aggregate

HBM, this should logically be reduced for a gravel aggregate in order to give the same load-transfer efficiency across cracks, gravel aggregate giving a higher expansion coefficient.

The point is that specifiers have to be careful, in stipulating both strength limits – sometimes upper as well as lower – and aggregate mineralogy, in order to ensure that the actual condition of the hydraulically bound layer is close to the design assumptions.

17.8.2 Avoiding damage during construction

A strong HBM layer is in a very fragile state during the first few days/weeks of its life. If the material is to perform well in service, this means it must not develop cracking other than deliberately formed transverse pre-cracks. Yet there is always pressure to expedite the placement of overlying asphalt layers, requiring loads from pavers, delivery wagons and rollers. There is no magic answer to this dilemma and, in reality, the likelihood of damage from asphalt construction plant depends on the thickness of the hydraulically bound base and the underlying foundation, as well as the strength of the HBM. UK specifications generally insist on 7 days before any sort of trafficking is permitted.

Weaker HBM layers (say $\leq$5 MPa compressive strength) generally avoid this problem. They are designed to be in a multiply cracked state long term and so moderate damage during construction presents no additional problem.

Slow-setting materials have an inherent ability to, at least partially, self-repair any early damage that occurs; however, the issue of damage during early life cannot be ignored. Laboratory testing indicates that the intensity of early-life trafficking can dramatically affect the final state of a slow-curing material but, unfortunately, the current state of knowledge means that it is impossible to make confident predictions. The advice contained in Table 17.1 should therefore be seen as relatively 'high risk'. It is extrapolated from experimental data on high-strength HBM obtained at the University of Nottingham and assumes that normal traffic levels are allowed onto the pavement within a few days of the layer being recycled. Actual behaviour will naturally also be a function of the details of pavement construction.

17.8.3 The HBM–asphalt interface

This author's experience is that a core through a composite pavement several years after construction will almost always reveal that the asphalt and HBM base are no longer bonded together. This is not a catastrophic problem. It will not significantly impact the stresses in either the HBM layer or the asphalt – but it does present an opportunity for water to accumulate and weaken both materials, particularly the asphalt.

Table 17.1 Early-life trafficking of strong slow-setting hydraulically bound layers

Time to 90% of design strength: days	Suggested design stiffness: % of fully cured undamaged		
	0.1 msa/year	0.3 msa/year	1 msa/year
28	60%	30%	12%
56	40%	20%	7%
90	30%	12%	4%
180	20%	7%	2.5%

So what should be done to combat this problem? The mechanisms of reflective cracking outlined in Section 17.4 suggest that layer separation either side of a crack through the HBM base is difficult to avoid. It is also clear that some sort of allowance for relative shear movement is helpful in controlling thermally driven strains in the asphalt. This leads to the following options.

- Don't worry about it too much; just use a standard bitumen emulsion bond coat.
- Go for a thick interlayer treatment, a material with a high bitumen content that allows relative shear movement whilst maintaining bond in the vertical direction. This is a stress-absorbing membrane interlayer (SAMI). It can take many forms, for example a thin (10–15 mm) paved layer or a form of surface dressing, often including polymer-modified bitumen and/or fibres, but the key properties are that it has to be tough, flexible and with plenty of bitumen to stick to. A SAMI is not cheap but it enhances reliability.

17.9. Summary

Design of composite pavements depends critically on having a correct understanding of the nature and likely long-term condition of the HBM base, which depends on a proper analysis of the stresses likely within the material. If the base can be considered as a continuous material, giving continuous support to the overlying asphalt, then asphalt layer design can proceed as if for a fully flexible asphalt pavement; if not, then the issue of reflective cracking cannot be avoided.

However, one question that has not really been addressed here is the degree to which cracking actually matters. On its own, a crack does not interfere with pavement usage at all, and this is particularly relevant in the case of reflective cracking in composite pavements. If such cracks are relatively few in number (e.g. where a strong hydraulically bound base is used without any pre-cracking), a pragmatic decision could be taken to ignore reflective cracking and to treat it as a maintenance issue; after all, it is only when a crack deteriorates and leads to more general break-up of the pavement that it becomes a real problem. And this point now forces consideration of the topic of how best to maintain this type of pavement, covered in the next chapter.

REFERENCES

AASHTO (American Association of State Highway and Transportation Officials) (2007) *Guide for Mechanistic-Empirical Design of New and Rehabilitated Pavement Structures.* AASHTO, Washington, DC, USA.

Bonnot J (1988) Fissuration de retrait des chaussées à assises traitées aux hants hydrauliques. *Bulletin de Liaison du Laboratoire Centrale des Ponts et Chaussées 156 and 157* (in French).

Cameron DA and Matthias CL (2001) Bonding of cement treated pavement layers. *Proceedings of the 20th ARRB Transport Research Conference, Melbourne, Australia.*

DoT (Department of Transport) (1996) TRH4: Structural design of flexible pavements for interurban and rural roads. DoT, Pretoria, South Africa.

Elsing A and Sobolewski J (1998) Asphalt-layer polymer reinforcement: long-term experience, new design method, recent developments. *Proceedings of the 5th International Conference on the Bearing Capacity of Roads and Airfields, Trondheim, Norway*, vol. 3, pp. 1523–1532.

Huhnholz M (1996) Asphalt reinforcement in practice. *Proceedings of the 3rd International RILEM Conference on Reflective Cracking in Pavements, Maastricht, the Netherlands*, pp. 456–463.

National Highways (2020a) *Manual of Contract Documents for Highway Works. Vol. 1: Specification for Highway Works*. Stationery Office, London, UK.

National Highways (2020b) CD 225: Design for new pavement foundations. *Design Manual for Roads and Bridges. Pavement Design*. Stationery Office, London, UK.

National Highways (2021) CD 226: Design for new pavement construction. *Design Manual for Roads and Bridges. Pavement Design*. Stationery Office, London, UK.

Thom NH (2000) A simplified computer model for grid reinforced asphalt overlays. *Proceedings of the 4th International RILEM Conference on Reflective Cracking in Pavements, Ottawa, Canada*, pp. 37–46.

Nick Thom
ISBN 978-1-83549-711-1
https://doi.org/10.1108/978-1-83549-710-420243021

Chapter 18
Maintenance of composite pavements

General maintenance of asphalt surfaces has been covered already in Chapter 12. This chapter will therefore concentrate on the specific issue for composite pavements, namely combatting reflective cracking.

18.1. Deterioration mechanism

No matter how good the design or how well a pavement is constructed, it will not last for ever and, in the case of composite pavements with a relatively strong hydraulically bound material (HBM) base, the first sign of deterioration is commonly reflective cracking from transverse cracks in the base.

These sorts of cracks are likely to widen relatively quickly due to the significant thermally driven expansion and contraction. If, for example, the HBM base changes average temperature by 6°C over a 24-h period, and if it is pre-cracked at 3 m centres, then, with an expansion coefficient of 10^{-5} per °C, the daily opening and closing of an average crack will be 0.18 mm, more at the surface due to asphalt expansion and contraction. That is quite a large movement and, of course, many individual cracks will move still more. This gives plenty of opportunity for detritus to enter the crack each time it opens and then to prevent full closure, resulting in an ever-increasing crack width, often affecting ride quality.

The wider the crack becomes, the easier it will be for water to enter and pass straight through the cracked base layer to the foundation. The result will be local softening of the foundation, allowing localised settlement either side of the crack. Each time a vehicle load passes, the base layer deflects and compresses the foundation, but the foundation doesn't quite recover, resulting in the development of a slight gap. Because an HBM is a rigid solid, it cannot follow the foundation down, so it begins to lose its support, increasing the bending stresses it experiences under load. In the end, the stresses will be too much and the base will crack, allowing it to settle locally to the level of the foundation and giving a distinct depression in the road surface around the original reflective crack location.

18.2. Reflective crack treatments

18.2.1 Crack sealing

Clearly the potential for water to significantly exacerbate pavement deterioration means that it is highly desirable to seal cracks before such damage occurs. But how to achieve it? The problem again is the large daily and seasonal opening and closing. A simple overband seal – refer to Section 12.3.1 – cannot be expected to last long. Indeed, the only sealing option that is likely to last for an acceptable time is one where a routed or saw-cut slot is filled with a high-flexibility, high-durability sealant, as shown in Figure 18.1(a).

Figure 18.1 Alternative reflective crack treatments

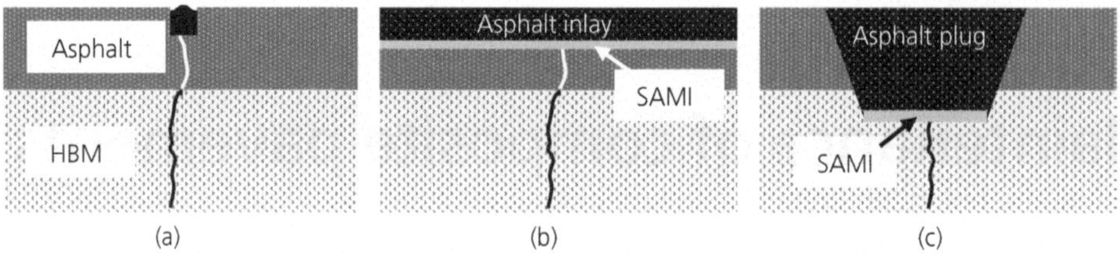

If a simple sealant treatment is insufficient, then localised use of a stress-absorbing membrane interlayer (SAMI) system, as described in Section 17.8.3, should be considered beneath the surface course. Figure 18.1(b) illustrates this. The SAMI could be as simple as a strip of geotextile impregnated with plenty of bitumen, preferably polymer-modified bitumen. The danger, if the SAMI strip is insufficiently wide, is that the single reflective crack is replaced by two, one either side of the strip, but at least it will be much harder for water to penetrate through to the foundation. Regarding the width of the SAMI, it is impossible to be definite because it is so dependent on the make-up of the SAMI itself and what grade of binder is used, but 200 mm either side of the original reflective crack is suggested.

18.2.2 Deeper treatments

Crack sealing and the use of SAMIs should usually be sufficient. However, in cases where the asphalt is thin, say 100 mm or less, a long-lasting repair may only be possible by local thickening, illustrated in Figure 18.1(c). In this case, a strip has to be planed across the road at the reflective crack location and planing has to extend down into the HBM base. This allows placement of a plug of asphalt, with or without a SAMI, and a locally thickened surfacing. In an extreme case, a trench can be milled right through the base and a deep asphalt plug formed.

Obviously such treatments are expensive. Also obvious is the fact that it is difficult to get good compaction in a narrow strip repair, which then introduces the danger of further asphalt densification under traffic and an uneven ride. The advice here is to try to avoid deep treatments by keeping cracks sealed.

18.3. Summary

Composite pavements are here to stay. The development of alternative cements, coupled with a likely future reduction in bitumen availability, means this type of pavement will surely remain an economically attractive choice for heavily trafficked roads and airfield pavements. This means that treatment of its distinctive mode of failure, reflective cracking, will continue to be important. The key, as with so many pavement issues, is to protect against water ingress, and for this the main point that has to be appreciated is that crack opening and closing will often be significantly greater than in most flexible pavements. To be successful, therefore, treatments must take this into account.

Part 6

Concrete pavements

Principles of Pavement Engineering, Third edition

Nick Thom
ISBN 978-1-83549-711-1
https://doi.org/10.1108/978-1-83549-710-420243022

Chapter 19
Pavement-quality concrete

Pavement-quality concrete (PQC) refers to the quality of concrete used as a surface material. However, PQC is a specific form of hydraulically bound material (HBM) and reference will therefore be made back to Chapter 16 in places to avoid any unnecessary duplication. Nor is it the purpose of this book to delve into the science of cements, a science that is rapidly changing, leading to ever-stronger concretes, increasingly effective shrinkage control and, of course, alternative low carbon dioxide cements. However, if sensible pavement design is to be achieved using PQC then it is important to fully understand the physical characteristics of such materials, the factors that affect their properties and the potential for them to become damaged during their lifetime.

19.1. Strength

Strength, the level of stress that can be withstood prior to fracture, is thought by most to be the key characteristic of PQC. Up until fracture, a PQC is a virtually non-deformable solid. After fracture, it is nothing more than a collection of pieces – basically an unbound material, albeit one with rather large particles.

As already noted for HBMs, the cause of fracture is generally considered to be *tensile stress* – although a good case could be made for *tensile strain* actually being the key quantity determining the point of failure. This may seem a fairly irrelevant detail, as stress and strain are linked directly via the *elastic modulus* of the material, which – in contrast to unbound materials and asphalts – is an approximately constant property, independent of stress conditions or temperature. However, this irrelevant detail does indeed become relevant as failure is approached, affecting the interpretation of different test types. For the moment, however, this section will follow convention and work in terms of stress.

The common modes of strength measurement are shown in Figure 19.1. In each case, the test is interpreted on the assumption that the material under test is a homogeneous linear elastic solid.

19.1.1 Tensile strength

Tension is the fundamental failure mode. However, the problem with tensile strength is that it is not easy to measure. Tests have been devised but they are not easy to carry out as a standard procedure, which means that alternative measures of strength are sought. Nevertheless, if tension really is the fundamental mode, it is important to take a look at the typical stress–strain relationship that a tensile test generates, and this is shown in Figure 19.2.

The explanation offered in Figure 19.2 acknowledges the fact that concrete is not really a homogeneous solid but is actually a collection of aggregate particles bonded together and, as these bonds begin to break, the material adopts some of the frictional characteristics of an unbound aggregate.

Figure 19.1 Modes of test configuration for concrete

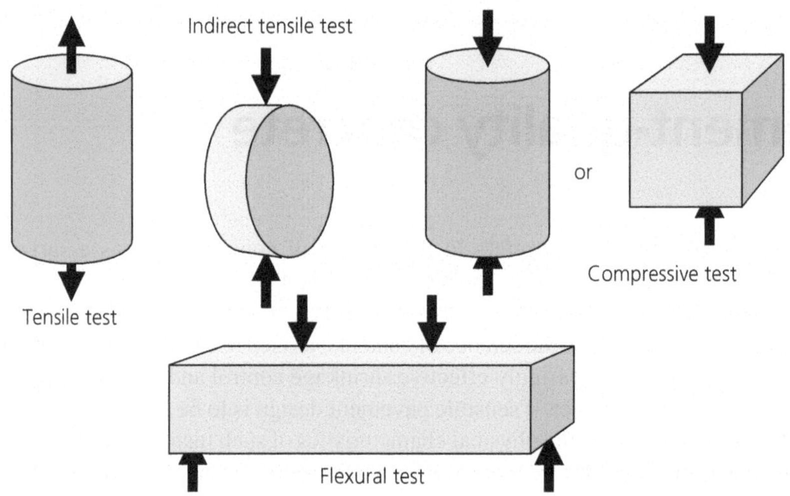

Figure 19.2 Typical tensile stress–strain relationship for a concrete

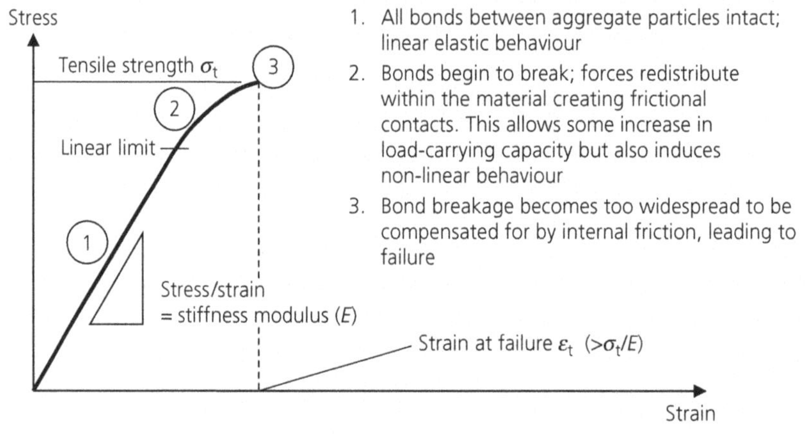

19.1.2 Flexural strength

Flexural strength is a much more convenient mode in which to test, and specifications include AASHTO (2018), ASTM (2022) and BSI (2019a). It is also reasonably representative of the way in which stress is experienced in a pavement layer under applied wheel load. Two different loading arrangements are possible, namely four-point (or 'third point' – see Figure 19.3) or three-point, in which the inner two loads are replaced by a single load. The advantage of the four-point arrangement is that it induces a zone of uniform bending moment in the central part of the beam, allowing the material to fail at a natural point of weakness (which will be dependent on the exact arrangement of aggregate within the mixture). A three-point arrangement has a single plane of maximum bending moment and therefore allows little freedom regarding the location of failure.

Figure 19.3 Stress distribution in concrete under bending

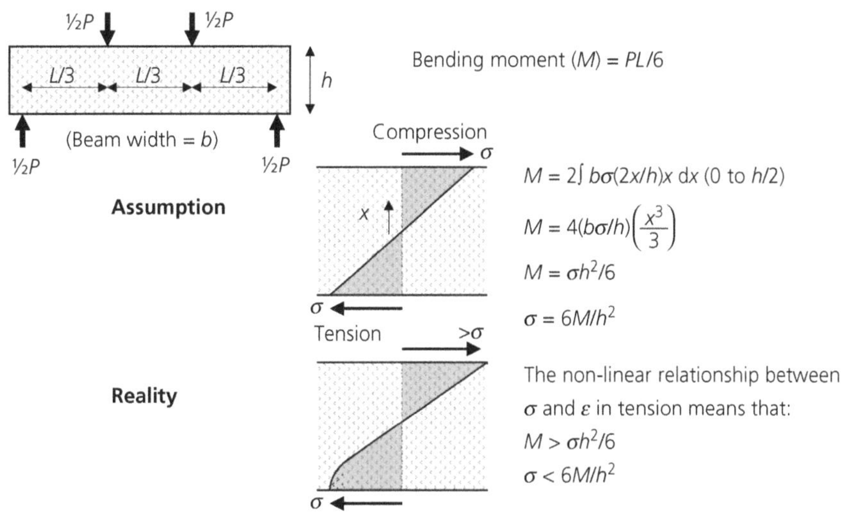

The failure load in either test arrangement allows the maximum bending moment to be calculated. This then has to be interpreted in terms of a maximum tensile stress (also known as a 'modulus of rupture'), assuming that the material is linear elastic. Figure 19.3 illustrates the assumptions and the consequent error. A given moment at failure M will be interpreted in terms of a *flexural failure stress* σ_f ($= 6M/h^2$). However, the equivalent tensile failure stress (σ_t) will actually be smaller because of the non-linear (and consequently more efficient) stress distribution that is present in reality. The difference between the two strength measures is typically 10–15%. Calculation Sheet 19.1 shows predictions for this difference based on the shape of the tensile stress–strain relationship.

Calculation Sheet 19.1. Concrete strength tests

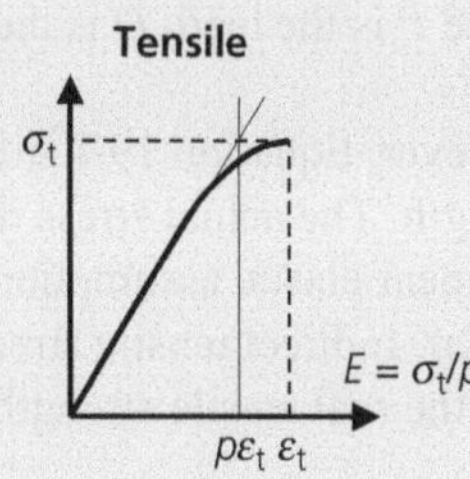

Flexural

Neutral axis z above mid-depth (at failure), beam depth h

From balancing compressive and tensile moment contributions (full working not shown):

$$z = \frac{\tfrac{1}{2}h(1-\sqrt[3]{1.5p-0.5p^3})}{1+\sqrt[3]{1.5p-0.5p^3}}$$

Moment $M = \sigma h^2/6$ (see Figure 19.3) $= (2\sigma_t/3p)(\tfrac{1}{2}h - z)^3/(\tfrac{1}{2}h + z)$

$\therefore \sigma_f = \sigma_t(4/p)(\tfrac{1}{2}h - z)^3/h^2(\tfrac{1}{2}h + z)$

Compressive (cube)

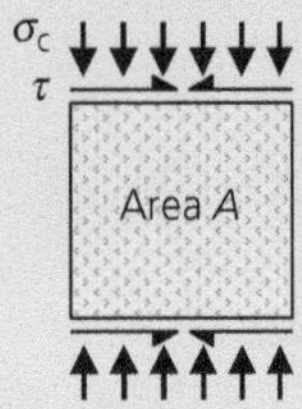

Friction $\tau = \sigma_c \tan\phi$ where ϕ is the angle of friction

Average horizontal stress $\approx \frac{1}{2}\tau \times 2 \times \frac{1}{2} A/A = \frac{1}{2}\tau = \frac{1}{2}\sigma_c \tan\phi$

Say failure when average horizontal strain reaches ε_t

$\varepsilon_t = \nu\sigma_c/E - \frac{1}{2}\sigma_c \tan\phi/E$

$\therefore \sigma_c = \varepsilon_t E/(\nu - \frac{1}{2}\tan\phi) = \sigma_t/[p(\nu - \frac{1}{2}\tan\phi)]$

Examples

If $p = 0.9, \nu = 0.20, \phi = 10°$:	$\sigma_f = 1.10\sigma_t$;	$\sigma_c = 9.94\sigma_t$
If $p = 0.9, \nu = 0.20, \phi = 5°$:	$\sigma_f = 1.10\sigma_t$;	$\sigma_c = 7.11\sigma_t$
If $p = 0.9, \nu = 0.15, \phi = 10°$:	$\sigma_f = 1.10\sigma_t$;	$\sigma_c = 17.97\sigma_t$
If $p = 0.8, \nu = 0.20, \phi = 10°$:	$\sigma_f = 1.20\sigma_t$;	$\sigma_c = 11.18\sigma_t$

19.1.3 Indirect tensile strength

The reason for carrying out indirect tensile (or 'Brazilian splitting') strength determination is that it can be conducted on cored specimens. It is therefore ideal for proving the in situ condition of a pavement layer. Specifications include ASTM (2017) and BSI (2023). Detailed analysis of the stresses induced has led to use of the following equation for the horizontal stress applying over most of the central zone of the specimen.

$$\sigma_{it} = 2P/\pi Dt \tag{19.1}$$

where P is the load, D is the horizontal displacement and t is the specimen thickness.

However, Equation 19.1 suffers from the same assumption of material linearity as that for flexural strength. The actual stress distribution as failure is approached will be slightly more efficient than the linear elastic assumption implies, just as was the case for the flexural test, and the result is also similar. Indirect tensile strength, interpreted using Equation 19.1, also tends to be 10–15% higher than the real tensile strength.

19.1.4 Compressive strength

Although much the *least meaningful* strength test, compressive strength is nevertheless the most common strength measure used in practice. The reason lies purely in the ease with which the test can be carried out. Setting up the test takes only seconds and there is no need for any complex grips or loading platens. For purposes of quality control, it is therefore an ideal test but it has serious limitations as a design measure. The test is illustrated in Figure 19.4. Specifications for compressive strength include AASHTO (2022), ASTM (2023) and BSI (2019b).

The problem here is that there are three different influences on the relationship between tensile and compressive strength. The first is the non-linear characteristic in tension, the effect which has

Figure 19.4 Stress distribution in concrete under compression

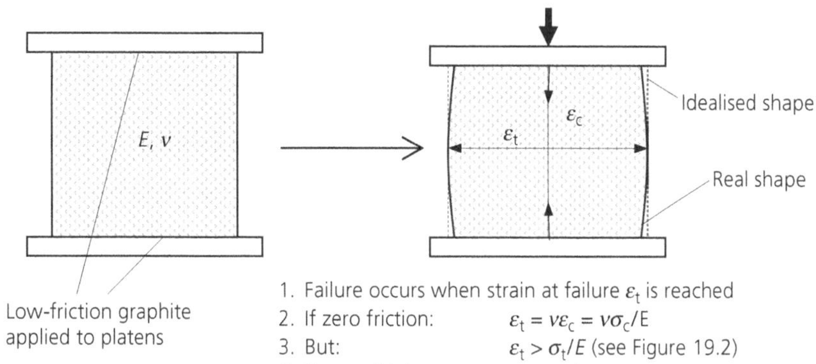

1. Failure occurs when strain at failure ε_t is reached
2. If zero friction: $\varepsilon_t = v\varepsilon_c = v\sigma_c/E$
3. But: $\varepsilon_t > \sigma_t/E$ (see Figure 19.2)
4. So, combining: $\sigma_c > \sigma_t/v$
5. If friction ≠ zero: $\sigma_c >> \sigma_t/v$

already led to differences between tensile, flexural and indirect tensile strengths. The second is the variation in Poisson's ratio (typically 0.15–0.20) and the third is the uncertainty regarding the frictional restraint offered by the end platens. This, unfortunately, means that the ratio of compressive to tensile strength can be anything from around 5 (almost linear tension characteristic; Poisson's ratio of 0.2) to 15. Poisson's ratio and the degree of non-linearity in the tension characteristic are affected by the aggregate type and also the strength of the mixture. Calculation Sheet 19.1 presents the numerical relationship between compressive strength and other strength measures and the examples given demonstrate the wide range possible.

Compressive strength, whether measured using a cube or a cylinder, is therefore *not a fundamental measure*. It should never be relied upon in pavement design. Although experience allows certain 'rules of thumb' to be applied, compressive strength is an inherently unreliable measure of actual quality.

Rules of thumb

- Strong mixes have low Poisson's ratio.
- Limestone aggregate results in low compressive/tensile strength ratio (e.g. 7).
- Granite aggregate results in moderate compressive/tensile strength ratio (e.g. 10).
- Gravel aggregate results in high compressive/tensile strength ratio (e.g. 12).

19.1.5 Strength gain with time

As for hydraulically bound bases and sub-bases, the rate of strength gain is a function of the chemical details of the cement, which is not the subject of this book. For PQC, the range is from rapid-hardening mixtures – designed for use in repairs to concrete pavements, and which can be trafficable within an hour or two – to standard ordinary Portland cement (OPC) bound materials that are designed on a 28-day strength. Figure 19.5 shows compressive strength data for an OPC concrete.

While design is usually based on a 28-day strength, preferably flexural for the reasons stated previously, 7-day or sometimes 3-day compressive strengths are used for quality control to give early warning of potential problems.

Figure 19.5 Typical strength gain of OPC concrete (Sampebulu, 2012)

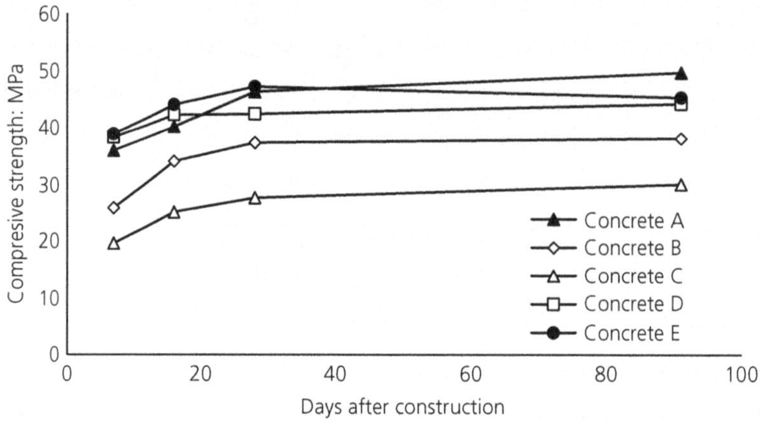

19.2. Fatigue

As is the case with all pavement materials, PQC layers rarely fail under a single load application. However, they can most certainly fail under multiple applications of a stress considerably less than the ultimate strength of the material. This is illustrated in Figure 19.6.

The data in Figure 19.6 are expressed in what has become the usual way for all HBMs, namely as a plot of numbers of load applications to failure against the ratio of applied stress to ultimate strength at failure. Note that the best-fit line suggests that a stress greater than the ultimate failure strength would be required to induce failure in a single load application – which cannot be correct. This is actually further evidence that the fundamental measure is the *strain ratio* rather than stress ratio,

Figure 19.6 Typical fatigue data for hydraulically bound material (Goel *et al.*, 2012; Thompson *et al.*, 1999)

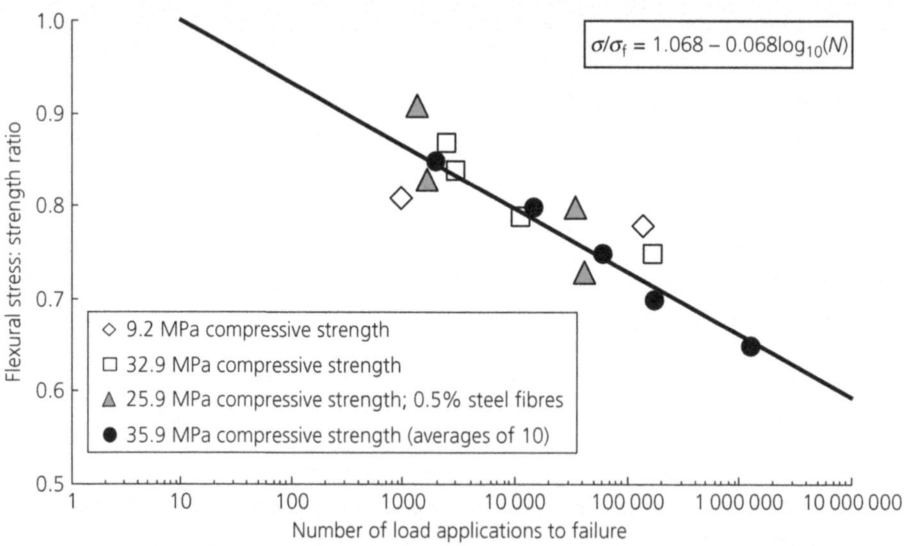

and the non-linearity in the tensile failure characteristic is reflected in a non-linearity in the fatigue characteristic at stress levels close to ultimate strength. The data in Figure 19.6 are compatible with a '*p* value' (see Calculation Sheet 19.1) of around 0.9.

In truth, this is a very complex area, and there are various subtle factors that affect fatigue life. However, experience is that a line approximately as shown in Figure 19.6 can be applied to all but the weakest materials. This leads to the rule of thumb that, as long as the stress is less than 50% of the failure stress, the life will be long. The form of the relationship, a straight line on a log–linear plot, is not really compatible with a power law, but the part that most usually concerns designers (stress ratios of 0.3–0.4) has an approximate exponent of −12 (or −0.083 depending on the way in which the equation is formulated). This contrasts strongly with the exponent found to apply to asphalt – around −4 or −5.

It is useful to consider for a moment just what this process of fatigue actually is. It is similar to that applying in asphalt (see Section 9.4.2) and consists of the progressive breakage of bonds between individual pieces of aggregate. As a tensile stress is applied, it will not be uniformly distributed within the mixture and there will inevitably be certain points, certain key bonds between aggregate particles, that approach their fracture strength even at relatively low overall stress levels. This is a function of the complex geometry of the aggregate matrix. As the first bonds break, this results in a redistribution of the loads within the matrix. Furthermore, it means that the detailed pattern of straining within the material will be slightly different following a bond breakage. This will result in other bonds, probably near to the broken ones, coming under excessive load and breaking in turn – and so the process continues until, eventually, macrocracks form and propagate.

19.3. Durability

A well-designed PQC should continue to increase in strength throughout its life, as chemical bonds continue to develop. The material should be almost impermeable and will not normally be susceptible to water damage – unless the water freezes. Frost damage to PQC can be seen as areas of surface flaking, giving a very rough ride quality to vehicles, and is therefore most definitely to be avoided. Measures to combat frost damage include the following.

- *Air entrainment.* This technique deliberately introduces tiny air bubbles (typically around 5% by volume) into the concrete. Different chemical additives have been used successfully to produce air entrainment, including natural resins, synthetic detergents and fatty acids. The entrained air content is also affected by the cement quality and the nature of any other fine materials used in the mixture. Air voids allow local distortion of the concrete to take place without any damage occurring. The downside is that the strength is reduced or, to put it another way, the cement content has to increase.
- *Maintain high strength.* Experience is that, if the strength is high enough (around 5 MPa flexural strength), then the expansive pressure from frozen trapped water is resisted. In part, this is due to the inevitably very low voids content in a strong mixture, reducing the possibility of water ingress.

19.4. Thermal properties

As for all HBMs, thermally induced expansion and contraction constitutes a significant factor in PQC behaviour and, in practice, this issue either demands the use of joints or of very heavy reinforcement, as explained in Chapter 20. The value of the coefficient of thermal expansion (α) depends chiefly on the aggregate used (refer back to Table 16.1).

Other thermal properties, not normally used in design but that do, in reality, affect the performance of a PQC pavement are the thermal conductivity (the measure of how quickly heat can flow through) and the specific heat capacity (the energy required to heat the material by 1K, equivalent to 1°C). These two properties, together with the density, determine the relationship between pavement surface temperature and temperature at depth, and this in turn influences the far-from-negligible thermally induced warping that inevitably occurs. Typical values are

- thermal conductivity: 2 W/m.K
- specific heat capacity: 880 J/kg.K.

19.5. Stiffness

PQC can reasonably be considered as *linear elastic*, although Figure 19.2 has illustrated the limitations of this assumption, namely that its validity reduces as the fracture strength is approached. There are two principal methods of measuring stiffness in the laboratory. The first is to use the same test types that have already been introduced for strength measurement, namely tensile, flexural, indirect tensile and compressive tests. The usual test procedure is simply to control the load and to record the load at failure, but the addition of a sensitive measurement transducer for strain allows the elastic modulus (E) to be calculated. Calculation Sheet 19.2 gives details of stiffness derivation from flexural beam testing.

Calculation Sheet 19.2. Four-point bending stiffness

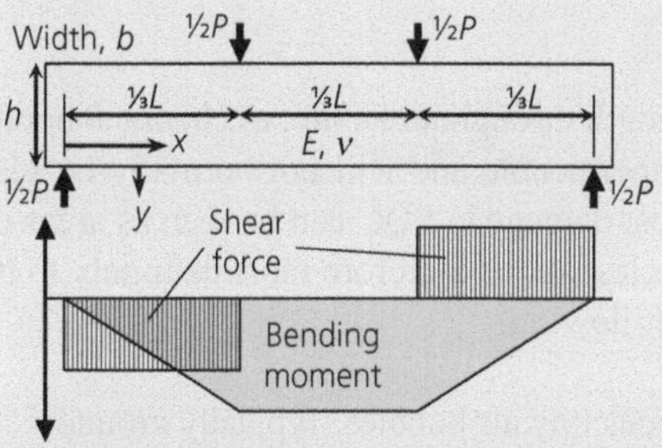

Shear effect

Shear affects the two side sections but not the centre

$$\delta_{shear} = \tfrac{1}{2} P/AG \times L/3$$
$$= PL/6AG$$

where A is the beam cross-sectional area and G is the shear modulus ($G = E/2(1+\nu)$)

Moment effect

$$\text{Curvature} = d^2y/dx^2 = -M/EI$$
$$= (-1/EI)[Px/2 - P(x - L/3)/2] = (-P/2EI)[x - (x - L/3)]$$

where M is the bending moment, E is the elastic modulus and I is the second moment of area ($= bh^3/12$)

Integrating:

$$dy/dx = (-P/2EI)[\tfrac{1}{2}x^2 - \tfrac{1}{2}(x - L/3)^2 + C_1]$$

when x = L/2, dy/dx = 0

$$\therefore L^2/8 - L^2/72 + C_1 = 0$$

$$\therefore C_1 = L^2/9$$

Integrate again:

$$y = (-P/2EI)[x^3/6 - (x - L/3)^3/6 - L^2x/9 + C_2]$$

$$\text{when } x = 0, y = 0 \therefore C_2 = 0$$

$$\delta_{\text{moment}} = y \text{ when } x = L/2$$
$$= (-P/2EI)(L^3/48 - L^3/1296 - L^3/18 + C_2)$$
$$= 23PL^3/1296EI$$

$$\delta = \delta_{\text{shear}} + \delta_{\text{moment}} = PL(1+\nu)/3bhE + 23PL^3/108bh^3E$$
$$\rightarrow E = (PL/bh\delta)[(1+\nu)/3 + 23L^2/108h^2]$$

One reason why such tests are uncommon is the difficulty (and time cost) in obtaining an accurate deflection reading. There are standards, notably the EU standard for obtaining stiffness from uniaxial (compression/tension) tests on HBM base layers (BSI, 2017), but such procedures are rare.

A simpler technique from the operator's point of view is *ultrasonic testing*, in which pulses of stress are applied to the end of a beam and the time taken for the pulse to travel down the beam, reflect from the end and return is measured. The direct measure obtained is the pulse wave velocity, which is a function of material density and stiffness modulus; so, if the density is known, the stiffness modulus can be deduced. This is quick and convenient; unfortunately, it is not necessarily correct. The ultrasonic test often produces significantly higher stiffnesses than those measured by direct mechanical means, perhaps by 30–50% (based on the author's experience). This is exemplified by comparing the range of ultrasonic stiffnesses quoted by the UK Transport Research Laboratory (Nunn, 2004) for cement-bound base materials: about 38 GPa for a 20 MPa compressive strength mix but typically around 25 GPa for those commonly measured in situ. Part of the problem may be that the level of stress applied in the ultrasonic test is just too small and that the first (normally invisible) part of the stress–strain relationship is actually steeper than the remainder, possibly a function of the non-homogeneous nature of the material (stiff particles in contact, less stiff paste surrounding). Therefore, the general advice is to use ultrasonic measurement for comparative purposes but not to base any analysis on such measurements.

In practice, the stiffness modulus of PQC varies from about 45 GPa down to 25 GPa – refer back to Table 16.2. Many design approaches use 35 GPa as a default value if no specific data are available. A key point to appreciate is that a material of lower stiffness is not necessarily inferior. Indeed, a high-stiffness material attracts high stress – inviting failure unless the strength is correspondingly high, which is not necessarily the case.

19.6. Mixture design

Practical mixture design is primarily concerned with strength. Material specification may also include a durability requirement, but fatigue life, stiffness and thermal properties will not normally be either specified or measured. These all affect pavement performance but to a far less significant degree than strength. The key, therefore, is to achieve the design strength.

19.6.1 Wet-formed mixtures

Traditional PQC is mixed to a consistency at which it can be poured and readily compacted by means of vibration. The water content should be adequate for the chemical reaction needed to convert the cementitious binder into hard-setting compounds (typically 0.22–0.25 g of water per gramme of cement); if not, then some of the binder will remain unused, reducing the strength. However, if the water content is too great (greater than around 0.45 g per gramme of cement), then the excess water will inhibit the formation of the most effective compounds, also reducing the strength. Thus, there is an *optimum water content* – or, in practice, there is a range of water contents that give an acceptably high mixture strength. Once a certain strength has been selected, requiring a certain cement content, limits are therefore placed on the range of possible water contents.

However, the material must also be sufficiently workable, a property most commonly specified by means of a 'slump' test, which is a highly practical site measure of the amount by which a truncated cone of concrete slumps once the conical mould is removed. If the desired combination of aggregate type and gradation plus cement type/content cannot achieve this slump with a water content in the permissible range, then the mixture has to be redesigned, which may involve adjustment to the fines content or even a change of aggregate type. Workability requirements therefore dictate that wet-formed mixtures are only suited to binder contents above a certain figure and, in practice, this means a PQC with a typical flexural strength of 3 MPa or over (compressive strength over about 20 MPa). Figure 19.7 illustrates the constraints applying.

To some extent, the design also depends on the paving equipment being used. Slip-form pavers (see Chapter 20) have become the machines of choice for large areas of concrete pavement because of the high productivity achievable, but they can only be successful if the concrete is able to retain its paved shape right from the moment it emerges from the paver. This demands lower workability, and therefore lower slump and lower water content, and places an upper limit on the appropriate binder content, because anything beyond this will not have sufficient water to activate it. Admittedly, binder chemistry can be adjusted to help overcome this and specialist cements (and additives) allow very high strength to be achieved at remarkably low water contents – but at a price.

Figure 19.7 Constraints on design: wet-formed and roller-compacted mixtures

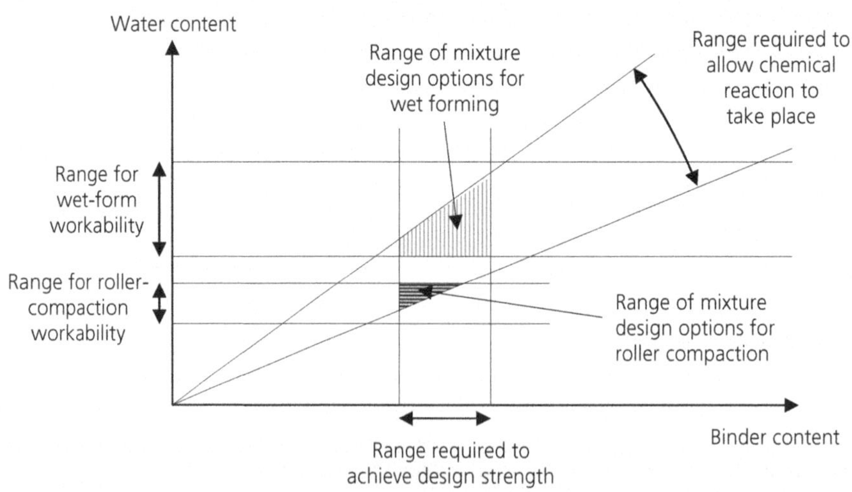

19.6.2 Roller-compacted mixtures

It is quite possible to use roller-compaction to construct a PQC. However, roller compaction introduces a much higher compactive effort and therefore requires a significantly *lower optimum water content* to achieve maximum compacted density. As the ratio of water to cement (or other hydraulic binder) is still of critical importance to the eventual strength of the mixture, this means that roller compaction is generally more suited to weaker materials with a relatively lower cement content. Nevertheless, roller-compacted PQC with up to about 6 MPa flexural strength, say 40 MPa compressive strength, can be achieved with certain mixtures, particularly those making use of limestone aggregate. The constraints that apply are illustrated in Figure 19.7.

19.6.3 Typical PQC mixtures

Typical proportions for PQC mixes are shown in Table 19.1. The proportions given should, of course, be seen as indicative only, as they are highly dependent on the aggregate and gradation used.

19.6.4 Relationship between laboratory and site

This is a very delicate area. Nevertheless, it is not one that can be dismissed. There are many reasons to expect a laboratory mixture to display a higher strength than the material actually in a pavement. For example, control of the mixture proportions may be better and curing may be better controlled. But the principal reason is that compaction will generally be better in a confined mould. Whatever the cause or combination of causes, many studies have shown that strengths of paving material compacted into moulds on site, and cured outside adjacent to the pavement, consistently outperform those measured on cores taken directly from the pavement layer, and this is found in both wet-formed and roller-compacted materials. The difference is often as much as 30% (e.g. Griffiths and Thom, 2007).

Where pavement designs are based purely on experience, this finding is of little importance. The difference is taken into account by the calibration process from which the design method was developed. However, in analytical design the difference cannot be ignored.

19.6.5 Fibre-reinforced mixtures

Fibres are introduced into PQC for two basic reasons – to control shrinkage cracking and to reinforce.

Table 19.1 Typical mixture proportions

Mixture type	Design strength at 28 days: MPa		Cement content: kg/m^3	Water content: l/m^3
	Flexural	Compressive		
Strong PQC	6	(45)	460	200
Standard PQC	4.5	40	420	180

Note: the value shown in brackets would not normally be used in material specification

- Shrinkage crack control is not normally achieved using fibres in pavement layers as it is more economical and more effective to use curing compounds on the pavement surface. However, if fibres were to be used, they should be of low-strength polymeric material.
- Reinforcement, on the other hand, demands that the fibre takes considerable stress – and this generally implies the use of high-strength polymer, glass, basalt, carbon or steel fibres.

The introduction of fibres into a mixture does not usually lead to any significant change in aggregate proportions or water or cement content. In effect, the fibres act as long, thin pieces of aggregate. They come in numerous shapes (Figure 19.8) and sizes, representing a compromise between mechanical efficiency, which is optimised by a long, thin shape and 'mixability'. Economics dictate that steel fibres are rarely used at over 0.5% by volume.

The effect of fibre reinforcement is most clearly seen in a tensile (or flexural) stress–strain plot, as illustrated in Figure 19.8. While the concrete is fully intact, the fibres do nothing other than act as stiff pieces of aggregate; however, as soon as local fractures between aggregate particles start to occur, the fibres begin to take control. The concept that frictional interaction between aggregate particles begins to play a part in behaviour as soon as the first fractures occur has already been introduced in Figure 19.2. With fibres present, this frictional phase is extended enormously because of the highly efficient shape of a fibre. Each fibre interlocks with numerous pieces of aggregate, requiring large energy input into the system in order to overcome it; hence the greatly increased failure strain shown in Figure 19.8. Fibres often give little or no increase in failure stress, but the high failure strain means that the concrete requires a lot more energy to fail it – the property known as 'toughness' – which is important in impact situations such as safety barriers, and also potentially useful in pavements.

Basically, the presence of fibres enhances the resistance of a concrete to fatigue. It takes energy to form a crack and it takes more energy if fibres need to be pulled out or broken.

Figure 19.8 Stress–strain behaviour of steel-fibre-reinforced concrete

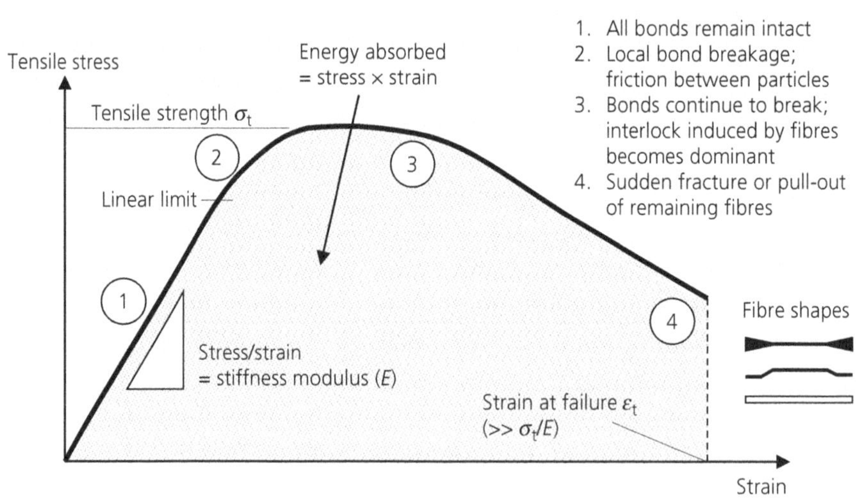

19.7. Summary

In some ways PQC, as an approximately linear elastic material, is relatively simple to understand and model. The main complication for the designer is that, as a solid, PQC is susceptible to thermally driven stresses and strains and Chapter 20 will bring out what this means for pavement design. The following are some of the key points from this chapter.

- The specifier should be particularly careful with regard to strength. Compressive strength is convenient to measure and is ideal for quality control, but is not fundamentally suited to characterising a material that is principally loaded in bending. Flexural strength is much better suited to giving inputs for design.
- Thermal expansion and contraction depends on the expansion coefficient, and this is affected by aggregate type.
- An appropriate water content is essential to achieving design strength and is also constrained by the needs of material compaction.

REFERENCES

AASHTO (American Association of State Highway and Transportation Officials) (2018) T97: Standard method of test for flexural strength of concrete (using simple beam with third point loading). AASHTO, Washington, DC, USA.

AASHTO (2022) T22M/T22: Standard method of test for compressive strength of cylindrical concrete specimens. AASHTO, Washington, DC, USA.

ASTM (2017) C496/C496M: Standard test method for splitting tensile strength of cylindrical concrete specimens. ASTM, West Conshohocken, PA, USA.

ASTM (2022) C78/C78M-22: Standard test method for flexural strength of concrete (using simple beam with third point loading). ASTM, West Conshohocken, PA, USA.

ASTM (2023) C39/C39M: Standard test method for compressive strength of cylindrical concrete specimens. ASTM, West Conshohocken, PA, USA.

BSI (2017) BS EN 13286-43:2003: Unbound and hydraulically bound mixtures. Test method for the determination of the modulus of elasticity of hydraulically bound mixtures. BSI, London, UK.

BSI (2019a) BS EN 12390-5:2019: Testing hardened concrete. Flexural strength of test specimens. BSI, London, UK.

BSI (2019b) BS EN 12390-3:2019: Testing hardened concrete. Compressive strength of test specimens. BSI, London, UK.

BSI (2023) BS EN 12390-6:2023: Testing hardened concrete. Tensile splitting strength of test specimens. BSI, London, UK.

Goel S, Singh SP and Singh P (2012) Flexural fatigue analysis of self-compacting concrete beams. *Proceedings of the Institution of Civil Engineers – Construction Materials* **165(6):** 367–376.

Griffiths G and Thom NH (2007) *Concrete Pavement Design Guidance Notes*. Taylor & Francis, London, UK.

Nunn ME (2004) *Development of a More Versatile Approach to Flexible and Flexible Composite Pavement Design*. Transport Research Laboratory, Crowthorne, UK Report, TRL615.

Sampebulu V (2012) Increase on strengths of hot weather concrete by self-curing of wet porous aggregate. *Civil Engineering Dimension* **14(2):** 92–99.

Thompson I, Peaston CH and Thom NH (1999) Fibre reinforced cement bound roadbase. *Proceedings of the 2nd Asia Pacific Specialty Conference on Fibre Reinforced Concrete, Singapore*, pp. 213–220.

Nick Thom
ISBN 978-1-83549-711-1
https://doi.org/10.1108/978-1-83549-710-420243023

Chapter 20
Rigid pavement – structural design

20.1. Construction practicalities

20.1.1 Placement and compaction

As noted in Section 19.6.2, pavement-quality concrete (PQC) can be roller-compacted, and the placement and compaction techniques are exactly the same as for hydraulically bound bases and sub-bases – and for unbound materials. The same restrictions on layer thickness apply (a limit of around 200 mm) because of the need to achieve density.

However, wet-formed PQC is a different matter entirely. Traditionally, concrete is laid in bays between strips of formwork, often in a chessboard pattern, so that half the bays (equivalent to the red squares on a chessboard) are formed first and the other half (the white squares) are then infilled. The concrete is wet and flows easily, hence the need for formwork. It is not stable enough for roller compaction and so vibrating pokers are used, often together with a vibrating screed. This brings the larger air bubbles to the surface and ensures that the concrete flows around any obstruction such as steel reinforcement bars.

For road construction, where a long thin length is required, the traditional approach has been to set up a so-called 'concrete train', a series of pieces of equipment that run on side rails, which are also used as side formers for the concrete. This allows a continuous high-efficiency pattern of work and a high-quality, well-regulated finish to the pavement.

However, for really high productivity, slip-forming is the technique to use (Figure 20.1). A slip-form paver is not unlike an asphalt paver in that it is a relatively short piece of equipment that spreads the material across the paving width and applies compaction by means of a vibrating screed. Side formers are attached to the paver and move continuously such that the formed PQC slab has to be able to support itself almost as soon as it is formed.

Figure 20.1 Plan view of slip-form concrete paving

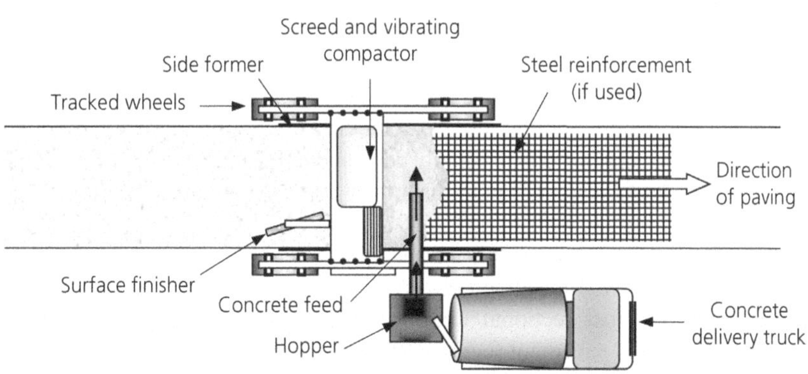

There is no limit on construction thickness for wet-formed PQC as a function of compaction. However, very thick lifts introduce the problem of differential set at different depths and consequent shrinkage cracking, and it is very difficult to lay PQC successfully in lifts greater than about 450 mm. This is irrelevant for roads, where 300 mm is as much as would ever be needed, but it becomes a constraint for airfield pavements.

20.1.2 Curing

If no special measures are taken and a PQC is simply left uncovered after placement, then shrinkage cracking is inevitable. The details will depend on the weather conditions as explained below, but it will certainly occur. The problems are as follows.

- The rate of set of the mortar is dependent on temperature and water supply.
- There is commonly a temperature differential through the depth of the slab, as setting of the body of the material generates heat.
- Heat generation can also mean that the temperature at the point of set is high, which will result in significant thermal shrinkage as the slab cools.
- Unless special measures are taken, the upper surface of the slab will dry quickest, especially if exposed to sunshine or wind.

If the body of the slab is at a much higher temperature than the surface, then it will tend to set first. The upper part will then try to set and, as it does so, to shrink. This shrinkage will be restrained by the already set material at depth and the result will be shallow cracking.

If, however, the surface is allowed to dry out quickly, it will be the first to set. This will then restrain the rest of the material as it tries to shrink and the result will be shrinkage cracks through the body of the material, and these will quickly propagate to the surface. This is a much more damaging situation, as the cracks will penetrate right through the slab.

To counter these problems, the surface must not be allowed to dry out prematurely. This is achieved either by physically keeping the surface wet (e.g. by means of a damp sheet of fabric) or by sealing in the moisture using a chemical spray. It is also important to prevent the upper part from cooling much more rapidly than the rest, which restricts the range of weather conditions suited to PQC construction.

20.1.3 Reinforcement

A perfectly satisfactory PQC pavement can be constructed unreinforced, relying on nothing more than the tensile strength of the concrete itself to prevent cracks from opening. However, reinforcement gives extra tensile strength. It does not stop cracks from forming but it does prevent them from opening, which can considerably slow down the process of deterioration in surface condition. In industrial pavements, for instance, it is common to provide a light mesh of reinforcing steel as an insurance policy, in case unexpected pavement cracking occurs. In highway pavements, where loads are often more certain and which are usually designed not to crack at all, it is common to omit the reinforcement because of the benefit in terms of cost and ease of construction.

In general, reinforcement does not overcome the need for joints, although the added tensile strength is sometimes used to extend the distance between them. However, if enough reinforcement is added (at least 0.4% of the cross-sectional area – see Section 20.5) then it is actually possible to provide a continuous jointless pavement (continuously reinforced concrete (CRC)), at least in one direction, because the stresses generated by thermal contraction of the concrete are insufficient to fracture the reinforcement. This is an attractive proposition, as joints are invariably a nuisance to maintain, but it does of course add considerably to the cost of the pavement and so is not usual except in the most heavily trafficked highway applications.

A final option, frequently used in internal pavement applications (in warehouses for instance), is to mix high-strength fibres into the concrete, thus forming a continuous randomly orientated reinforcement within the material. The added tensile toughness often allows joints to be either omitted or else very widely spaced. The UK Concrete Society report TR34 is an example of a design method that includes fibre reinforcement as an option (Concrete Society, 2018). However, external applications, subject to much greater and more rapid temperature variation, render fibre reinforcement less useful.

20.1.4 Joints

In practice, almost all PQC pavements have joints. Even CRC pavements have joints in the longitudinal direction. Wherever there is a tendency for significant expansion and contraction or warping, the brittle nature of concrete dictates the use of joints. Without joints, the concrete would be so highly stressed, as a consequence of temperature variation, that it would crack anyway, so the obvious course of action is to make sure that these cracks are pre-provided in the form of straight, well-controlled joints.

Joints come in three basic types (Figure 20.2), namely *expansion, contraction* and *warping*, allowing these three types of pavement deformation to take place. They all place a burden on the construction process because, whatever else is provided, it is necessary somehow to ensure that a straight crack is formed at the chosen location. As the pavement will crack anyway as it shrinks due to cement hydration and then due to thermal effects, the trick is to persuade it to crack where the designer wants it to, and this is achieved by making sure that it is significantly weaker along the line of the desired joint than it is elsewhere in the body of the material. Figure 20.2 illustrates possible techniques.

Figure 20.2 Joints in PQC pavements; bar-type dowels shown

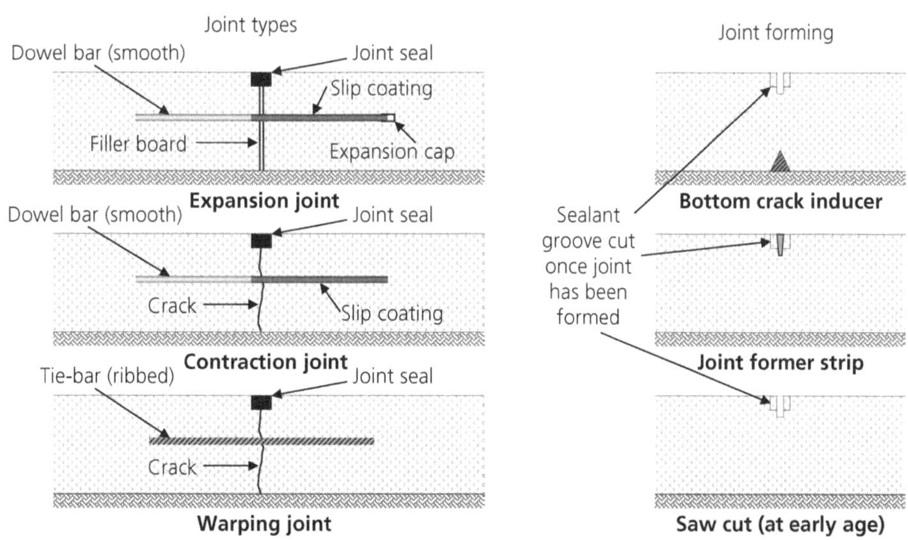

In the case of an expansion joint, a further complication arises in that it is not straightforward to fix the compressible 'board' of material that has to be incorporated in order to allow the concrete either side of the joint to expand; it has to be fixed sufficiently well for it to remain in place during concrete placement.

The other major issue is that a joint represents a serious weakness in the pavement. In cold weather, when the concrete shrinks and the joints open up, there will be a loss of continuity across the joint; the concrete slab each side will behave as if it had an unsupported edge. One way around this is to provide *dowels* (for expansion and contraction joints) and *tie-bars* (for warping joints) – see Figure 20.2 – which means that these either have to be inserted into the wet concrete or they have to be accurately and securely fixed in place beforehand. Both are disruptive processes and represent potential sources of construction defects. Dowel bars in particular will fail to perform well if they are misaligned more than a fraction. For example, the UK National Highways (2020) places an absolute limit on dowel bar misalignment of 6 mm over a 300 mm length (just over 1°). The incentive to avoid joints wherever possible is therefore a strong one, and it is also quite possible to avoid the use of dowels as long as the slab is thick enough and strong enough.

It should also be noted that dowels do not have to be bars. Plate dowels (see Figure 20.3) have the advantage that the stress is transferred to the surrounding concrete over a larger area and so localised concrete failure is less likely. They are also more forgiving with regard to precise alignment, allowing some lateral movement between slabs.

Finally, joints need to be sealed, otherwise they are easy passages for water to access the often moisture-susceptible subgrade soil. This requires the formation of a groove and the application of a sealant. Furthermore, it often necessitates regular maintenance of the seal, as even the most advanced sealant materials have a finite life in such an aggressive environment.

20.2. Current design practice

All the guides and methods referred to in Section 10.2 for flexible pavements also apply to rigid pavement design, particularly the Interpave design guide (Knapton, 2007). The descriptions will not be repeated here, but the following additional documents are highlighted.

Figure 20.3 Alternative dowel: plate type

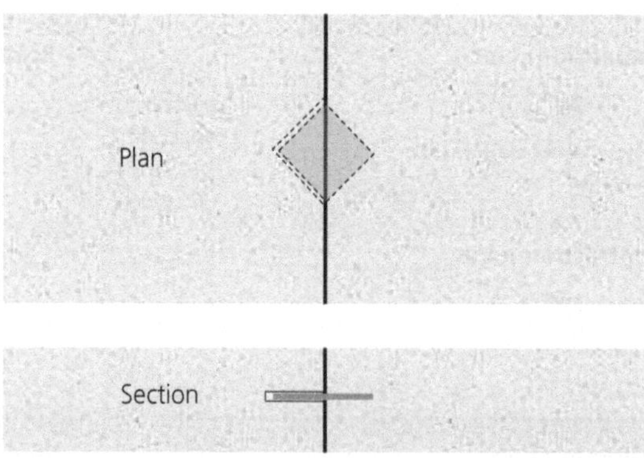

20.2.1 Britpave *Concrete Hardstandings Design Handbook*

This method (Britpave, 2023) refers back to UK National Highways documentation for thickness design. However, it includes much useful detail on joint design and construction, joint arrangement in awkward areas of pavement, edge details and practical advice on specification.

20.2.2 UK Concrete Society *Concrete Industrial Ground Floors: A Guide to Design and Construction*

The basis for the designs in this document (Concrete Society, 2018) is the analysis approach known as 'limit state', described in Section 20.3.3. This method accounts for the energy needed to induce cracking failure, and so is particularly suited to designs including fibre reinforcement. However, the document is principally aimed at indoor pavements (e.g. warehouses) and caution is advised if applying to external locations, particularly with regard to joint spacing and detailing.

20.3. Unreinforced concrete pavements

Chapter 19 introduced the fracture behaviour of PQC and the different ways in which it is measured, and the point was made that the fundamental mode of failure is in *tension*. Every PQC has a tensile strength and, if that is exceeded, the material will fail. An unreinforced PQC layer is therefore in danger from the moment it sets as a solid. So, the first point to make is that joints are absolutely required. The issue of what type of joint should be used in which location will be tackled in Section 20.3.6, but first the problem of stress analysis under traffic load will be considered.

20.3.1 Westergaard analysis

Since the 1920s, the equations developed by Westergaard (1926) have represented the most widely used approach to analysis of concrete pavements under load. The original equations were derived assuming a concrete pavement to act as a slab in pure bending, and several subsequent modifications have been made over the years to increase the accuracy with which a real pavement can be modelled. The equations are in terms of the maximum tensile stress in the concrete due to slab bending, and they are for three load locations: (*a*) internal (i.e. distant from a joint); (*b*) edge and (*c*) corner. The load is assumed to consist of a uniformly stressed circular area. Recommended versions of the equations (based on Ioannides *et al.* (1985)) are shown in Box 20.1.

Box 20.1. Westergaard equations for stress in a concrete slab

Internal loading; stress at base of slab

$$\sigma_{\text{tensile}} = [3p(1+\nu)/2\pi h^2][\ln(2L_s/a)+0.5-0.577]+[3p(1+\nu)/64h^2][(a/L_s)^2]$$

Edge loading; stress at base of slab

$$\sigma_{\text{tensile}} = [3p(1+\nu)/\pi(3+\nu)h^2] \times [\ln(Eh^3/100ka^4)+1.84-4\nu/3+(1+\nu)/2+1.18(1+2\nu)\times(a/L_s)]$$

Corner loading; stress at top of slab

$$\sigma_{\text{tensile}} = (3p/h^2)[1-(2a/L_s)^{0.72}]$$

where p is the load, a is the radius of the loaded area, h is the slab thickness, ν is Poisson's ratio, L_s is the radius of relative stiffness, $L_s = \{Eh^3/[12k(1-\nu^2)]\}^{0.25}$, E is the concrete stiffness modulus and k is the modulus of subgrade reaction

A few words of explanation are required here. The radius of relative stiffness (L_s) is a convenient parameter expressing the ease with which the slab bends. It has no fundamental meaning on its own but depends on other quantities. The modulus of subgrade reaction (k), however, is a fundamental input to the equations. It is defined as the pressure required to cause unit distance of vertical deflection, and it represents a computationally convenient way of expressing foundation stiffness. Traditionally, it is measured using a large (762 mm diameter) plate loading test (see Section 6.3.10). It can also be estimated from the combination of foundation layer stiffness moduli expected (see Section 1.5.2.5), but the problem is that no Poisson's ratio effect is present when a modulus of subgrade reaction is used, and so conversion from E and ν values is imprecise.

Clearly, once the tensile stress has been calculated, it can be compared with the tensile strength, preferably derived from a flexure test. If the computed stress is lower than the strength, then the slab should survive loading; if not it will crack.

This simplistic evaluation can then be extended for cyclic loading with reference to the typical fatigue behaviour of PQC, as illustrated in Section 19.2. The following equation was given, derived from laboratory fatigue testing of various concretes and other hydraulically bound materials, allowing determination of a limiting stress for design.

$$\sigma/\sigma_f = 1.068 - 0.068\log_{10}(N) \tag{20.1}$$

where σ is the tensile stress due to load, σ_f is the flexural strength and N is the number of load applications to failure.

The slope of this fatigue characteristic is a shallow one, which means that a small increase in stress can lead to a large reduction in life. Designs should therefore be appropriately conservative, bearing in mind the effect of locally reduced concrete thickness or loss of slab support.

A common problem, however, is the difference between the idealised load cases covered by the Westergaard equations and real life. In a concrete highway pavement, for example, the joints represent points of weakness and, therefore, locations of increased stress. But a joint is not the same as an edge. Dowels are generally used to ensure a degree of load transfer but, on the other hand, the situation is definitely worse than the internal load case. This is a real dilemma, and it limits the usefulness of the equations. A common approach would be to calculate both the Westergaard internal and edge load cases and then to make a judgement, depending on the type of joint present, as to just how much better than a free edge a real joint is likely to be. It always has to be appreciated that joint efficiency changes with temperature as concrete expands and contracts, so a conservative, low-temperature case should be taken. The author's experience is that a multiplier between 1.2 and 1.5 applied to the stress from the internal load case is usually appropriate, whereas the edge loading stress typically works out at about 1.8 times the internal stress.

A particular issue is how much attention should be paid to the corner loading case, as this is always the most critical in terms of the predicted stress. However, free corners do not occur. Even at joint intersections there is always a degree of load transfer to adjacent slab sections. It should also be borne in mind that (*a*) joint intersections are not generally in the wheel-paths on highways (although they may be for airfields or industrial areas) and (*b*) corner cracking is a much less serious phenomenon than other forms of cracking, as a corner crack is of limited extent. Once again, engineering judgement is needed, supported by approximate analysis.

A final problem lies in the lack of flexibility in load position. Most real load situations involve interacting effects from more than one wheel, for example aircraft wheel gears, and the Westergaard equations make no allowance for this. Of course, they can still be used – but only as approximate guides.

20.3.2 Multi-layer linear elastic analysis

Multi-layer linear elastic analysis, introduced in Section 7.3.2, represents an alternative to Westergaard analysis. As in the Westergaard approach, the relevant output from the analysis is the maximum tensile stress in the concrete. However, this type of analysis has one major advantage over Westergaard, which is that load combinations can be included, such as dual tandem or dual tridem wheel sets. Unfortunately it also has a major disadvantage, which is that all layers have to be infinite in extent; there is no way of analysing an edge or corner situation.

A reasonable compromise, attempting to extract the best from each method, is to combine the two approaches. The Westergaard equations give the relative effect of an edge or a corner compared with the internal case, and this can be translated into a stress multiplier and applied to the stress derived from multi-layer linear elastic analysis – which always relates to the internal case. This is a sound, common-sense, approach and is to be recommended. The value of the multiplier should be around 1.2 for the case of joints with good load transfer, becoming as much as 1.5 for poor joints.

20.3.3 Limit state analysis

A conceptual problem with a purely elastic analysis such as Westergaard's, or layered elastic theory, is that concrete cannot simply crack at a single point. If cracking is to occur, then this will involve a 'mechanism' of cracks. Thus, even if, in theory, the stress at a particular point is sufficient to induce a crack after a certain number of load applications, the reality is that this cannot occur until adjacent sections of concrete are also ready to crack. What will actually happen is that the point of theoretical failure will simply reduce in stiffness locally as the first interparticle fractures begin to occur (refer back to the explanation in Sections 19.1 and 19.2), shedding stress onto neighbouring regions. The upshot of this is that conclusions based on an elastic analysis will generally be *conservative*. The alternative form of analysis is termed 'limit state'. It is a plastic analysis rather than elastic, and the different way in which stress–strain behaviour is idealised is shown in Figure 20.4. This type of analysis was pioneered by Meyerhof (1962). Concrete Society report TR34 (Concrete Society, 2018) is an example of a current design manual based on this type of calculation.

It is clear from Figure 20.4 that, while the linear elastic assumption may be conservative, the limit state assumption is *optimistic*. Only in the particular case of fibre-reinforced concrete, for which the stress–strain behaviour genuinely does include a large post-peak stress-carrying capacity, can it be considered as realistic.

Nevertheless, the limit state approach opens the door to solving problems that would otherwise be intractable. No matter how complex the interaction between load locations and joints, all that is required is that a crack mechanism is postulated and an energy balance carried out. The relevant equation is

$$\text{Work done by loads} = \text{energy absorbed by the foundation} + \text{energy dissipated at cracks} \quad (20.2)$$

Figure 20.4 Idealisations of stress–strain behaviour

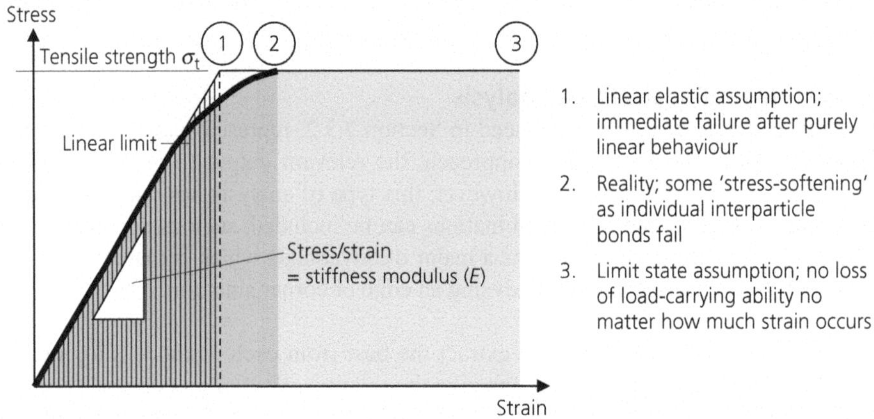

The crack mechanism (see Figure 20.5 for examples) is assumed to move by a small amount δ (vertically in Figure 20.5). The work done by the loads is then given by the magnitude of each load multiplied by the vertical movement, which is related to δ. The energy absorbed by the foundation has to be estimated; it is the product of stress and deflection, summed over the entire deflected area. The energy dissipated at cracks is the bending moment at failure multiplied by the imposed angle of rotation, summed over all lengths of all cracks. It is the *bending moment at failure* per linear metre of crack (M_f) that is the desired output from this equation. The *flexural stress at failure* (σ_f) is then (see Section 19.1.2)

$$M_f = \sigma_f h^2/6 \tag{20.3}$$

where h is the slab thickness.

Figure 20.5 Possible crack mechanisms for use in limit state analysis

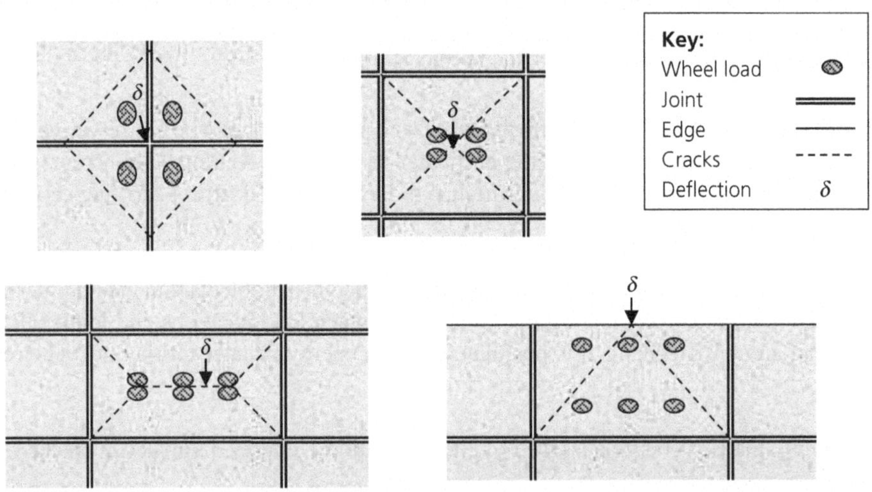

Once a mechanism has been postulated, it is relatively simple to work out the deflection at each load location and the rotation angle (in radians remember) at each crack. The uncertain element is always the distribution of foundation stress. In some cases, it is worth carrying out a multi-layer linear elastic calculation in advance to predict this; however, a reasonable assumption would be that the sum of all the applied wheel loads is distributed over the deflected area in proportion to the magnitude of deflection. A simpler but conservative assumption would be to assume a uniform distribution over the deflected area. An example of the use of limit state analysis is given in Calculation Sheet 20.1.

Calculation Sheet 20.1. Limit state analysis

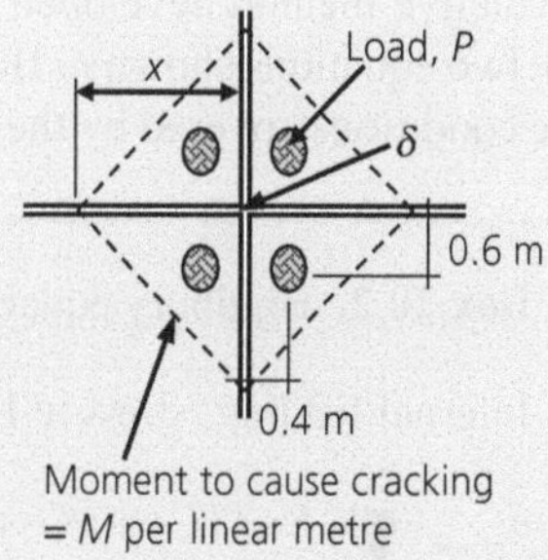

Simplifications

(*a*) load uniformly distributed over area of foundation enclosed by cracks

(*b*) square wheel pattern with 0.5 m offsets everywhere

Work done by loads $= 4P\delta[1-(0.5\sqrt{2})/(x/\sqrt{2})] = 4P\delta(1-1/x)$

Energy dissipated at cracks $= 4\sqrt{2}Mx[\delta/(x/\sqrt{2})] = 8M\delta$

Energy absorbed by foundation $= 4P(\delta/3)$

Energy balance: $4P\delta(1-1/x) = 8M\delta + 4P\delta/3 \rightarrow M = P(1/3 - 1/2x) \approx P/3$ when x is large

Since M also equals $\sigma_f h^2/6$ (h is the slab thickness), so $\sigma_f = 2P/h^2$

Example

$P = 200$ kN; limit σ_f to 3 MPa $\rightarrow h = 0.36$ m

Limit state analysis is less frequently carried out for pavements than is elastic analysis, but the advice here is not to ignore it. The technique can be used to address specific cases such as the effect of a joint with no load transfer, unusual jointing patterns or load arrangements such as those from specialist pieces of industrial equipment or aircraft wheel gears. However, it must be remembered that the method is inherently 'unsafe' in that the assumption made regarding stress–strain behaviour is optimistic unless designing with fibre-reinforced concrete. A conservative treatment of foundation stress distribution is therefore recommended in order to balance this out.

20.3.4 Warping stress analysis

Thermally induced warping stresses, resulting from a temperature difference between the top and the bottom of a slab, can be highly significant. In most design methods, they are not explicitly taken into consideration, but the calibration upon which the method is based implicitly takes them into account. If using an analytical approach, the usual technique is to build in extra conservatism when considering stresses due to traffic, and an example of this is the use of the so-called 'Packard line'

(Packard and Tayabji, 1985) for fatigue assessment, an ultra-conservative fatigue characteristic commonly used by designers. The equation of the Packard line is

$$\sigma/\sigma_{\mathrm{f}} = 0.96 - 0.0799\log_{10}(N) \tag{20.4}$$

where σ is the tensile stress due to load, σ_{f} is the flexural strength and N is the number of load applications to failure.

However, while this may be satisfactory for standard cases, it is not recommended for non-standard applications, other than as a rough check. For such cases, it is appropriate to attempt a thermal stress prediction, which may either be carried out from first principles or use may be made of the predictive method developed by Bradbury (1938). The Bradbury approach can be summarised in the two equations shown in Box 20.2, one for internal stress and the other for edge stress, mirroring the conditions covered by the Westergaard equations.

Box 20.2. Bradbury equations for maximum warping stress in a concrete slab

Internal loading; stress at base of slab

$$\sigma_{\mathrm{tensile}} = \tfrac{1}{2}E\alpha T(C_x + \nu C_y)/(1-\nu^2)$$

Edge loading; stress at base of slab

$$\sigma_{\mathrm{tensile}} = \tfrac{1}{2}E\alpha T C_x(\text{or } C_y)$$

where E is the stiffness modulus, α is the coefficient of thermal expansion, T is the top–bottom temperature difference, C_x (or C_y) depends on x (or y)/L_s as shown on graph, $L_s = \{Eh^3/[12k(1-\nu^2)]\}^{0.25}$, h is the slab thickness, k is the modulus of subgrade reaction and v is Poisson's ratio

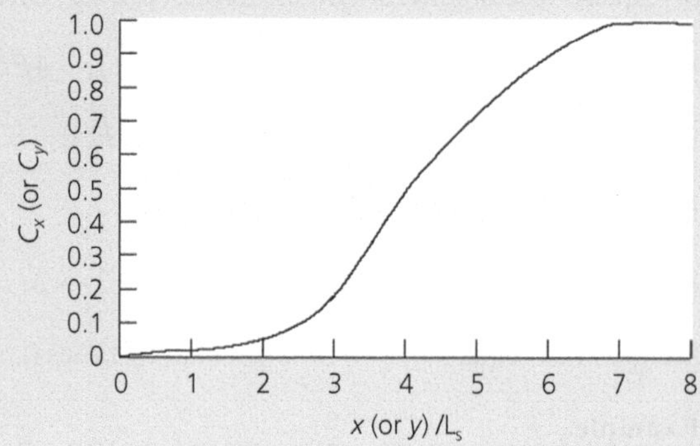

Approximate analysis from first principles is also possible, simplifying the situation to just two dimensions rather than three. Figure 20.6 illustrates the basis of the computation and presents theoretical solutions for the idealised case where a concrete slab is founded on a fully rigid support.

Putting typical numbers to the symbols in Figure 20.6, the length of the unsupported section of slab is likely to be between 1 and 2 m under night-time conditions (say 5°C temperature difference), and potentially up to around 3 m under the most severe daytime conditions (say 8°C temperature difference). This gives worst-case stresses of around 1 MPa at night and nearly 2 MPa during the day (depending on slab thickness etc.). However, the most serious problem comes when a vehicle passes over the unsupported section of slab, leading to an overall stress significantly higher than would have been predicted under wheel loading with full support conditions. The interaction is complex; simply summing the maximum thermally induced and (internal) load-induced stresses will give an exaggerated estimate of the real worst case.

Figure 20.6 Stresses due to slab warping

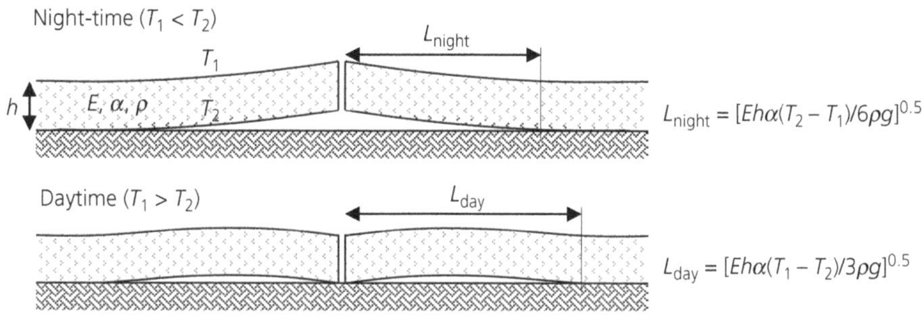

Stress calculation (temperature only)

If unresisted, curvature due to temperature = $\alpha\Delta T/h$ where $\Delta T = |T_1 - T_2|$

In flat part of slab, moment exactly negates curvature due to temperature difference; that is $M/EI = \alpha\Delta T/h$; but $M = \sigma h^2/6$ and $I = h^3/12$ (per metre width);

$\therefore$ Stress (σ) = $\alpha\Delta TE/2$

- This is the maximum night-time stress (at surface)
- The maximum daytime stress (at bottom of slab) is slightly greater

This is not the end of the story however; for a start, the hardness of the support is clearly a major issue, particularly in the case of the daytime state. It would only require a fraction of a millimetre of foundation deflection to dramatically reduce the unsupported length. Some design methods therefore recognise that the use of an unbound sub-base layer beneath the concrete reduces warping stresses.

The final point to bear in mind if an analytical assessment of thermally induced warping stresses is carried out is that the worst-case scenario does not happen very often. Reference to the fatigue characteristic for concrete means that it is theoretically possible to convert a small number of severe stress applications (i.e. combination of full traffic load and unfavourable thermal conditions) to a much larger number of 'standard' stress applications – or vice versa. However, because temperature variation within the slab will not be known with any certainty, any such conversion can only result in an estimate for the number of design traffic loads, but it at least represents a rational approach to solving the problem. Calculation Sheet 20.2 presents an example computation.

Calculation Sheet 20.2. Combining traffic and thermal stresses

Say

(*a*) maximum stress due to traffic loading = 1.7 MPa

(*b*) design traffic = 150 000 load applications

(*c*) maximum additional stress due to thermal effects = 0.9 MPa

(*d*) thermal stress has equal probability of being anywhere in the range 0–0.9 MPa

Question: How many equivalent load applications at full thermal stress?
Replace fatigue characteristic such as Figure 19.6 by a power law with exponent 12.

$\therefore$ Damage = $k\sigma^{12}$ (k is a constant)

Integrate between 1.7 and 2.6 MPa for **average damage**

$$\int k\sigma^{12}\, d\sigma/(2.6-1.7) = k(2.6^{13}-1.7^{13})/(13\times 0.9) = 21122k$$

This is equivalent to the damage due to N_{eq} load applications at 2.6 MPa $= N_{eq}k2.6^{12}$

$\therefore N_{eq} = 21122/2.6^{12} = 0.221$; that is, take $0.221 \times 150000 = 33150$ at full thermal stress

20.3.5 Foundation choice

In general, a stiff foundation will result in reduced traffic-related stress in the concrete slab. This is, of course, desirable, and its effect can be included in the stress computations introduced earlier in this section. However, when the computation is carried out it is generally found that the theoretical benefit gained is slight, often too slight to justify the extra cost of a bound sub-base. Furthermore, as presented in the previous section, a stiff foundation actually increases warping stresses, making it even less easy to justify. On the other hand, there are other less readily quantifiable advantages, namely

- a bound surface beneath the concrete makes construction of the concrete slab easier, and is therefore likely to raise quality
- a bound sub-base gives continuity of support at joints, reducing shear force across the joint, and so reduces the rate of joint deterioration.

This issue is one that is far from resolved. On major highways and airfield pavements it is nowadays most common to specify a hydraulically bound sub-base, recognising that it is usually the safest option. On industrial pavements and lower category roads it is less usual, as the extra cost is that much harder to justify. However, this is an area of continuing debate, and this book cannot advise strongly one way or the other.

20.3.6 Joint type

The different types of joint were introduced in Section 20.1.4. But the question of which type of joint to use is one with no simple answer. Dowelled transverse joints are common in highway pavements, and dowels are often used for all joints in a large expanse of concrete pavement such as an airfield apron. In most cases these are *contraction joints*, allowing the concrete to contract relative to its dimension at the time of set – but not to expand. If construction takes place in cool weather, then it may be necessary to include *expansion joints* at intervals (e.g. every tenth joint or so). This avoids the possibility of blow-up failures in very hot weather, when slabs in compression literally buckle under pressure. The problem with expansion joints, however, is that they can never achieve interlock; they rely solely on dowels for load transfer. An expansion joint is therefore a point of weakness – a step nearer to the 'edge loading' condition. Expansion joints present two choices: either ignore the weakness and accept an accelerated failure locally; or strengthen (usually thicken) the pavement locally. The first option is the more common, but serious consideration should really be given to the second.

If the total linear extent of the pavement is no more than about 20 m (notably transversely across a highway), then freedom for expansion and contraction is not really required. Thus, the standard longitudinal joint type in a highway is a *warping joint*, in which the dowel is replaced by a tie-bar, thus ensuring good interlock across the joint under all circumstances, but at the same time allowing a degree of rotation at the joint.

Note that neither dowels nor tie-bars have to be used. Some organisations have taken the view that they are a nuisance and a likely source of future trouble, and that a pavement is safer without them! They accept that load-transfer efficiency is reduced and use designs that recognise this by increasing slab thickness accordingly.

20.3.7 Joint spacing

While the joint pattern is routinely ignored by users of elastic analysis, the section on limit state analysis made the point that load-carrying capacity is integrally tied to the pattern of joints present. And, although joint separation was not explicitly mentioned under 'warping analysis', it may have been noted that the length of unsupported slab introduced in Figure 20.6 supposes a semi-infinite slab length; if the actual joint spacing is less than twice the calculated unsupported length, the warping stress will clearly be reduced.

However, the primary reason for the choice of joint spacing is simply related to linear thermal contraction and the maximum resulting joint opening, which in turn relates to *load-transfer efficiency*. Although joints may be equipped with dowels to assist in carrying the load from one side to the other, these will only be of limited use unless there is a considerable degree of *interlock* across the joint between the two concrete faces – and this presumes that joints will never open excessively. Studies have shown that, once an opening exceeds around 1 mm, interlock across the joint becomes minimal, and it is the need to control joint opening that really controls spacing. For example, with a 20°C temperature reduction, giving a strain of around 2×10^{-4}, a 5 m joint spacing would give a 1 mm opening. In fact, experience suggests that, for most climates, joint spacings between 3.5 and 6 m represent the range of reasonable compromise between the cost and nuisance of joint construction and maintenance and the need to maintain load-transfer efficiency; less joint spacing and the cost becomes too great, more and joint-related problems become increasingly likely.

20.4. Jointed reinforced concrete pavements

Concrete reinforcement is a subject that extends way beyond pavement engineering. With a stiffness of 210 GPa, steel is around five times as stiff as PQC, while the tensile strength of high-yield steel (about 450 MPa) is nearly a hundred times that of PQC; basalt and glass fibre bars are both much stronger still – although less stiff. Reinforcement, therefore, has something to offer to concrete, in pavements as well as in other applications.

Light reinforcement is reasonably common in PQC, usually as a means of controlling shrinkage cracking. It is also used to cut back significantly on the number of joints. The argument goes

- joints are a nuisance and their number should therefore be minimised
- with greater joint spacing, warping stresses increase and load transfer at joints decreases
- therefore, there is a much greater likelihood of cracking
- but, if reinforcement is present, cracks are 'controlled'; they will remain narrow and the slab will not break up.

Basically, in this sort of pavement, with moderate levels of reinforcement, 'hairline' cracking is accepted, but there is sufficient reinforcement to hold the slab together, ensuring excellent interlock across each hairline crack and therefore no significant loss of performance. And such pavements work. They are particularly common in industrial applications where a slight loss of ride quality due to cracking is unimportant. Many road authorities have also permitted jointed reinforced pavements, although long-term maintenance difficulties caused by increased joint opening and closing mean that they have now become less popular on highways.

Thickness design is an interesting issue here. In a pavement that is designed to crack, it is unclear how the thickness is to be selected. Logically, while stresses sufficient to cause cracking are acceptable, crack deterioration is not. Once cracks begin to spall, load-transfer efficiency will be lost no matter how strong the reinforcement, and this type of deterioration will then accelerate. Unfortunately, however, this type of distress is hard to model and current practice, perhaps wisely, is to base thickness selection on experience. For example, standard practice in the USA (AASHTO, 1993, 2007) is to adopt the same thicknesses as for unreinforced concrete, while in the UK a slight thickness decrease is permitted (Mayhew and Harding, 1987), adopted by the UK National Highways design standard CD 226 (National Highways, 2021).

The next question is the degree to which joint spacing can be opened up. There is still a limit to the length of unjointed pavement possible, and this stems from the tensile stress induced in the slab due to contraction (during cooling) and the frictional restraint provided by the pavement foundation. Figure 20.7 illustrates the situation and outlines the analysis involved. For example, with an angle of friction between the concrete slab and the foundation of 30°, an intact 4 MPa tensile strength concrete could theoretically sustain a length of about 540 m between joints – a massive distance. However, at crack locations, the entire tensile force has to be taken by the steel reinforcement. Thus, for example, with 12 mm diameter bars at 250 mm spacing, giving 4.5 cm^2 of steel per metre width, and assuming high-yield steel with a strength of 450 MPa, the maximum sustainable tensile force would be just over 200 kN/m. In a 250 mm-thick pavement of density 2500 kg/m^3, and with a 30° friction angle, the limiting length reduces to 114 m. Anything more and the reinforcement would fracture.

In practice, the angle of friction between the slab and the foundation is hard to predict. It also has to be remembered that steel is subject to fatigue failure under repeated loading at stresses less than its ultimate strength. For these reasons it is prudent to impose a factor of safety of at least 3 and, in fact, joint separation distances of 15–30 m are commonly used. With greater separation, joint sealing becomes very difficult due to the magnitude of daily joint opening and closing.

Figure 20.7 Tensile stresses due to slab shrinkage

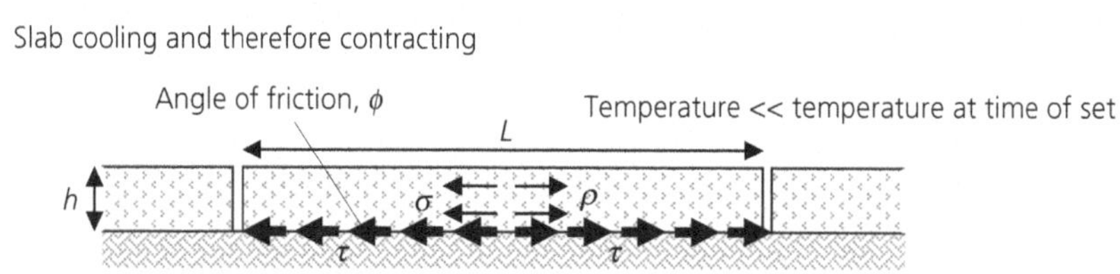

Calculation

Shear stress $\tau = \rho gh \times \tan \phi$

$\therefore$ Force across centre of slab = $\tau L/2 = \rho ghL/2 \times \tan \phi$ (per metre width of slab)

If the slab is intact

→ Tensile stress at slab centre ≈ $(\rho ghL/2 \times \tan \phi)/h = \rho gL/2 \times \tan \phi$

In a cracked reinforced slab

→ Stress in steel reinforcement = $\rho ghL/2A \times \tan \phi$

in which A is reinforcement area per metre width

20.5. Continuously reinforced concrete pavements

Continuously reinforced concrete (CRC) is the ultimate concrete pavement. The principle is simple. The reinforcement has to be strong enough to resist concrete contraction due to cooling, giving reinforcement quantities typically between 0.4% and 0.8% of the concrete area. Calculation Sheet 20.3 suggests a theoretical basis for these percentages, although the real justification is practical experience.

Calculation Sheet 20.3. Continuous reinforcement design

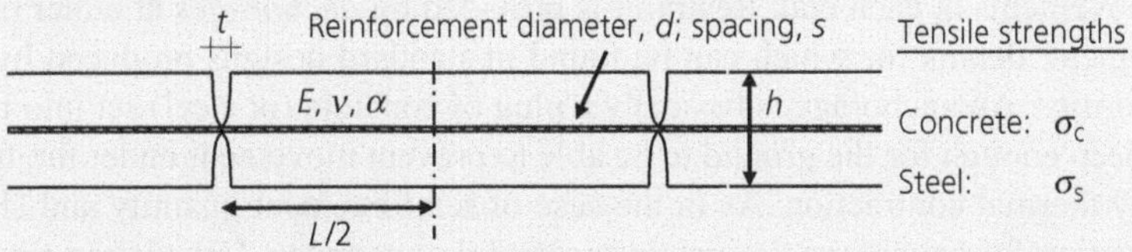

Extension over length $L/2$ with force F in reinforcement bar = extension due to stress in mass of concrete + local extension around bar (Boussinesq equation – see Figure 6.3)

$$= L/2 \times (F/hs)/E + F(1-v^2)/dE$$

balanced by thermal contraction under temperature drop T: $L/2 \times \alpha T$

Equate: $L/2 \times \alpha T = L/2 \times (F/hs)/E + F(1-\nu^2)/dE$

Design steps

(*a*) Propose d and s; compute L for likely combinations of T and F (assume $= \sigma_c hs$) during the curing process

(*b*) Select L for design; suggest use 75% of lowest L from (*a*); default = 1.5 m

(*c*) Compute new d with selected L, worst-case T and $F/(\pi d^2/4) = \sigma_s$ (this gives a quadratic equation for d)

(*d*) Select d to be larger than value from (*c*); revise s and start again if necessary

Thickness design of CRC is not straightforward. Cracking is accepted; indeed it is inevitable, with a typical spacing of around 1.0–1.5 m. This means that the pavement is relatively flexible, as each crack acts as a hinge, allowing a small degree of rotation. Yet the pavement must be thick enough to prevent excessive crack deterioration – and this balance is not an easy one to achieve. Where crack deterioration is excessive, ride quality is severely compromised and maintenance is very difficult indeed. It is suggested here that, with a concrete flexural strength of 4 MPa, the shear stress across each crack should be kept to less than about 500 kPa if significant crack deterioration is to be avoided and that, under commercial truck loading, this typically demands a slab thickness of at least 150 mm. A similar thickness is required to prevent longitudinal cracking.

This suggestion of at least 150 mm assumes many tens of millions of standard axles of loading, including axle loads up to about 120 kN, but it also assumes continuous slab support. Problems can occur where this support is lacking in specific locations due to differential settlement, which induces increased flexure of the slab, spalling, crack deterioration and longitudinal crack development. This combination leads to the failure mode known as 'punching', where discrete blocks of pavement concrete are 'punched' down, forming distinct steps in pavement level and a very poor ride quality. Bearing in mind the high-value nature of a CRC pavement, it is advised that thicknesses should be rather greater than the theoretical requirement. The UK National Highways standard CD 226 (National Highways, 2021) calls for a minimum of 200 mm, reducing to 150 mm where a 100 mm asphalt overlay is to be applied.

One of the key additional requirements for CRC pavement is that the concrete has to be restrained in order to avoid large movements at each end. Restraint is provided by *anchorages* at either end of a length of CRC pavement, details for which can be found in standard designs produced by a number of highway authorities. An anchorage is basically a plug of concrete (or steel) set into the ground, and it has to be deep enough for the ground to be able to prevent movement under the full tensile force generated by thermal contraction. As in the case of reinforcement quantity and slab thickness, the real justification for anchorage designs is successful experience. Calculation Sheet 20.4 provides theoretical backup – to be used cautiously!

Calculation Sheet 20.4. Anchorage design

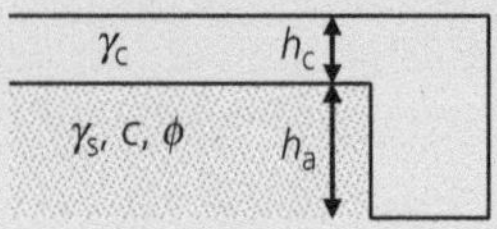

Say tensile force in concrete pavement is sufficient to cause cracking (i.e. tensile strength × area: $F = \sigma_{tc} h_c$ per metre)

Strength parameters c and ϕ define a limiting relationship between horizontal and vertical stress in the foundation material, as explained in Section 6.2

$$\sigma_h = [2c\cos\phi + \sigma_v(1+\sin\phi)]/(1-\sin\phi)$$

Average σ_v at anchorage $= h_c\gamma_c + 0.5h_a\gamma_s$

$\therefore F$ to cause anchorage failure $= h_a\sigma_h = h_a[2c\cos\phi + (h_c\gamma_c + 0.5h_a\gamma_s)(1+\sin\phi)]/(1-\sin\phi)$

Example

$c = 10$ kPa; $\phi = 50°$; $\gamma_c = 24$ kN/m²; $\gamma_s = 18$ kN/m²; $h_c = 0.2$ m; $\sigma_{tc} = 3$ MPa

$\rightarrow F = 600$ kN/m requires

One anchorage with	$h_a = 2.4$ m
Two anchorages with	$h_a = 1.5$ m
Three anchorages with	$h_a = 1.2$ m
Four anchorages with	$h_a = 1.0$ m etc.

20.6. Fibre-reinforced concrete pavements

Fibre reinforcement (see Section 19.6.5) is quite different from continuous bar reinforcement. If 0.5% of the pavement cross-section is steel bar, this is sufficient to make a CRC pavement; however, 0.5% steel fibre is barely enough to increase the tensile strength of the composite material.

There are various reasons for this. One is the random orientation of the fibres; another is the very modest anchorage available to each individual fibre. Whereas a crack through a pavement reinforced with continuous bars has no choice but to go right through the bar, in a fibre-reinforced material the crack can work its way around most of the fibres. Fibre reinforcement is therefore a very inefficient means of increasing overall strength – but what it can do is increase the toughness significantly (i.e. it takes a lot more energy to break it).

But what difference should this make to pavement design?

Basically, it means that the assumptions of limit state analysis become much more realistic than is the case for unreinforced concrete where, in Section 20.3.3, it was suggested that limit state analysis was inherently unsafe. In the author's experience a design thickness based on limit state is often 5–10% less than one based on elastic analysis (Westergaard and/or multi-layer linear elastic), so this is logically the order of magnitude of the potential saving when fibres are used. The problem is that the cost penalty is likely to be considerably greater than this, which means it isn't worth using fibre reinforcement unless there is some additional benefit.

Additional benefit is possible, however. Cracks may still form in fibre-reinforced concrete but the fibres will tend to hold the material together, giving a longer post-cracking phase before the pavement genuinely fails. Cracks should therefore remain almost invisible and still give good load transfer since the fibres will limit crack opening.

20.7. Maximising reliability

Rigid pavements are inherently unforgiving. Failure soon leads to very poor ride quality and expensive maintenance, as detailed in Chapter 22. It is therefore particularly important for this pavement type to make sure that resilience is not compromised.

20.7.1 Specification

The key advice here is to specify flexural strength rather than compressive strength since it is much more closely related to the actual mode of failure. It therefore gives much greater assurance that the as-designed material will actually perform in the pavement.

A secondary point is that, if pavement analysis is being carried out, then the expected difference between the strength of a specimen compacted into a mould and the material on site needs to be taken into account.

And a final point is that a generous safety factor should be used to account for the uncertainty present. The Packard fatigue line, introduced in Section 20.3.4, is a suitably safe relationship between the stress/strength ratio and the number of load applications to failure, but only if the strength (i.e. the flexural strength) is known with some confidence.

20.7.2 Curing

In practical terms, curing is probably the area where problems are most likely to occur and curing problems can be very serious, resulting in the formation of full-depth cracks. This simply means that every effort should be made to avoid rapid drying of the surface, through specification, supervision and, preferably, friendly partnership with the contractor. The weather is very difficult to specify out of the equation and so the combination of temperature, humidity and wind speed is best countered through agreement since it is never in anyone's interest to create problems for the future.

The back-up plan, of course, could be the use of fibres, not necessarily strong reinforcing fibres, but thin polymeric fibres that have been proven to aid in combatting shrinkage cracks.

20.7.3 Joint sealant

This may sound like a trivial element compared to the PQC itself or to other joint components like dowels, but the truth is that a good and long-lasting sealant is essential to ensuring a resilient pavement. The design issues are as follows.

- The smaller the daily/seasonal strain cycle, the happier the sealant will be, the better it will stick to the concrete and the longer it will last.
- A small strain cycle demands a wide sealant strip, for a given crack opening and closing.
- But a wide strip means an unpleasant 'thump' as each tyre crosses it.

So what should the designer do? To be honest there is no perfect solution to this. Most joints in PQC are chased out to widths between 20 and 40 mm to keep strains in the sealant to acceptable levels, but even then they don't last for ever. The advice here is not to compromise on sealant life. Naturally, not all sealants are the same, and those that tolerate strain better and are more adhesive are, of course, more expensive. But a joint that leaks a lot of water is a much more serious problem in the making, as Chapter 22 will make clear, and the highway management problems caused by having to renew sealants are far from trivial, leading to leaky joints commonly being left unrepaired. They say 'the devil is in the detail' and with rigid pavements this is certainly true.

20.7.4 Sub-base material

It is hard to justify the use of a bound sub-base from the standpoint of pure design theory. The calculated effect on stress in the PQC slab is minimal. Nevertheless, organisations such as UK National Highways and several airport authorities have reached the conclusion that the long-term pavement durability benefits of a bound foundation are significant and outweigh the cost. In part, it is due to the greater ease of PQC slab construction, but there is also a real and demonstrable benefit in reducing joint deterioration. The point is that load transfer across joints is provided by a combination of dowels (or tie-bars), aggregate interlock and sub-base continuity. An unbound sub-base has no continuity; it can readily deform, allowing relative vertical movement of the slabs either side of a joint. A hydraulically bound sub-base, on the other hand, is a rigid solid – albeit probably a cracked one – and it will therefore provide significant resistance to relative vertical movement. It provides a valuable contribution to overall pavement resilience.

20.8. Summary

The key to design of rigid pavements is that the PQC must not crack – except at joints, which are really just cracks in a controlled location. If the material is well specified and constructed, if the design has taken sensible account of overloaded vehicles and if joint weakness has been compensated for in the slab thickness specified, then the life should be practically indefinite – as long as timely maintenance is carried out. Conceptually, a rigid pavement is a simple enough structure, but please bear in mind how disastrous it is once it starts to fail. Failure is not an option!

REFERENCES

AASHTO (American Association of State Highway and Transportation Officials) (1993) *Guide for Design of Pavement Structures*. AASHTO, Washington, DC, USA.

AASHTO (2007) *Guide for Mechanistic-Empirical Design of New and Rehabilitated Pavement Structures*. AASHTO, Washington, DC, USA.

Bradbury RD (1938) *Reinforced Concrete Pavements*. Wire Reinforcement Institute, Washington, DC, USA.

Britpave (2023) *Concrete Hardstanding Design Handbook*, 3rd edn. Britpave, Wokingham, UK.

Concrete Society (2018) *Concrete Industrial Ground Floors: A Guide to Design and Construction*, 4th edn. The Concrete Society, Camberley, UK, Report TR34.

Ioannides AM, Thompson MR and Barenberg EJ (1985) Westergaard solutions reconsidered. *Transportation Research Record* **1043**: 13–23.

Knapton J (2007) *The Structural Design of Heavy Duty Pavements for Ports and Other Industries*, 4th edn. Interpave, Leicester, UK.

Mayhew HC and Harding HM (1987) *Thickness Design of Concrete Roads*. Transport and Road Research Laboratory, Crowthorne, UK, Report RR87.

Meyerhof GG (1962) Load carrying capacity of concrete pavements. *Journal of the Soil Mechanics and Foundations Division, ASCE* **88(3)**: 89–116.

National Highways (2020) *Manual of Contract Documents for Highway Works. Vol. 1: Specification for Highway Works*. Stationery Office, London, UK.

National Highways (2021) CD 226: Design for new pavement construction. *Design Manual for Roads and Bridges. Pavement Design*. Stationery Office, London, UK.

Packard RG and Tayabji SD (1985) New PCA thickness design procedure for concrete highway and street pavements. *Proceedings of the 3rd International Conference on Concrete Pavement Design and Rehabilitation, Lafayette, IN, USA*, vol. 9, pp. 225–236.

Westergaard HM (1926) Stresses in concrete pavements computed by theoretical analysis. *Public Roads* **7(2)**: 25-33.

Nick Thom
ISBN 978-1-83549-711-1
https://doi.org/10.1108/978-1-83549-710-420243025

Chapter 21
Concrete surfaces

21.1. Achieving good ride quality

One must never forget that excellent ride quality is not always necessary (such as on many industrial pavements), but on high-speed roads it most definitely is. Typical tolerance limits specified at the surface of such a road would be ± 3 mm in 3 m; in order to achieve this, an appropriate 'system design' is needed.

21.1.1 Wet-formed pavement-quality concrete (PQC)

There should be no problem achieving the required tolerance in a machine-laid wet-formed concrete pavement (although this is less true of hand-laid pavements). Therefore, concrete should logically be well suited to high-speed applications. Unfortunately, one of the few things that virtually all highway users can agree on is that it is not a pleasure driving on a concrete road. The problem, however, is not failure to achieve construction tolerance. It is the hard, unforgiving nature of concrete itself, unable to absorb much energy from the tyres. This results in high *tyre vibration*, which is felt within the vehicle as well as being associated with relatively high noise (see Sandberg and Ejsmont, 2002). It is true that surface texture also plays a key role here, and that both longitudinal grooving and an exposed aggregate finish (see Section 21.2) provide reasonable surfaces to drive on – comparable to dense asphalt. Nevertheless, concrete suffers from a fundamental problem due to its lack of energy absorption, although this is only a problem at high speed and only for road vehicles – aircraft noise drowns out anything due to the pavement.

A secondary problem is that of joints. Perhaps surprisingly, even a 20 mm gap between bays is readily felt as a mild impact, despite the excellence of modern car suspension systems. The problem is that the frequency range of this impact is vastly higher than that for which the suspension (including car-seat damping) is designed. It is not a major problem in reality, but it is a perceived nuisance. Again, it is not a significant issue for airfield pavements or for low-speed applications, and of course it is overcome in the case of continuously-reinforced concrete.

Naturally, concrete is still an option for high-speed roads, but this will depend on the relative importance assigned to ride quality against other issues, most notably economics.

21.1.2 Roller-compacted concrete

Unfortunately, it's quite impossible to achieve the same tolerances in dry roller-compacted mixtures as in wet-formed concrete, which effectively rules out such materials as surface layers on major roads. The problem is that layer thickness would generally be at least 150 mm and this is much too thick for tight level control. As a surface material, roller-compacted concrete is, therefore, only suited to low-speed roads or industrial applications.

21.2. Surface finish

A further complication when constructing a wet-formed PQC pavement is the surface finish. Whereas an asphalt can be designed such that its natural surface gives adequate skid resistance without any special treatment, this is not generally the case for wet-formed concrete, which would be very smooth indeed if no special measures were taken. The micro-texture of the surface mortar on the concrete would generally be adequate but the texture depth, which is so important in the provision of wet skid resistance (refer back to Chapter 11), has to be specially formed. Figure 21.1 shows the options available.

Surface finish is undeniably one of the major issues affecting the popularity of concrete-surfaced pavements. As an intrinsically hard material that dissipates little of the energy applied by tyre loading, anything the surface texture can do to ameliorate the problem would be most welcome.

21.2.1 Transverse grooving or brushing

This is a much-used traditional finish. It can readily be achieved while the concrete is wet, so cost is low. It gives rapid drainage of surface water to the side and so is good from a safety point of view. But it can be really noisy! Figure 21.2 illustrates the problem. A rotating vehicle tyre would form a continuous contact strip with a smooth pavement surface; there would be no impact between tyre rubber and the surface, just a smooth 'landing' as each element of tyre gently makes contact. An asphalt is far from smooth, but the random arrangement of stones means that even if one tyre

Figure 21.1 Concrete surface finish options

Grooving		A set of steel tines is dragged across the surface of the fresh concrete immediately after paving (can also be achieved by sawing hardened concrete)
Brushed finish		The fresh concrete surface is brushed using an appropriately heavy-duty brush to form a ridged finish
Exposed aggregate	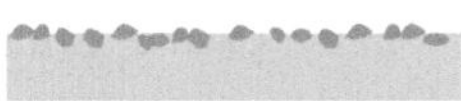	A retarder is sprayed onto the finished concrete and loose mortar is brushed away from around the larger aggregate pieces about 12 hours afterwards
Burlap drag		A sheet of rough fabric is dragged over the surface of the wet concrete, leaving a rough finish

Figure 21.2 The tyre–groove impact problem

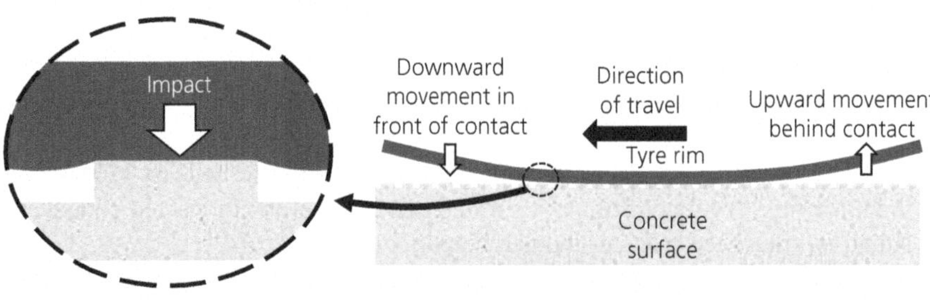

element lands in a gap between stones, its neighbours land securely on a stone and disruption to the tyre is minimised. But if there is a gap in the pavement going right across the wheel-path then all elements of the bit of tyre that lands in the gap find themselves without any support, and the next bit of tyre impacts on the concrete slab beyond the gap; the wider the gap, the greater the impact velocity and consequently the noise and vibration level.

Many variations on transverse grooving have been tried in an attempt to solve this problem. One theory was that the problem was the regularity of the grooving and so quasi-random spacings were used; although this produced some change in the nature of the noise it didn't really solve the problem – because the gaps were still there. If Figure 21.2 genuinely represents the principal noise/vibration-inducing mechanism then the only ways to mitigate the problem are to reduce the width of the individual grooves – which is not easy – or to break them up in some way, mimicking the more random nature of an asphalt surface.

21.2.2 Longitudinal grooving or brushing

This is the most obvious alternative. It is equally convenient and low-cost and it immediately changes the nature of the tyre–surface contact. Now the majority of the tyre has continuity of support and so no impacts are induced. Strips of tyre have zero support, but it is continuous zero and so still no impact. The result is a nice smooth ride – well, as smooth as concrete can be without any energy absorption.

But the problem is surface drainage. Rainwater is trapped and has no escape path to the sides of the road, although the total texture depth (refer back to Chapter 11) can be adequate so skid resistance is still good under most conditions. Extensive studies have been carried out over the last 20 years into the safety of longitudinal grooving, prompted by reports of motorcyclists finding the surface dangerous; the consistent finding is that there is no statistically significant loss of safety.

21.2.3 Exposed aggregate

An exposed aggregate finish directly mimics an asphalt. Roller-compacted concrete naturally produces an exposed aggregate finish and, if it weren't for the surface tolerance problem mentioned in Section 21.1.2, it would be a great surface to drive on. A similar finish can be induced in wet-formed concrete – but it comes at a cost. The procedure is

- finish the wet concrete as usual, leaving a smooth surface
- spray on a retarder, which will delay the set of the uppermost few millimetres of mortar
- come back the next morning and brush away the loose (not yet set) mortar; this exposes the large aggregate particles.

It isn't complicated, but it involves an additional construction process and therefore added cost. Furthermore, the intrinsic skid resistance, the micro-texture, is now given by the large aggregate, which means it has to have a suitably high polished stone value (see Chapter 11), which adds to the cost of the PQC material itself. It will also not last forever. The aggregate will abrade under traffic loading, reducing the texture depth and therefore the wet skid resistance, and the treatments that can readily be used when this happens to asphalt are either ineffective or much more difficult on concrete.

Nevertheless, despite the extra cost, an exposed aggregate finish is a proven method of generating a concrete surface that is almost equivalent to an asphalt with regard to ride quality.

21.3. Use of an asphalt surface

Some highway authorities, for example National Highways in the UK (National Highways, 2021), specify an asphalt surface over a CRC pavement for any new concrete pavement construction, and asphalt surface properties have been discussed in Chapter 11. Bond to the underlying concrete is achieved by means of a suitably tough bitumen bond coat.

One point of relevance is the relative structural contribution of such an asphalt surface which, it was suggested in Section 11.3.1, is approximately related to the square root of the base/surface stiffness ratio. In the case of an asphalt overlay to concrete, this ratio is often nearly 20, giving a ratio of 4–5 in equivalent thickness; that is, a 40 mm surfacing becomes structurally equivalent to about 10 mm of PQC.

21.4. Summary

There is no perfect solution to the problem of concrete surfaces on major highways. It would seem that the road authority has to choose between (*a*) a noisy surface, (*b*) a lower-noise surface that may be slightly dangerous to motorcyclists, (*c*) a lower-noise but costly surface with limited life, (*d*) a lower-noise but uneven surface or (*e*) a high-cost asphalt overlay. Experiments have been carried out on various alternatives over the years, for example diagonal grooving, but in the current state of practice no single solution can yet be recommended.

REFERENCES

National Highways (2021) CD 226: Design for new pavement construction. *Design Manual for Roads and Bridges. Pavement Design*. Stationery Office, London, UK.

Sandberg U and Ejsmont JA (2002) *Tyre/Road Noise Reference Book*. Informex, Kisa, Sweden.

Nick Thom
ISBN 978-1-83549-711-1
https://doi.org/10.1108/978-1-83549-710-420243026

Chapter 22
Maintenance of rigid pavements

22.1. Deterioration mechanisms

The designs described in Chapter 20 basically addressed one thing: the pavement must not crack, or at least not in an uncontrolled manner. That's OK, but that doesn't mean that nothing can go wrong, and if problems are not dealt with as they arise then the dreaded uncontrolled cracking could indeed occur.

22.1.1 Surface texture deterioration

Whatever form of texture has been provided, it will not endure forever. Abrasion through tyre action is a fact of life, applying to both stone and cement mortar, so over the course of time the texture depth will decrease, potentially to the state where something has to be done to maintain adequate skid resistance.

22.1.2 Spalling

Spalling refers to the flaking off of small pieces of surface concrete, usually around a joint or crack. It is not hard to understand why this might happen. Figure 22.1 illustrates what happens as a vehicle tyre travels over a joint or crack. As it approaches or leaves, it induces a slight step between the slabs either side of the crack, the magnitude of which depends on the load-transfer efficiency of the joint. The step itself is no problem; it is only a few micrometres in magnitude and it recovers fully once the tyre has passed on. The problem is the shear stress across the joint, which could easily reach around 200 kPa on average as a heavy goods vehicle tyre passes over. The shear at the joint will induce a tensile stress in the concrete of similar order of magnitude, say 200 kPa, and will try to break off a small triangle of concrete. And of course the shear stress across the joint is not uniform. It is transmitted through individual pieces of coarse aggregate, which means local stresses may be many times greater, sometimes great enough to cause a local fatigue failure and to remove a small piece of concrete.

The other cause of spalling is the large horizontal compressive stress across a tightly closed crack, induced when the tyre is immediately above, also potentially inducing failure of a small triangular element of concrete.

Figure 22.1 Development of spalling at joints

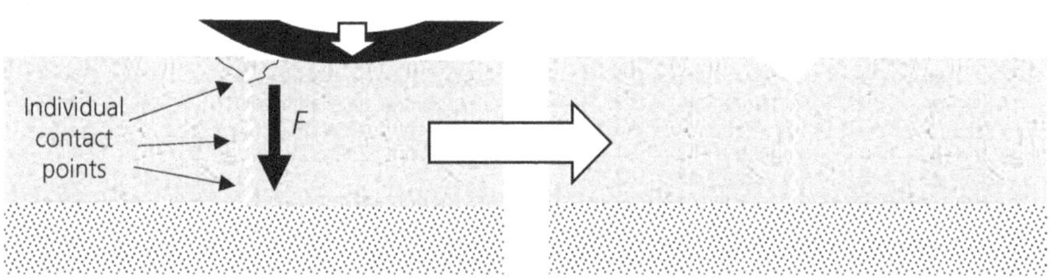

Spalling is most likely in cases where concrete placement results in an excessively thick layer of mortar on the surface. But is it a real problem? It certainly isn't a structural failure, but it contributes to a poor ride quality and also, where it occurs at joints, to failure of the joint sealant.

22.1.3 Joint deterioration

Whether or not spalling occurs, the repeated shear stresses experienced across a joint can lead to (*a*) a deterioration in the aggregate interlock efficiency and (*b*) failure of concrete immediately adjacent to the dowel (or tie-bar). With regard to reduction in interlock efficiency, Figure 22.2 presents best-fit lines from data obtained from a series of beam tests in which the rate of reduction in load-transfer efficiency (expressed as a stiffness–shear stress/differential vertical movement) was measured during multiple applications of load across a crack in the beam. The crack opening (w) was varied up to a maximum of 1 mm and the influence of this parameter is clear. Unsurprisingly, the other key factor is stress, expressed as the ratio of shear stress across the joint to flexural strength.

The data in Figure 22.2, combined with experience from various sites, suggest that the deterioration rate is, like fatigue, a function of the ratio of stress to material strength. If the 200 kPa shear stress quoted in the previous section is accepted (although note that it can be calculated approximately using multi-layer linear elastic analysis, using the stiffness expected from intact concrete), then the ratio of shear stress to flexural strength might typically be 0.04, less than any of the ratios in the figure, in which case this type of deterioration should be very slow.

The greatest deterioration rate in Figure 22.2 is seen when the crack is wide, which occurs in cold weather as joints open up. But as a joint opens, the load transfer is increasingly taken via dowels or tie-bars, reducing aggregate interlock stress, so problems should only occur if the efficiency of the dowel or tie-bar is compromised. Unfortunately, this can happen. If a large load is transferred via a dowel or tie-bar, this will induce very high stresses in the surrounding concrete, potentially

Figure 22.2 Crack deterioration under repeated shear load (Thompson, 2001)

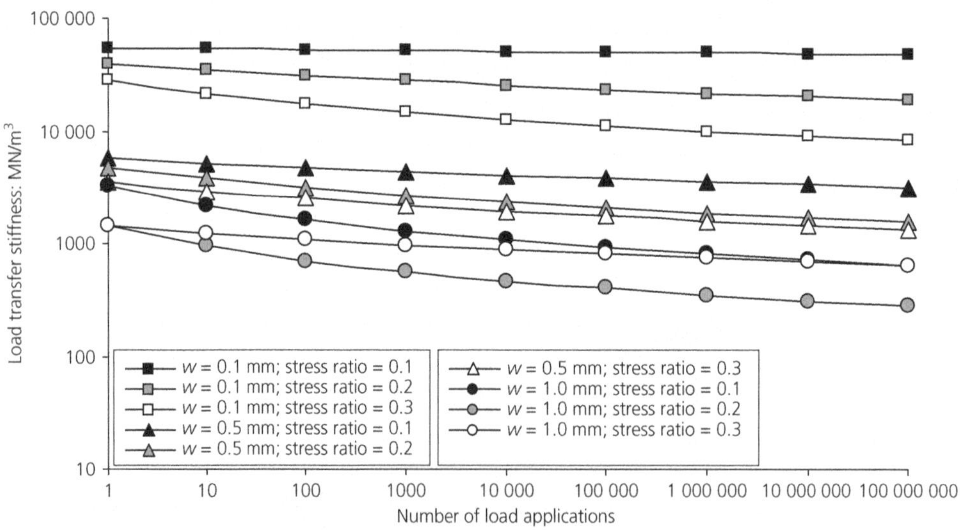

leading to localised internal failure and the effect will be that a length of dowel or tie-bar either side of the joint will lose its support. This then allows the dowel or tie-bar to flex, which reduces its load-transfer efficiency, thus throwing a greater proportion of the load onto the aggregate interlock stress.

It's all very complicated, particularly because the temperature fluctuations mean that every day is different, but the end result is that a slow loss of load-transfer efficiency is likely to occur. The stronger the concrete, the thicker the slab or the wider the dowel, the slower this deterioration will be.

In an extreme case, loss of load-transfer efficiency can lead to the distress type known as *faulting*. This is where a permanent 'step' develops between the two slabs either side of a joint. It will usually be associated with foundation damage and consequent loss of slab support, more on one side of the joint than the other.

22.1.4 Water damage

PQC will not be damaged by water directly; problems are indirect and they start with leaking joint sealant. Modern sealants can accommodate amazing amounts of movement over an enormous range of temperatures while continuing to adhere to the surrounding concrete, but they still have finite life, and the greater the stresses and strains they are asked to withstand, the shorter that life will be. Even the best joint sealant will eventually leak.

As soon as the seal begins to fail, rainwater will be able to make its way to the location of any dowel or tie-bar connecting adjacent slabs of concrete and, given the chance, it will initiate corrosion. This will increase the local damage to the surrounding concrete as steel corrosion is an expansive reaction, accelerating loss of load-transfer efficiency.

The other problem is that water reaching the foundation will accelerate permanent deformation and the loss of slab support either side of a joint, particularly if an unbound sub-base has been used. Thus, a leaky joint will lead to more rapid loss of load-transfer efficiency and a greater likelihood of faulting.

22.1.5 Transverse cracking

Transverse cracks, often at mid-distance between joints, indicate thermally induced stress, assisted by traffic loading. The pavement is making up for an excessive joint spacing. A joint spacing about 20 times the slab thickness very approximately marks the maximum spacing at which joints in unreinforced concrete can be expected to function properly. Fortunately, there is a limit to how far transverse cracks can go; they may progress to the edge of the pavement or they may stop at an adjacent longitudinal joint and go no further.

The other common location for transverse cracking is about a metre from a joint. The cause this time is upward curling of the slab end during the night when the surface is cooler than concrete at depth (refer back to Figure 20.6) combined with traffic load. It is not an inevitable occurrence; it is more likely in a long slab than a short one and would also logically be encouraged by poor joint load-transfer efficiency.

In the case of a continuously reinforced concrete (CRC) pavement, transverse cracks are expected of course. They should form at a spacing of 1–1.5 m but, in a properly functioning pavement, they will remain narrow and there should be no spalling. The occurrence of spalling implies excess relative vertical movement across cracks, which means a damaged load-transfer capability for some reason.

22.1.6 Longitudinal cracking

This almost certainly indicates traffic-induced damage. Cracks may not be exactly in the wheelpath and may wander around either side, but the likelihood is that they will jump across transverse joints and so may propagate rapidly along the pavement. Longitudinal cracking is a sure sign of overloading and a very serious mode of distress in concrete pavements; significant pavement rehabilitation is likely to be required. The cause may be inadequate PQC strength or thickness, but it is also highly likely that poor joint load transfer contributes. In Section 20.3.1 the relatively higher concrete stress adjacent to a joint compared to the centre of the slab was mentioned, suggesting that the ratio might be around 1.2 for joints in good condition rising to 1.5 if in poor condition; if correct, this means that poor load transfer could increase the tensile stress in bending by 25% which, bearing in mind the sensitivity of all concretes to overloading, would reduce the life by a factor of over 10.

22.2. Restoring surface texture

There is really only one treatment that actually restores concrete texture and that is physically saw-cutting a new set of grooves. It works, but the costs are significant. It also incrementally eats into the overall PQC slab thickness.

The alternative is to apply a surface dressing, as described in Section 7.6.1. This has been used successfully on the most heavily trafficked concrete highways and can definitely restore skid resistance. It does, however, generate a very rough and noisy surface.

22.3. Joint repairs

22.3.1 Resealing

Resealing of concrete joints (chasing out the old sealant, cleaning and applying a replacement seal) is strongly recommended as soon as they begin to show signs of deterioration and leakage; as already noted, delay really can prove fatal to the pavement.

However, it has to be acknowledged that delay is not always fatal. The world is full of concrete pavements with poor or non-existent joint seals that remain in perfectly serviceable condition. It may be the case that slab support has disappeared either side of each joint and that load transfer is negligible, but if the concrete itself is strong enough then it will still not break. So, while it is good practice to reseal defective joints, it is worth assessing the structural capacity of the concrete before panicking.

22.3.2 Shallow spall repairs

The repair must make use of suitably strong material, be well bonded to the existing PQC and joints have to be re-formed and sealed (Figure 22.3). Nowadays, rapid-setting compounds allow trafficking as little as one hour after placement (depending on ambient temperature), although the cost of such materials is, of course, considerably greater than that of standard PQC. Nevertheless, on busy highways speed is vital and, when user-delay costs are factored in, it becomes worthwhile.

Figure 22.3 Local patch repairs in concrete pavements

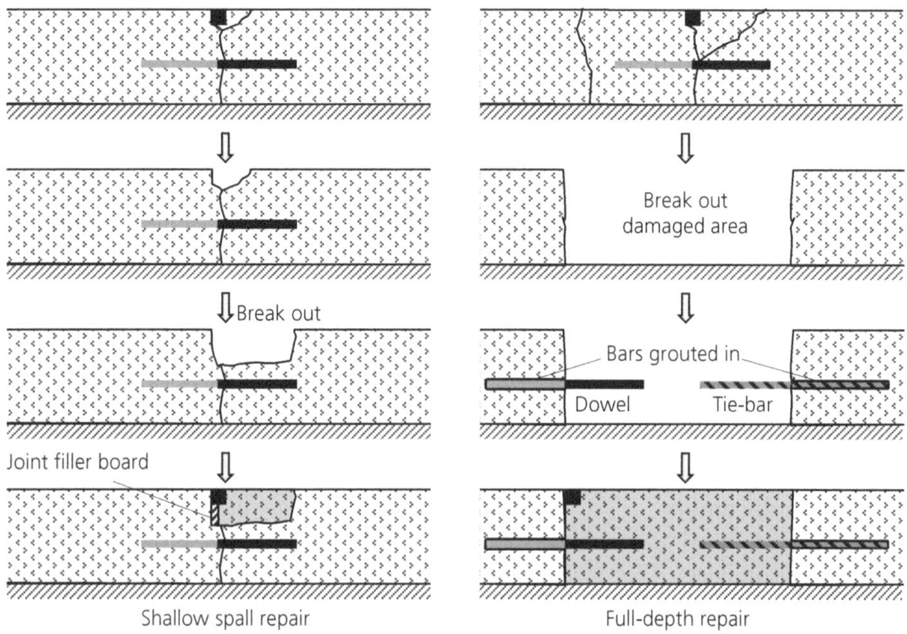

Repairs to concrete pavements are not without cost in terms of ride quality. Concrete is much more sensitive to minor surface imperfections than asphalt – as witnessed by the fact that even transverse joints are easily noticeable to the motorist – and a patch repair cannot possibly be finished to the same standard of evenness as is achieved by a concrete paving machine.

22.3.3 Full-depth joint repairs

A shallow repair is useless if the defect penetrates as far as a dowel bar. In this case (termed 'deep spalling'), the whole PQC slab thickness needs to be replaced, as shown in Figure 22.3. This is obviously a much more time-consuming procedure, even when a rapid-setting repair material is used. Joints have to be re-formed; other interfaces require that a tie-bar is inserted to avoid a crack forming easily between old and new materials. Both tie-bar and dowel installation mean that holes have to be drilled into the faces of the existing concrete and bars grouted in. This is clearly not an operation that can be undertaken lightly.

22.3.4 Under-slab grouting

Faulting is a sure sign of slab support problems, where permanent deformation of the foundation has led to the opening up of a small void between the PQC slab and the upper surface of the foundation, thus allowing the 'fault' to develop. This is illustrated in Figure 22.4. Under-slab grouting is a very effective treatment for this and several specialist processes have been developed, either using cementitious grout or an expansive geopolymer. The technique is to simply drill small holes in the slab adjacent to the joint, insert the injection lance, apply a measured quantity of grout and fill the drill holes.

Figure 22.4 Faulting in a concrete pavement

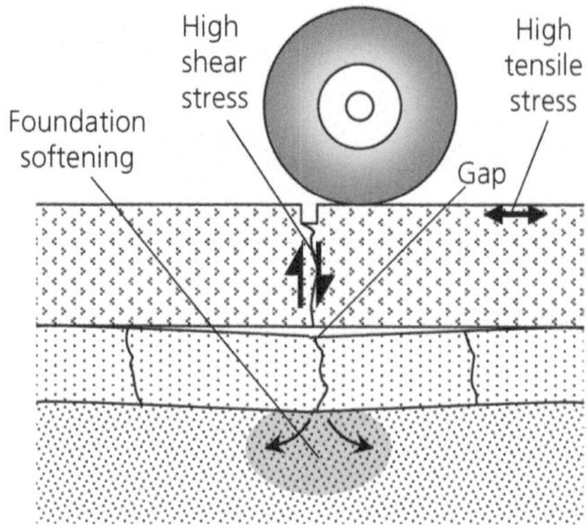

22.4. Slab repairs

22.4.1 Bay replacement

In many cases, individual concrete bays (i.e. lengths between joints) suffer cracking, while adjacent bays remain intact. This occurs as minor differences such as joint defects lead to minor cracking, and the consequent lack of concrete continuity then places ever greater stress on remaining materials, leading to a more general fragmentation of the bay. The enclosing joints form barriers to the progression of cracking beyond the individual bay, although it must be appreciated that these barriers will eventually be breached. In such a case, it is logical to carry out a whole-bay repair, following exactly the principles shown in Figure 22.3 for full-depth repairs. Care must be taken to ensure that joints are replaced and, where they are displaced longitudinally (as shown in Figure 22.3), the mismatch at adjacent longitudinal joints must not be allowed to inhibit transverse joint performance. The length of 'stagger' at longitudinal joints between new and old transverse joints should therefore be treated to allow slip (using bituminous primer or similar).

One further point is that the possibility of thickening a replaced bay should be considered. If the local deterioration was due to factors that are likely to continue (e.g. foundation weakness or high loads from cornering), then it makes sense to increase the pavement strength.

22.4.2 Crack stitching

An alternative to expensive and time-consuming full-depth repairs is the technique known as crack stitching. It does not represent as complete or long-lasting a treatment, but it can be effective at slowing down the rate of deterioration. The idea is to tie the two sides of a crack together. Thus, although the crack is not repaired, it is at least prevented from widening further, preserving load transfer and making crack sealing more likely to be effective. Figure 22.5 illustrates two different crack stitching options. In both cases, tie-bars are grouted (into slots or holes) such that they prevent further crack movement. The sealant layer will usually be recessed into a shallow channel milled into the surface, and should preferably be an asphaltic material with high deformation resistance; a cementitious product would be too brittle to withstand the expected flexure at the crack.

Figure 22.5 Crack stitching techniques

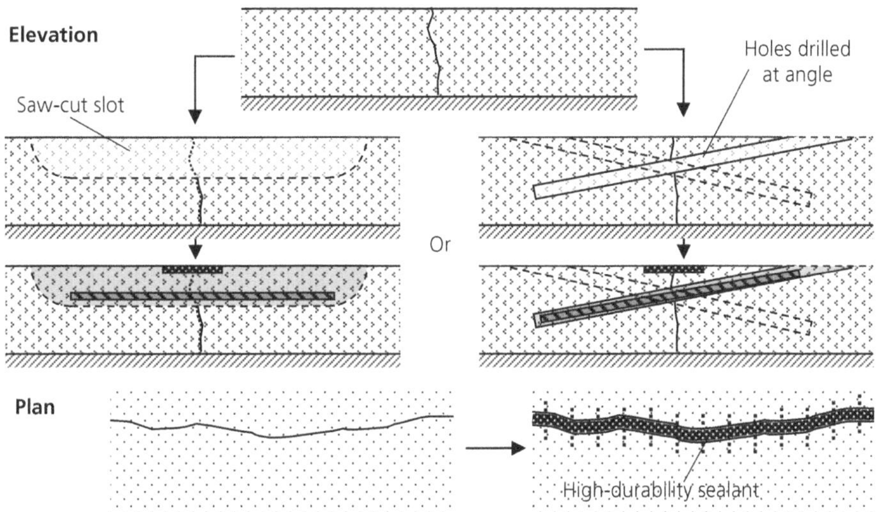

22.5. Summary

One point worth repeating is that rigid pavements are unforgiving in nature. If something goes wrong it can go very badly wrong. Getting the design (and construction) right in the first place is obviously plan A. Timely proactive maintenance such as resealing of joints and under-slab grouting is plan B. Crack sealing and full-depth repairs constitute plan C but, to be honest, once plan C is in operation the costs are becoming unsustainably high and a nightmare scenario for the highway authority is rapidly developing. Chapter 27 will describe how it may be possible to bring such a failing pavement back from the dead, but costs will not be trivial.

REFERENCE

Thompson I (2001) *Use of Steel Fibres to Reinforce Cement Bound Roadbase*. PhD thesis, University of Nottingham, Nottingham, UK.

Part 7

Water and drainage

Nick Thom
ISBN 978-1-83549-711-1
https://doi.org/10.1108/978-1-83549-710-420243027

Chapter 23
Water movement

Water content has been identified several times as a key parameter influencing material performance. Previous sections have shown that a moderate water content generates suction within an unbound material and is therefore beneficial to shear strength, stiffness and resistance to deformation under repeated load. Conversely, a high water content leads to positive pore pressures and a loss of performance in all three departments. Yet water is commonly added to unbound pavement materials in order to facilitate compaction (as explained in Chapter 7), leaving the layer in a near-saturated state – which is highly undesirable in the longer term. Permeability is, therefore, a key property.

23.1. Permeability

A highly permeable material will never be adversely affected by water as long as a drainage outlet is available. On the other hand, it will not derive any significant benefit from suction. A fine-grained low-permeability material may develop an impressive suction – but this can only be maintained if further water is prevented from entering. The first option is a low-risk, low-profit scenario; the second involves higher risk but offers the promise of high reward. Permeability must therefore be understood.

23.1.1 The fluid mechanics of permeability

The coefficient of permeability (the ratio of head loss per metre to flow velocity) depends on the size of the gaps joining void spaces within a granular matrix, and this in turn is clearly a function of particle size – but which particle size? In a single-sized material, it is possible to analyse this problem geometrically, at least approximately. However, in a graded aggregate, the difficulty is considerable.

In general, fluid flow may be either *laminar* (i.e. smooth and steady) or *turbulent* and, for most purposes, flow through paving materials and soils may be considered laminar. This means that the so-called Reynolds number (Re) (see specialist fluid mechanics texts for details) – the dimensionless quantity used to describe the state of fluid flow – will be relatively low. Equation 23.1 for Re applies to laminar flow in pipes.

$$\mathrm{Re} = DV/\nu < \text{about } 2000 \tag{23.1}$$

where D is the pipe diameter, V is the flow velocity and ν is the kinematic viscosity (1.14 mm²/s for water at 20°C).

Laminar flow also means that the quantity known as the friction factor (λ) can be directly (although approximately) related to Re number. This approximation is derived from the so-called Poiseuille equation for laminar flow.

$$\lambda \approx 64/\mathrm{Re} = 64\nu/DV \tag{23.2}$$

If flow through an unbound material is assumed to resemble flow through a circular pipe of diameter D, then the head loss H over a length L can be expressed theoretically as (see standard texts on fluid flow)

$$H \approx 16\lambda LQ^2/2g\pi^2 D^5 \tag{23.3}$$

where Q is the volume flow rate ($= V \times \pi D^2/4$) and g is acceleration due to gravity (9.81 m/s²).

Substituting for λ and Q gives

$$H = 64\nu LV/2gD^2 \tag{23.4}$$

Inserting values for ν and g and rearranging

$$V = 269\,000D^2(H/L) \text{ (in m/s)} \tag{23.5}$$

The trick now is to know what value of D is applicable within the matrix of an unbound material and also to make an appropriate conversion from a regular cylindrical pipe to a tortuous path between particles. Any estimate will therefore be approximate.

23.1.2 Predicting permeability – single-sized material

Permeability is expressed as a coefficient k (in m/s) and the key equation is Darcy's law (Darcy, 1856).

$$V = k(H/L) \tag{23.6}$$

where all the parameters are as defined earlier. H/L is known as the *hydraulic gradient*.

Clearly the form of Darcy's law is similar to the equation derived above for pipe flow and the permeability k therefore equates to $269\,000D^2$ – but what value should be assigned to D? A simple procedure is to assume that the water has to flow through tubes shaped as shown in Calculation Sheet 23.1. This ignores the fact that the void available for water flow opens up between the constrictions illustrated in the figure, giving a greater average cross-section, but it also ignores flow path tortuosity. The two simplifications are therefore opposite in effect, meaning that the prediction may be acceptably close to the truth. The result, derived in Calculation Sheet 23.1, is that $k = 2300r^2$ m/s.

23.1.3 Predicting permeability – graded aggregate

Although the prediction shown in Calculation Sheet 23.1 may be about right for a single-sized material, such materials are few and far between in real pavement engineering. The question, therefore, is just how to choose a representative particle size within a graded aggregate – the particle size that dictates resistance to water flow. And it is suggested here that the concept of *critical particle size*, the particle size where the actual gradation just touches that of a Fuller curve (as described in Section 6.1.3), also applies here. This concept is outlined in Calculation Sheet 23.2 and a reasonably practical methodology is suggested.

Calculation Sheet 23.1. Predicting permeability

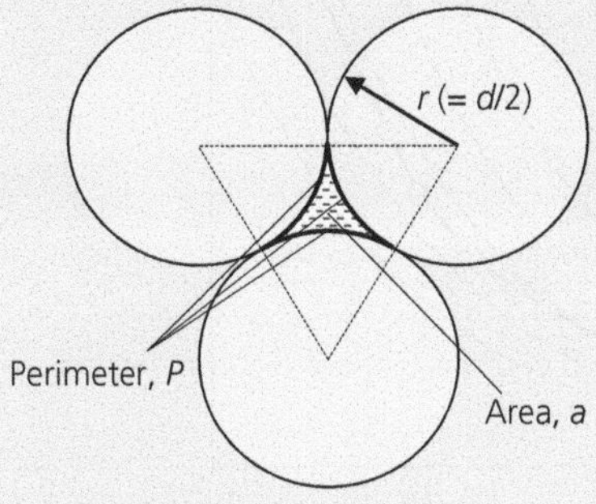

Area, a

$= \text{area of triangle less } 3 \times 60° \text{ segments}$

$= \sqrt{3} \times r^2 - 0.5\pi r^2$

$\approx 0.16r^2$

$= \text{equivalent to circle of diameter } 2\sqrt{(0.16/\pi)}r$

Void as proportion of total area $= 0.16/\sqrt{3} = 0.092$

Perimeter, P

$= 3 \times 60° \text{ sectors} = \pi \times r$

$= \text{equivalent to circle of diameter } r$

Equivalent to flow in circular pipe (see text):

$$V_{pipe} \approx 269\,000D^2(H/L)$$

where $D = 2\sqrt{(0.16/\pi)}r$ – except factor V_{pipe} by $\pi D/P$ due to drag on increased perimeter

$\therefore V_{pipe} \approx 25\,000r^2(H/L)$

Darcy's law (see text): $V = k(H/L) = V_{pipe} \times 0.092$

$\rightarrow$ Permeability $k \approx 25\,000r^2 \times 0.092 \approx 2300r^2$ m/s

Calculation Sheet 23.2. Critical particle size

The series of Fuller curves shown in the figure represents idealised gradations giving maximum packing density, but with differing maximum particle sizes. If a real grading curve is steeper than the Fuller curve, particles will be loose; if it is less steep, larger particles will become separated from each other in a matrix of smaller-sized particles.

The tangent point shown is the **critical size** (d_{crit}), representing the smallest matrix through which water must flow.

Initial estimate

$$k(d_{crit}) = 2300r^2 \text{ (Calculation Sheet 23.1)} = 575d_{crit}^2$$

Adjust because particles larger than d_{crit} form impermeable barriers

$\rightarrow \times \%_{crit}/100$ where $\%_{crit}$ = percentage passing d_{crit}

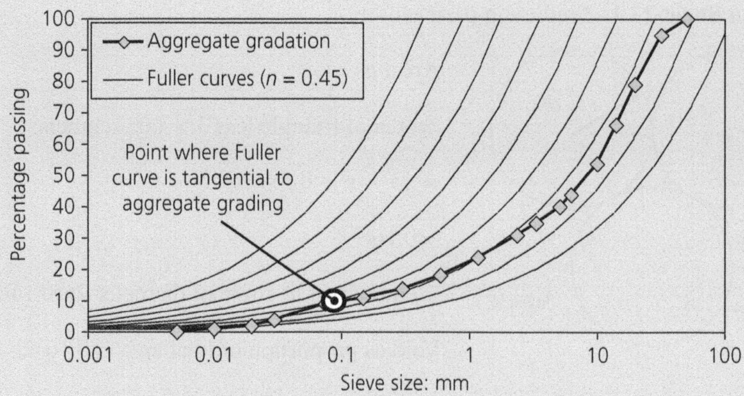

Adjust because sizes $< d_{crit}$ also represent additional flow restrictions

$$\rightarrow \times (1-\text{AR})$$

where AR = area ratio = (area under aggregate gradation; $< d_{crit}$)/(area under Fuller curve; $< d_{crit}$)

Improved estimate

$$k \approx k(d_{crit}) \times \%_{crit} / 100 \times (1-\text{AR})$$

Example

$d_{crit} = 0.09$ mm; $\%_{crit} = 10\%$; AR = 0.7

$\rightarrow k = 575 \times 0.09^2 \times 10^{-6} \times 10/100 \times (1-0.7) = 1.410^{-7}$ m/s

An alternative relatively widely used and much simpler method is Hazen's formula (Hazen, 1892).

$$k \text{ (in m/s)} = cd_{10}^2 \qquad (23.7)$$

where c is a coefficient, usually taken to be 0.01, and d_{10} is the sieve size (in mm) through which just 10% of the material passes.

Variations on Hazen's formula have been proposed using either d_5 or d_{15}, but it has to be acknowledged that such methods cannot possibly cater for the full range of possible aggregate gradations. For this reason, many engineers simply fall back on assumptions of permeability based on material description. Table 23.1 contains such commonly used assumptions.

As a check on these different approaches, consider the four different materials with the gradations shown in Figure 23.1. The first is a silty clay soil; the second is a clean sand; the third and fourth are typical of sub-base layers in road construction. In each case, the critical particle size according to the procedure in Calculation Sheet 23.2 is shown, as is the d_{10} value in three cases. Permeability estimates using the three techniques are given in Table 23.2.

Table 23.1 Typical coefficients of permeability. The soil descriptions are based on the Unified Soil Classification System (USCS) given in ASTM (2020); the permeability ranges quoted are simply intended as broad estimates

Material	Typical permeability range: m/s
Well-graded gravels	10^{-5} to 10^{-3}
Poorly graded gravels	5×10^{-5} to 10^{-3}
Silty gravels	10^{-8} to 10^{-4}
Clayey gravels	10^{-8} to 10^{-6}
Well-graded sands	5×10^{-6} to 5×10^{-4}
Poorly graded sands	5×10^{-7} to 5×10^{-6}
Silty sands	10^{-9} to 10^{-6}
Clayey sands	10^{-9} to 10^{-6}
Low-plasticity silts	10^{-9} to 10^{-7}
Low-plasticity clays	10^{-9} to 10^{-8}
High-plasticity silts	10^{-10} to 10^{-9}
High-plasticity clays	10^{-11} to 10^{-9}

Figure 23.1 Example materials for estimation of permeability

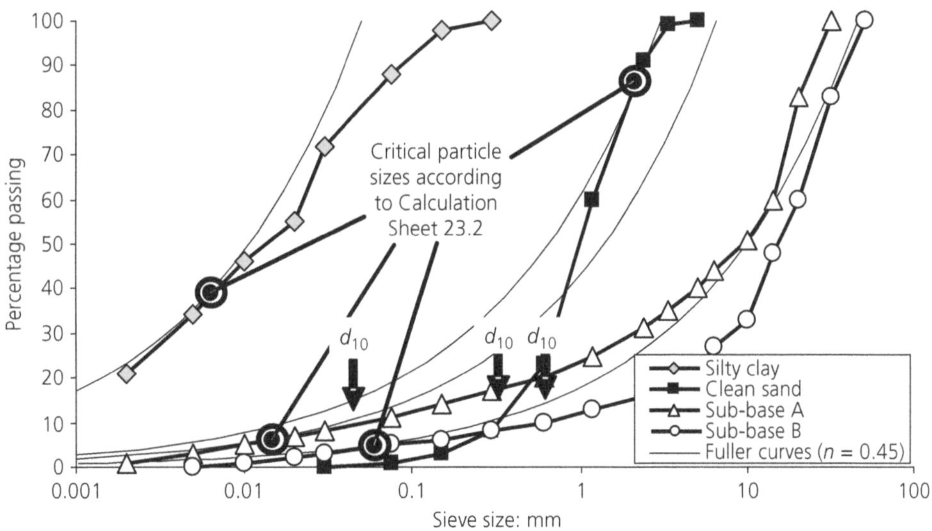

The chief discrepancies in Table 23.2 are in the sub-base – and they are very big discrepancies indeed. This illustrates the point that it is extremely difficult to predict the permeability of a graded aggregate. Such materials are highly sensitive to small variations in gradation, particularly in the fraction smaller than 75 μm, the details of which are not generally known. This is reflected by the

Table 23.2 Comparison of permeability predictions

Material	d_{10}: mm	Calculation Sheet 23.2			Predicted permeability: m/s		
		d_{crit}: mm	%crit	Area ratio (AR)	According to description	Hazen's formula	Calculation Sheet 23.2
Silty clay[a]	0.0005	0.006	39	0.9	10^{-9} to 10^{-8}	2.5×10^{-9}	3.4×10^{-9}
Clean sand	0.32	2.05	87	0.4	5×10^{-6} to 5×10^{-4}	1.0×10^{-3}	1.3×10^{-3}
Sub-base A	0.044	0.014	6	0.6	10^{-8} to 10^{-4}	1.9×10^{-5}	1.1×10^{-8}
Sub-base B	0.60	0.06	4.5	0.7	10^{-8} to 10^{-4}	3.6×10^{-3}	1.1×10^{-7}

[a]d_{10} value for silty clay extrapolated from data in Figure 23.1

large range given for silty gravel (taken to represent the sub-base) in Table 23.1, and seemingly inexplicable scatter in test data is common. Note also the fact that the Fuller curve nearly touches the grading curve for sub-base B a second time – at 30–40 mm. A slightly greater percentage of large particles, meaning insufficient fine material to fill the space available, would have resulted in a very different prediction indeed.

However, a key practical point is that all such materials are segregated to some extent, which means that relatively permeable paths will exist alongside almost impermeable zones. In practice, therefore, the average permeability of a graded aggregate may be one or two orders of magnitude greater than theoretically expected. It should also be noted that vertical permeability is often lower than horizontal permeability because of the dense closed surface formed under roller compaction.

23.1.4 Measuring permeability

This is a fairly straightforward procedure when carried out in a laboratory and there are numerous test specifications to choose from. The key point is that a specimen of material of uniform cross-section has to be enclosed by walls that do not allow preferential drainage paths to develop, and it is then subjected to a small difference in water pressure between the two ends of the specimen. Figure 23.2 illustrates the concept.

The arrangement shown in Figure 23.2 is particularly appropriate for unbound pavement layers as it is the horizontal permeability that is of most significance to the performance of a pavement. However, it is extremely important that the lid of the test box is pressed down into the material, generally via a layer of rubber, otherwise water will pass easily over the surface and lead to a false result. The other common arrangement, particularly for soils, is a vertical cylinder of material, and the easiest form of test to carry out involves allowing water to permeate vertically under gravity. However, results should always be treated with as much caution as predictions because (*a*) repeatability is often poor due to sensitivity to small variations in local gradation, (*b*) the hydraulic gradient used is often unrealistically large for pavement conditions and (*c*) vertical permeability is less important than horizontal permeability in most pavement situations. On the subject of variability, Deb and Shukla (2012) reviewed numerous published studies and found that a factor of 10 difference between minimum and maximum measurements is not uncommon, and that different types of test can give rise to very different results.

Figure 23.2 The principle of permeability testing in the laboratory (Jones and Jones, 1989)

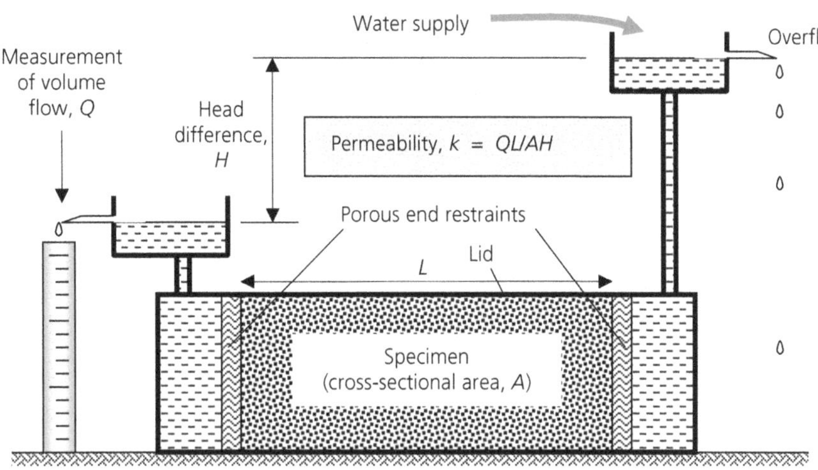

23.2. Partial saturation and suction

This subject is highly relevant to pavement foundations. They are constructed wet, often at approaching 90% saturation, in order to aid compaction (see Chapter 7), but it is expected that their long-term condition will be much drier than this. However, this depends on the propensity of the material to retain water, which in turn depends on particle size, very much as permeability does. The point is that the surface-tension effect at an air–water interface generates a pressure difference between the water collected near to a particle contact and the surrounding air, and this acts to resist water drainage. Calculation Sheet 23.3 presents an approximate prediction of the suction applying in a wet material. Figure 23.3 shows a plot of maximum predicted suctions (i.e. negative pore pressures) against particle size, together with the degree of saturation at which the maximum suction value occurs.

Calculation Sheet 23.3. Predicting suction

Surface tension T (N/m) acts along the edge of the air–water interface, at an angle ϕ to the face of the particle (0° if wet; up to 30° if dry). The water pressure is p below the air pressure.

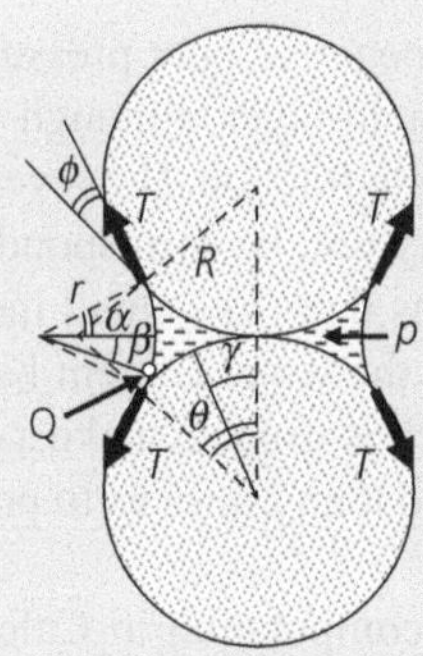

Geometric relationships

$$\alpha = (\pi/2) - \theta - \phi$$

$$r = R(1 - \cos\theta) / \sin\alpha$$

Perimeter at edge of wetted area $= 2\pi R \sin\theta$

$\therefore$ Total horizontal surface tension force $= 2 \times T \times 2\pi R \sin\theta \times \sin\alpha$

Perimeter at general point Q on the air–water interface $= 2\pi(R\sin\theta + r\cos\alpha - r\cos\beta)$

$\therefore$ Horizontal force due to $p = 2p\int 2\pi(R\sin\theta + r\cos\alpha - r\cos\beta)\times r\cos\beta\, d\beta$ from 0 to α

$$= 4\pi\, pr[(R\sin\theta + r\cos\alpha)\sin\alpha - \tfrac{1}{2} r\sin\alpha\cos\alpha - \tfrac{1}{2} r\alpha]$$

Equate forces: $p = TR\sin\theta\sin\alpha / r[(R\sin\theta + r\cos\alpha)\sin\alpha - \tfrac{1}{2} r\sin\alpha\cos\alpha - \tfrac{1}{2} r\alpha]$

Vertical inter particle force due to $p = \int p\times 2\pi R\sin\gamma\times R\, d\gamma\times\cos\gamma$ from 0 to θ

$$= \pi pR^2\sin^2\theta$$

$\therefore$ Effective pressure due to suction given by: $\pi pR^2\sin^2\theta/\pi R^2 = p\sin^2\theta$

Figure 23.3 The effect of surface tension in an idealised single-sized aggregate

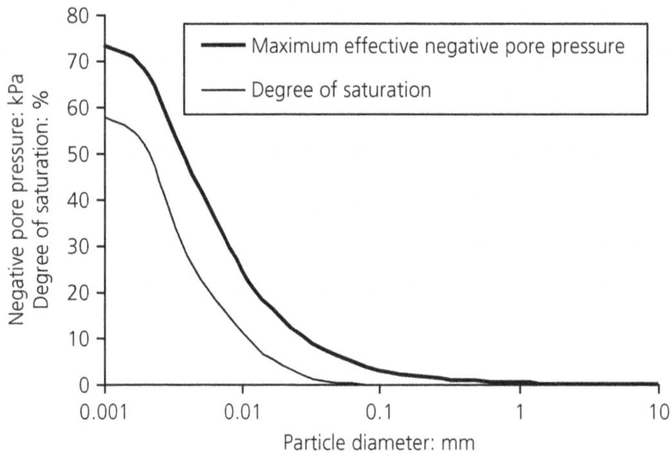

- Surface tension of 73 mN/m assumed
- Single-sized aggregate, 40% void space
- Drained condition (i.e. equilibrium)

The negative pore pressures shown in Figure 23.3 are highly beneficial to the performance of an unbound layer, although these benefits will only accrue if the material contains a significant fine particle fraction. However, there is a price to pay. If the particle sizes are too small, then the permeability will be correspondingly small and the equilibrium degree of saturation will be high, which means that (*a*) a small increase in water content can drastically affect material properties and (*b*) once the water content has increased, it will take a long time to drain. For sensible design, therefore, it is usual to make a compromise, allowing enough fines for some suction to be generated, but not so many fines as to prevent drainage.

The computation in Calculation Sheet 23.3 and Figure 23.3 relates to an idealised single-sized material. For a graded aggregate, it is suggested that predictions should be carried out for the *critical particle size*, which would give effective suctions of about 10 kPa and 3 kPa, respectively, for sub-bases A and B in Figure 23.1 at low water contents under drained conditions, and indirect laboratory measurement of effective suction tends to give values in this range for a fine-graded aggregate.

One result of all this is that suction will tend to draw water up into a pavement foundation from below until the water state reaches equilibrium. This is termed 'capillary rise'. The extent to which this happens depends on the connectivity of the water phase and is also complicated by the movement of water in vapour form. However, if it is assumed that connectivity is present at above about 30% saturation, then the approach in Calculation Sheet 23.3 suggests that capillary action could lift liquid water by 700 mm and 150 mm in sub-bases A and B, respectively. Vapour would then create a high-humidity atmosphere within the air phase, allowing water to condense out at higher levels than this. These effects self-evidently become important if an overlying bound material is potentially susceptible to water attack, and they should be considered in overall pavement and drainage design and related material specification.

23.3. Summary

This chapter has prepared the essential background to controlling the water content in a pavement, and the detrimental effect of water on all pavement materials has been emphasised many times throughout this book. The key points are

- permeability of unbound materials is highly sensitive to small changes in particle size distribution
- nevertheless, it is possible to make sensible predictions
- the effects of suction, both in enhancing material properties and in generating capillary action, need to be understood.

The next chapter will take this forward into design of sub-surface pavement drainage.

REFERENCES

ASTM (2020) D2487-17: Standard practice for classification of soils for engineering purposes (Unified Soil Classification System). ASTM International, West Conshohocken, PA, USA.

Darcy H (1856) *Les Fontaines Publiques de la Ville de Dijon*. Dalmont, Paris, France (in French).

Deb SK and Shukla MK (2012) Variability of hydraulic conductivity due to multiple factors. *American Journal of Environmental Science* **8(5)**: 489–502.

Hazen A (1892) *Some Physical Properties of Sands and Gravels with Special Reference to Their Use in Filtration*. Massachusetts State Board of Health, Boston, MA, USA, 24th Annual Report.

Jones RH and Jones HA (1989) Horizontal permeability of compacted aggregates. In *Unbound Aggregates in Roads* (Jones RH and Dawson AR (eds)). Butterworths, Sevenoaks, UK, pp. 70–77.

Nick Thom
ISBN 978-1-83549-711-1
https://doi.org/10.1108/978-1-83549-710-420243028

Chapter 24
Drainage

This chapter is intended as a reasonably complete review of drainage issues. The topic is enormously important for all pavement types since all paving materials are to some extent vulnerable to the presence of excess water.

24.1. Practical drainage options

Fundamentally, drainage has two roles – to remove surface water as rapidly as possible and to drain the internal structure of the pavement. The first is partly driven by safety issues, maximising skid resistance and minimising spray, but also in order to minimise the time during which surface water has the opportunity to seep into the underlying structure. The second recognises that no pavement surface is ever perfectly sealed and so water will always access the pavement layers to some extent. This section explains the techniques by which these two roles are generally achieved.

24.1.1 Crossfall and camber

Maintaining a slope on the pavement surface for surface drainage reasons is a key element in alignment design. But an important question is: what should that slope be? For surfaced roads, most highway design standards specify 2–3% as a minimum crossfall, which is obviously a compromise between the competing demands of drainage and vehicle control. On airfields, where vehicle control demands are more stringent, 1.5% is the norm, even less across wide apron areas where planes are refuelled.

For roads that are paved by a paving machine with a rigid screed, the crossfall across each paved width will be approximately constant. Unless superelevation is applied due to cornering, the crossfall will be outward either side of the road centreline. In cases where the pavement is not paver-laid, typically unbound pavements or those with only a surface dressing, the most convenient shape from a construction point of view will be more curved with zero gradient at the centre and an increasing gradient toward the edges of the road. This is known as a 'camber'.

24.1.2 Types of drain

Both water coming off the pavement surface and water from within the pavement have to be removed somehow and transported to an outfall of some sort. This book will not cover drainage system design, but the immediate means of collecting water from the pavement can affect pavement performance and so will be covered here. Figure 24.1 illustrates the types of drains commonly used.

Side ditches would be the default first-choice drainage structure. The reason is their large capacity for water storage, meaning that the system has a good chance of coping with an intense rainfall event. They are also relatively easy to clean and maintain. The downside is land take and, in an urban environment, ditches may not be feasible. Note the need to connect the road surface to the ditch, often by means of so-called 'grips' (small lateral channels). Also note that if ditches are to

Figure 24.1 Drainage options

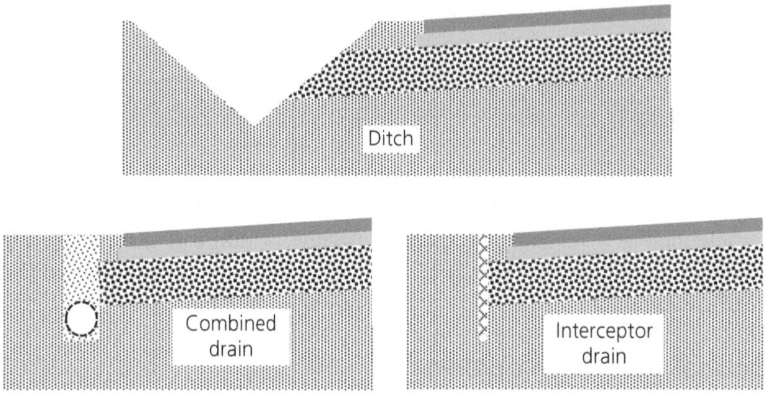

be effective at sub-surface drainage (see Section 24.3) – then a permeable layer of sufficient water-carrying capacity has to extend from the pavement foundation to the ditch. Unfortunately, this is something that is not always done well, for example relying on a series of small-diameter pipes that can easily become clogged and, depending on the nature of the subgrade, this can lead to a build-up of water within the pavement structure and consequent degradation in material properties.

In urban environments, surface water would typically be transported to a sub-surface carrier drain, either via gullies or drainage channels set into the surface.

Also illustrated in Figure 24.1 is a combined surface and sub-surface drain. Fundamentally, this does the same job as a side ditch but with much reduced water storage capacity (typically 10% or less) and much reduced land take (typically 0.5 m against 4 m). It can be located immediately next to the carriageway since it does not present the same hazard to motorists as an open ditch, making connection to road foundation layers relatively easy. However, combined drains are undeniably hard to maintain. Ideally, the upper part of the drainage medium will be designed as a filter, allowing water to pass but trapping detritus, which means it is only necessary to replace the upper part, perhaps 150 mm, from time to time. However, this ideal scenario is not always achieved and it is not uncommon for clogging throughout the drainage medium to impair functionality – but of course this isn't easy to observe from the surface. Radar systems exist that allow remote monitoring and these may become more commonly used.

The final type of drain illustrated in Figure 24.1 is an interceptor drain. This has virtually zero land take but very little capacity, which means it is suited to slow seepage-type water ingress, either from the surrounding terrain where a road is built into a hillside or from a pavement foundation. It can therefore be effective at reducing the sub-surface water table but cannot play any part in removing surface water. Maintenance access is generally impossible so the system has to be designed to develop a natural filter and so not to clog.

24.1.3 Depth of drain

One of the key design decisions is the depth of drain installation. It clearly has to be deep enough to allow water to flow from the pavement foundation, but there is also the issue of the underlying water table. Section 23.2 described the issue of capillary rise and, if the water table is too close to the pavement layers, there is the danger that water will be drawn up to the extent that it compromises the performance of the pavement materials. Naturally, cost has to be borne in mind but it is advised that the potential harm done by capillary rise effects, particularly if this may extend up into bound layers, should be taken seriously.

24.1.4 Drainage layers

Road sub-base is commonly described as free-draining – and sometimes it is. But Chapter 23 illustrated the extreme sensitivity of a dense-graded aggregate to small changes in gradation and this means that, if a permeable layer is required, then it should be specified to be a little more open-graded than a typical dense-graded sub-base. This is a significant problem with many current pavement design standards and specifications and, as Section 24.3 will illustrate, it can mean that the pavement foundation reaches high levels of saturation. A genuinely free-draining layer, either sub-base, capping or a drainage geosynthetic, connected through to a side drain, should ensure optimal water conditions and maximised material properties.

24.2. Surface water flow

For both surface water and flow in drains the key equation is the Manning formula. Calculation Sheet 24.1 goes through usage of this equation for different scenarios including surface flow. It is an empirical equation rather than fundamental, but it gives a pretty good estimation of water flows over a very wide range of scales, from small pipes to rivers. For the particular case of surface flow, the A/P ratio simplifies to water depth d, which means that the flow velocity is proportional to $d^{3/2} \times s^{1/2}$, which in turn means that the flow volume is proportional to $d^{5/2} \times s^{1/2}$. For example, if the slope is halved, for a given water flow volume, the water depth will only increase by about 13%, revealing relative insensitivity to slope.

Calculation Sheet 24.1. Calculating water flows

Manning formula: $v = [(A/P)^{2/3} s^{1/2}]/n$

where v = water velocity (m/sec)

A = flow cross-sectional area (m^2)

P = wetted perimeter (m)

s = slope (no units)

N = Manning coefficient (m$^{-1/3}$.s)

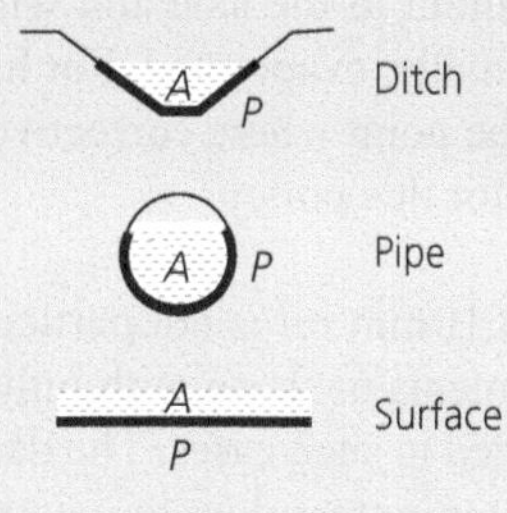

Typical Manning coefficients are:

Concrete	0.012
Asphalt	0.016–0.020
Earth	0.022
Vegetated channel	0.030

During rainfall on a cambered road, the water flow volume will be proportional to distance from the centreline. To achieve an even water depth across the road and therefore even skid resistance in both wheel-paths, this means that the slope has to be proportional to the square of the distance from the centreline, which as near as makes no difference means the shape should be a circular arc. This may be impossible to achieve with a paver, but it is a natural shape for a road shaped by a dozer (i.e. an unsurfaced road).

24.2.1 Effect of texture

Surface texture affects water depth because

- it will accommodate a certain rainfall depth, typically around 1 mm
- it will affect the Manning coefficient and therefore the water depth during rainfall, but not by much
- interconnectivity of voids within the texture potentially allows flow without surface water ever rising above the level of the stones in contact with the tyres.

The last point is particularly important. For example, a surface dressing may allow a flow calculated with water depth around 30% of the texture depth before the water reaches the level of the tyre–stone contact. With a texture depth of 1.2 mm, a Manning coefficient of 0.02, a crossfall of 2.5% and a rainfall intensity of 20 mm/h, a lane width of 2.5 m can be drained in this way. On the other hand, for a dense asphalt surface it may be that only 10% of the texture depth is available, meaning that only 0.4 m width can be drained before the water covers the surface stones, reducing skid resistance as explained in Chapter 11.

24.2.2 Effect of surface deformation

The crossfall slope may have only secondary influence on water depth during rainfall when the value is in the normal design range, say 1.5–5%, but if it falls to zero then there will be no water flow at all and ponding will occur. Most obviously this happens if ruts form. For example, in the case of a 10 mm-deep rut across a wheel-path of width 0.8 m, and assuming a sinusoidal surface profile across the rut, water will pond over a width of about 350 mm to a depth of 2 mm. This water will never drain to the side and will only disappear as it is splashed to the side by the action of tyres and, finally, evaporated. For high-speed roads, a 10 mm rut is often seen as an 'intervention' level (i.e. the point where corrective maintenance is needed), and the danger of ponding provides a rationale for this policy.

However, a 10 mm rut is not particularly large in the case of low-traffic roads, particularly unsurfaced pavements or those with only a surface dressing, and so a greater design crossfall is to be recommended in such cases. The danger here is not only a matter of traffic safety but also the damage that water ingress does to material properties.

24.3. Sub-surface water flow

Chapter 6 has made very clear the effect, both positive and negative, that water can have on material properties. A low (but non-zero) water content gives rise to suction and an increase in strength and stiffness; a high water content leads to positive pore pressures under traffic load and decreased strength and stiffness. Given that a certain amount of seepage through the surface is inevitable, the trick is to try to use gravity to maintain the water content at each level at an acceptably low level – and then somehow to persuade the water to make its way to a side drain. In reality, the subject is enormously complex and the reader is referred to Dawson (2008) for a detailed study. However, subsequent sections will describe a simplified approach.

24.3.1 Vertical flow under gravity

Darcy's law was introduced in Chapter 23. It states that flow velocity = permeability × hydraulic gradient ($V = k(H/L)$). In the case of a surface layer with water available at the surface and atmospheric pressure at the bottom of the layer, the hydraulic gradient (H/L) works out as 1 and, once the layer is saturated, the inflow velocity is (approximately) equal to the permeability. The question is: what happens next?

Figure 24.2 presents four scenarios depending on the relative permeability of the different layers.

In case A, the previous paragraph doesn't apply because the pressure at the bottom of the surfacing will not be atmospheric. The inflow rate will be dictated by the permeability of the less permeable base, which will itself also become saturated. The water will pass easily through the very permeable sub-base, which will be at a relatively low level of saturation, and down through the fairly permeable subgrade with an intermediate level of saturation until it reaches the water table. In this scenario, the base will have practically zero suction and will be very vulnerable to positive pore pressures under traffic, and the surfacing will also be vulnerable to whatever damage prolonged exposure to water might do.

In case B, both the foundation layers and the subgrade are all more permeable than the surfacing. The base will remain unsaturated and with a small amount of beneficial suction, increasing again once the rain stops; similarly the sub-base and subgrade.

Case C reflects the fact that the subgrade commonly includes a clay content and is of low permeability. The problem now is that the water will 'back up'. The subgrade will only allow a small rate of ingress (approximately equal to its permeability) and if the rain continues for long enough all the overlying layers will become totally saturated, obviously an undesirable situation.

The solution is case D, where water is removed from the system at sub-base level.

24.3.2 Transverse flow to a drain

If a path is available to a side ditch, a combined drain or an interceptor drain then it is possible to calculate (approximately of course) the rate at which water can be removed from the system, as in case D in Figure 24.2. Calculation Sheet 24.2 explains how this can be done.

Figure 24.2 Relative permeabilities of different layers

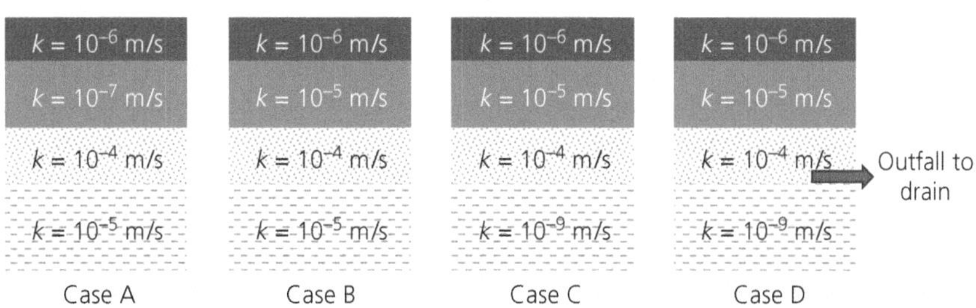

Calculation Sheet 24.2. Calculating sub-surface water flow

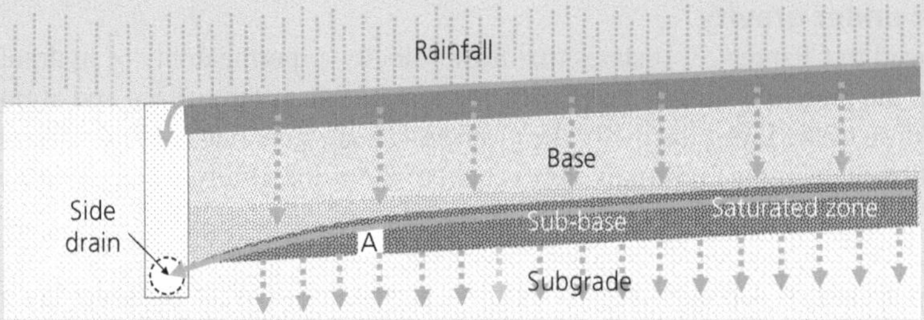

1. Water penetrates the asphalt according to its permeability
2. Surplus rainwater travels straight to the side drain
3. Water continues to travel vertically through the pavement
4. A saturated zone builds up at the bottom of the pavement
5. Some water seeps from the saturated zone into the subgrade
6. The remainder flows into the side drain

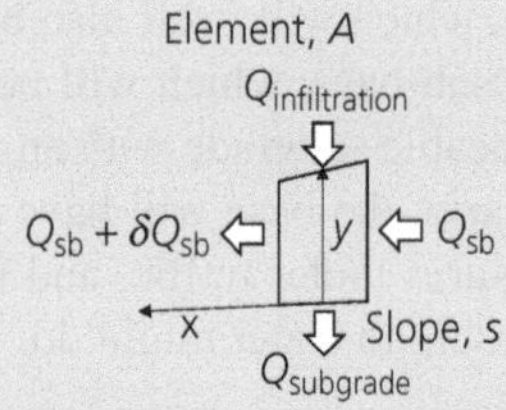

The following equations apply:

$$\delta Q_{sb} = Q_{infiltration} - \delta Q_{subgrade}$$

$$Q_{sb} = ky(dy/dx + s) \text{ per metre}$$

In a real scenario, some of the transverse flow may be forced to take place through the base or even the surfacing, depending on the permeability and thickness of the sub-base, but that's a matter of design; in a well-designed structure the sub-base will have enough capacity. The sequence of events during rainfall is therefore as follows.

- Water percolates into the surfacing for as long as the rainfall event lasts.
- In theory, the time between the water's arrival at the surface of the subgrade and its stopping is the same as the rainfall duration – but non-uniformity in material permeabilities will inevitably extend this to some extent.
- Inflow into the subgrade occurs throughout this extended duration and for a further period while remaining water in the sub-base makes its way to the side drain.

For example, with the permeabilities shown in case D in Figure 24.2, a crossfall of 3% over a pavement width of 4 m and 200 mm thickness each for surfacing, base and sub-base, under a 1.5 h rainstorm of intensity 5 mm/h, it is predicted that the water would temporarily saturate the sub-base and most of the base, and that water would be available to percolate into the subgrade for just over 4 days. However, if a much more free-draining sub-base was used, with a permeability of 10^{-3} m/s, then saturation would only ever occur within the bottom 70 mm of the sub-base and free water would only be present on top of the subgrade for about 18 h. Thus, while the method in Calculation Sheet 24.2 is far from precise, it clearly gives useful information to a designer.

24.3.3 Effect on pavement performance

This topic has been stressed repeatedly. The previous section has outlined how a successful drainage system should function. Basically

- keep water from getting in
- if it ever does get in, make sure it gets out quickly.

A key principle is that materials should be increasingly permeable with depth down to a drainage layer of some sort, either the subgrade itself if permeable or – more likely – a sub-base or capping connected through to a side drain, but this is not easy to ensure. But what happens if this principle is violated? Although the answer is obvious, it is worth spelling out the consequences since these can be very severe.

24.3.3.1 Surface course more permeable than underlying course

In wet weather, the surface course becomes saturated, the bitumen–aggregate bond is weakened, water-filled voids are pressurised by vehicle tyres and the layer is vulnerable to frost. The result is a much-increased damage rate for the surface course, leading to ravelling and, eventually, potholes.

24.3.3.2 Asphalt layers more permeable than the underlying base/sub-base

This circumstance can happen if the uppermost unbound layer is too fines-rich and compaction creates an almost impermeable top surface. In this case, the entire asphalt thickness can become temporarily saturated, feeding water into any accessible void, for example voided zones at the bottom of individual asphalt courses. This again promotes asphalt degradation as well as interlayer debonding.

24.3.3.3 Inadequate permeability in unbound layers/lack of drainage connection

Clearly, in this case, the foundation layers will be too wet for too long, also allowing too much water to infiltrate into the subgrade. The result is a reduction in modulus (poorer support to the asphalt) and strength (possible internal deformation – so-called structural rutting).

24.4. Effect of water on pavement material properties

This section will concentrate on how to select long-term material properties that reflect the effectiveness or otherwise of the drainage and will draw attention to problems that can occur if drainage is poor.

24.4.1 Subgrade softening

Experience suggests that a softening of the upper 100–500 mm of any low-permeability soil beneath an unbound foundation should be accepted as inevitable, with the actual degree of softening being seasonal, as more or less water makes its way through the pavement and into the foundation, and of course being drainage dependent. This loss of performance will of course worsen if the side drains fail to function properly. But how can this effect sensibly be predicted?

The key effect is that shown in Figure 6.4 with its attendant equation relating suction to position on the scale from plastic to liquid limit, as follows.

$$\text{Suction}(\text{kPa}) \approx 63\,[1-(w-\text{PL})/(\text{LL}-\text{PL})]^3 \tag{24.1}$$

This means that a soil at its plastic limit (PL) will suck in any free water that is available with a suction pressure of about 63 kPa but that as its water content (w) then increases, the suction pressure will reduce. If it ever reaches the liquid limit (LL) there would be no more suction and so no more increase in water content. Figure 24.3 illustrates for the case of a medium-plasticity, low-permeability (10^{-9} m/s) clay.

The time shown in Figure 24.3 is the time for which free water is available just above the subgrade. If the surface layer is of very low permeability then this may never happen; if the layer immediately above the subgrade is free draining and well connected to a side drain, then any water that does get through will quickly disappear. And it has to be remembered that the percentage of the year when it is raining (or snow is melting) is relatively small, even in fairly wet climates (maybe around 5% in the UK for example), and while there may be no water inflow for long periods there will probably still be continual, if very slow, flow down to an underlying water table.

Nevertheless, Figure 24.3 does warn of the danger. If the upper layers are relatively permeable for whatever reason and the foundation layers above the subgrade do not get rid of the water efficiently, then within a few years the upper part of the subgrade may approach its liquid limit – in theory at least. Figure 6.17 illustrated the effect that such a change in water content is likely to have on the subgrade modulus. And while strength and stiffness modulus are not uniquely related, a similar proportional change in shear strength is likely.

None of this actually tells the reader what long-term assumptions to make of course. The advice here is to take a conservative but not overly pessimistic approach. The author's experience from evaluation of pavements in situ is that a zone of around 150 mm of softened soil should be anticipated, but that good drainage and good pavement maintenance can limit further softening.

24.4.2 Unbound pavement layer properties

A well designed, constructed and maintained pavement will consist of layers of increasing permeability down to a free-draining layer (either sub-base, capping or the subgrade itself) connected

Figure 24.3 Predicted water content after 4 h inflow

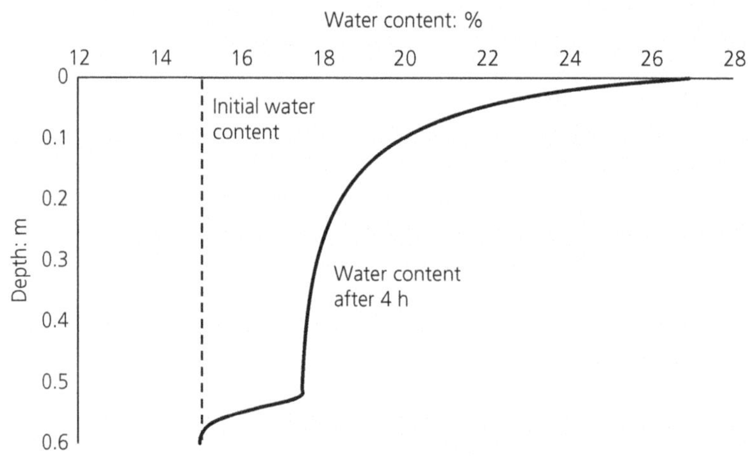

to an outfall (side drain or water table in a permeable soil). This means that there is never enough water in unbound bases, sub-bases and cappings to cause saturation, and during periods without rain the layers will slowly reduce in water content. In such cases, a near-full suction effect can be assumed and the full modulus of the material applies, for example as suggested in Table 6.4.

However, if drainage is poor – as is often the case in urban areas where space is restricted – then this is unrealistic. The 'effective' modulus (i.e. the modulus to use in any pavement analysis) may be as low as half that under good conditions.

24.4.3 Bound pavement layer issues

The issues of asphalt and hydraulically bound material durability have been covered in earlier parts of this book. The key point to be aware of here is that neither material is immune from water damage while free water is available. The surface itself is a specific problem; it will be exposed to free water during rainfall of course, but proper surface profiling and edge drainage should limit the duration. But it is essential that water is not trapped within bound materials due to low-permeability layers underneath – hence the statement above that permeability should always increase downward to a designated drainage layer. If drainage is poor and water 'backs up' within an unbound foundation, then expect the life of bound materials to be significantly shortened, the exception being pavement-quality concrete.

24.4.4 Frost heave

A particular problem of fine-grained materials (i.e. many soils) is frost heave. It is not a subject that will be discussed at any length here, but the reader should be aware that any pavement layer (including the subgrade) that becomes frozen will tend to undergo a volume change due to the growth of ice crystals, and that this volume change can be much greater than that expected simply as a result of water volume being replaced by ice volume. This is because, as ice crystals grow, they push soil or fine aggregate particles apart, creating void space. This void space, lying on the water–ice boundary, invites further water ingress if there is surplus water available; this then freezes and the result can be a large volume increase, leading to a raised and uneven road surface. Worse, when the ice melts the water content remains much higher than it was before freezing – which leaves a very weak subgrade (or unbound layer).

Qualitatively the solutions are (*a*) do not use frost-susceptible materials within the frost zone, commonly assumed to be 450 mm in the UK and (*b*) do not allow water to accumulate (i.e. ensure proper drainage). If the second point can be confidently adhered to then the first becomes irrelevant; however, if there is a danger of high water content within the frost zone, then it is sensible to ensure that none of the materials within this zone are frost-susceptible. A *frost heave test* (BSI, 2009) has been used in the UK for many years in order to exclude undesirable materials. However, the simplest method of overcoming any danger of frost heave is to use a relatively open-graded (and therefore free-draining) material. Of course, there are climatic zones where frost penetration is so deep that frost heave cannot be prevented; in such cases the overlying pavement has to be designed such that it can accommodate large heave (and subsequent settlement) without excessive damage. Traffic weight restrictions may have to be applied during the 2–3 weeks of spring thaw.

24.5. Use of geosynthetics

In addition to the reinforcement and separation roles of geosynthetics outlined in Chapter 7, they can also form part of a drainage and subgrade protection system. Two possible roles are as follows.

- *Drainage layer*. This basically replaces the permeable sub-base or capping referred to previously. The material itself comprises a relatively thick mat with sufficient connected voids to facilitate lateral flow. An advantage is that a large lateral flow is possible with a small thickness; a disadvantage is that such geosynthetics tend to be rather expensive. And as with all pavements, recyclability has to be borne in mind.
- *Impermeable membrane*. This would be placed immediately over the subgrade, preventing water ingress. This sounds like a good idea, and it probably is in many cases. But on railways, problems have been noted where water has been trapped with no possibility of escape via the foundation layers. Logically, this should not be a problem so long as the water table is deep enough to maintain a draw-down effect, but if any rise in water table occurs, perhaps due to downslope water flow through surrounding terrain, then the danger is real.

24.6. Drainage pavements

Having said so much about the need to exclude water, drainage pavements are the exception. The concept is shown in Figure 24.4.

Figure 24.4 The principle of drainage pavements

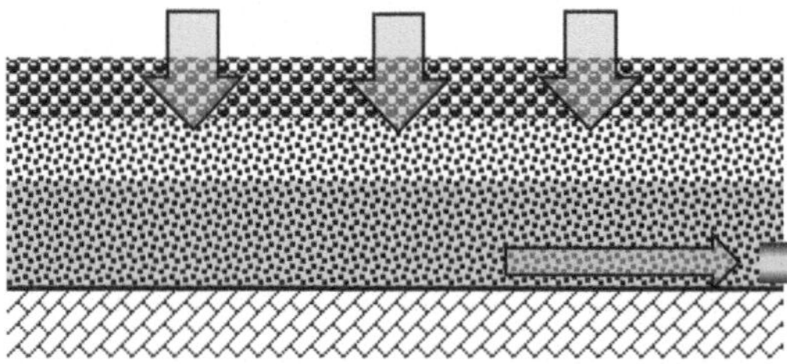

Drainage pavements are deliberately designed to accommodate water as part of a so-called 'sustainable drainage system' (SuDS), delaying the flow of water into watercourses that might be in danger of flash flooding. Basically, the pavement is turned into a large reservoir. The surfacing has to consist of either porous asphalt or porous blocks. The open-graded base layer, consisting of near single-sized aggregate, will have a voids content of around 30%, which means that a 300 mm-thick layer can accommodate 90 mm of rainfall over the area of the pavement. Unless the subgrade is non-water-susceptible (e.g. a sand), it will be protected by an impermeable (or low-permeability) membrane. The outlet or outlets are designed to allow only slow flow into the local drainage network, thus preventing overload.

Drainage pavements (see Woods Ballard *et al.* (2015) for full design guidance) are suited to large areas that take relatively light load, such as car parks. Open-graded unbound material can develop no suction, as would be the case in a continuously graded aggregate, which means it should be assigned a lower stiffness modulus in any pavement analysis, particularly if saturated conditions are expected for significant periods of time, perhaps 100 MPa (commonly assumed for railway ballast) rather than the 150 MPa often assumed for unbound sub-base.

24.7. Summary

This has been an enormously important chapter. We can be as clever as we like in calculating stresses and strains and so predicting the course of pavement damage due to traffic loads, but if the underlying assumptions of material properties are wrong then the prediction cannot be trusted. It is essential (*a*) to take every possible measure to protect the pavement structure from the damaging effects of water and (*b*) to be realistic about the long-term properties of materials since no protection or drainage system is perfect.

REFERENCES

BSI (2009) BS 812-124: 2009: Testing aggregates. Method for determination of frost heave. BSI, London, UK.

Dawson AR (ed.) (2008) *Water in Road Structures: Movement, Drainage and Effects*. Springer, Berlin, Germany.

Woods Ballard B, Wilson S, Udale-Clarke H *et al.* (2015) *The SuDS Manual*. CIRIA, London, UK.

Part 8

Pavement management

Nick Thom
ISBN 978-1-83549-711-1
https://doi.org/10.1108/978-1-83549-710-420243029

Chapter 25
Pavement monitoring and evaluation

Pavements cannot simply be subjected to a fully pre-planned, pre-programmed maintenance schedule. Unfortunately, their rates and modes of deterioration cannot be predicted with that degree of confidence, and so a genuine engineering appreciation is needed in order to enable the right decisions to be taken. And, in order to achieve that engineering appreciation, it is necessary to gather condition data in a scientific way. It is a little like diagnosing an illness in a human patient. If insufficient tests are carried out, then even the best medical professional is relying on intuition rather than genuine knowledge – and that can be decidedly dangerous. The same is true of a pavement. Inspired guesswork is unlikely to prove an effective decision-making strategy. In this chapter, the different methods of gathering data are presented and, at the end, the issue of planning a pavement evaluation is tackled.

25.1. Visual condition surveys

Minor maintenance treatments are carried out entirely on the basis of visual surveys, either by an inspector or a vehicle-mounted video camera. A visual survey is the most basic, and yet often the most useful, survey type of all. Put simply, not all cracks are the same; nor are all ruts or surface defects. An experienced engineer, potentially also a well-trained artificial intelligence system, can deduce a great deal about the internal health or otherwise of a pavement just by inspecting the surface.

It is not the intention here to describe directly any of the many dozens of methodologies, procedures, condition indexes and so on that have been devised by highway and airport authorities all over the world. For example, the so-called pavement condition index (PCI), developed by the US Army Corps of Engineers, is one of the most commonly used measures of visual condition, and is based on a summation of the contributions from different distress types (ravelling, cracking, rutting etc.). Effective pavement management demands that visual condition data are treated in a statistical way and that scores are awarded according to overall condition, perhaps divided into certain key categories. However, this section will concentrate on identifying the detailed engineering knowledge that can be obtained. The reader will then be in a position to judge the relevance or effectiveness of any given system, or indeed to devise his or her own if necessary.

25.1.1 Crack patterns – asphalt surfaces

Cracks can be longitudinal or transverse, close or widely spaced, wide or narrow, short or long – and every crack has a meaning. There will often be a combination of transverse and longitudinal cracks present because, unsurprisingly, more than one mode of damage may be taking place at the same time. Deterioration of the base may be masked by top-down cracking, for example. In old pavements that have already been subject to previous strengthening treatments, these issues become ever more complicated. Cracks may simply be reflective cracking from an older pavement construction. Crack patterns can tell the engineer a lot – but not everything. Here are some pointers.

- If wheel-path cracking predominates, then it is the traffic that is responsible, whatever the cracking may look like.
- If cracking extends more or less equally over all areas of the pavement, then the traffic is irrelevant and the cause is environmental or due to a general material defect.
- A single well-developed longitudinal crack in the wheel-path indicates traffic-induced fatigue of a thick asphalt pavement, the crack direction almost certainly being top-down, and crack depth is often no more than 50 mm or so.
- Multiple cracking (crazing) in the wheel-path implies a shallow failure, although this may overlie a deeper problem. Either the asphalt is thin or the upper layers have become debonded and are cracking independently.
- Regular transverse cracking is probably reflective cracking from an underlying layer of hydraulically bound material (HBM).
- Fine, usually white-looking, material around cracks indicates 'pumping' – water has been ejected by traffic loading after becoming trapped within the pavement, either within the bound materials (e.g. at an interface) or else in the upper part of an unbound foundation.

25.1.2 Crack patterns – concrete surfaces

It is probably fair to say that the range of options is a little more limited with concrete. Cracks can be transverse, implying that they are thermally generated, or longitudinal, implying traffic-induced.

25.1.3 Rutting

As with cracks, not all ruts are the same. A rut is a function of deformation taking place in all the various layers of a pavement and, depending on which layer or layers are dominant, the shape will be different. For example, if it is the surface course that is primarily to blame, the rut will typically be narrow with marked shoulders either side. It is possible to judge from the width of a rut, taking into account the likely lateral distribution of wheel locations, as well as the overall shape, approximately where the problem lies. Beware, however, of making the assumption that any narrow rut must be due to surface-course problems; in reality it is often due to the binder course, which is where the greatest shear stresses are found.

25.1.4 Surface defects

Besides cracks and ruts, several other surface features can be observed visually and may be important in diagnosing pavement problems.

25.1.4.1 Chipping loss

Individual aggregate particles may be lost from the surface for two principal reasons. The first is a loss of adhesive properties in the binder, an indication that the bitumen has aged significantly. The second reason is that cracks are beginning to develop, although they may as yet be invisible. A developing crack is associated with a concentration of curvature in the pavement under load, with associated high surface strain, and it is this high strain that leads to an eventual loosening of surface aggregate. Either way, chipping loss gives an early indicator of developing problems.

25.1.4.2 Ravelling

When loss of aggregate particles is more widespread, the term 'ravelling' is commonly used. This is an indication of an aged binder at the surface. Ravelling is eventually likely to lead to the development of potholes.

25.1.4.3 Bleeding

Otherwise known as 'fatting up', this is the phenomenon whereby excess bitumen in the pavement works its way up to the surface. It is seen as a smooth black sheen on the surface and is a significant safety hazard.

25.1.5 Other visual survey data

A visual survey will also be able to pick up

- edge distress
- concrete joint spalling and sealant loss
- condition of construction joints
- road ironwork problems
- marking defects
- drainage defects.

All of these are important to the decision-maker, who has to balance the various needs of the pavement and come up with the most cost-effective treatment or set of treatments. However, there is a limit to how far a visual survey can reach. Where large sums of money are at stake, the engineer needs more quantitative information.

25.2. Profile surveys

The two primary input parameters used in the discipline known as 'pavement management' (see Chapter 28) are visual survey data and profile data. Automated profile surveys are a highly cost-effective means of obtaining quantitative condition data.

25.2.1 International roughness index (IRI)

The IRI is one of the few measures of pavement condition with genuine international acceptance. It was developed from trials carried out in Brazil in the early 1980s under World Bank sponsorship (Paterson, 1987) and is primarily a measure used on highways rather than other sorts of pavement. It is defined by the amplitude of motion of a standard vehicle suspension system as it travels along the road, measured in cumulative metres of suspension system movement per kilometre of travel (m/km or mm/m). However, there are a large number of vehicles on the market (commonly known as *bump integrators*) that are purported to measure IRI, and they vary significantly in their details (wheels, axles, suspension systems and vehicle mass); calibration is therefore absolutely vital. Most bump integrators are car or van sized, and their speed of travel can be that of normal highway traffic. Figure 25.1 illustrates the principle.

Bump integrator surveys have many advantages: the equipment is not particularly expensive and it is therefore suited to use in developing countries; the measurement (IRI) is a direct indication of ride quality and is therefore directly related to 'serviceability' as felt by both vehicle and driver; data can be rapidly obtained. These qualities mean that IRI has become the primary indicator of pavement condition in many highway pavement management systems, usually supplemented by a visual condition index (VCI) of some sort – a combined score for visual condition from all different distress modes. The IRI is a 'blunt instrument' – a high IRI may be due to many different actual distress types, including rutting, ravelling, potholes, surface texture issues and cracking. However, it is ideal for monitoring changes in condition over the years. The trend can be observed and, when the IRI reaches a certain level, a more detailed survey can be carried out to determine the actual deterioration mode and thus the most appropriate maintenance strategy. It will only be at this detailed investigation stage that it becomes apparent just how much money needs to be

Figure 25.1 Bump integrators and the IRI

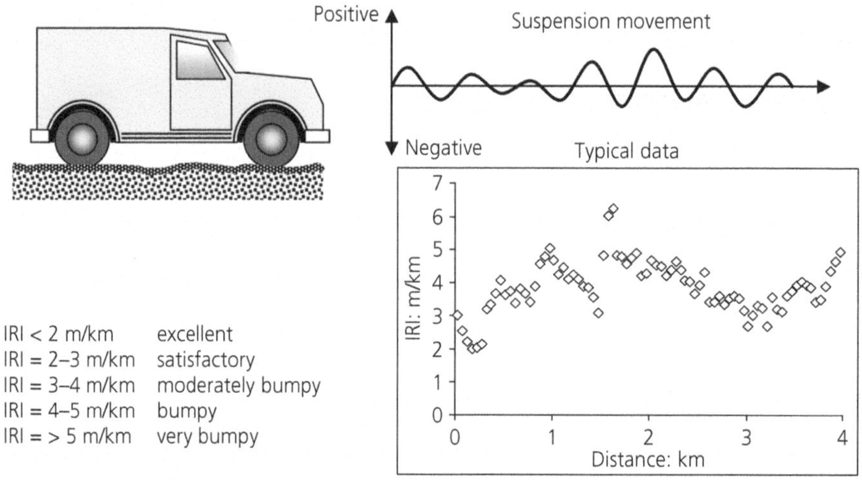

spent, which means that, until then, the management system will simply have to assume average spend quantities based on experience. However, for network-level budgeting purposes, this is quite sufficient.

25.2.2 Laser profile surveys

At the high-technology end of profile measurement are a string of laser-based systems on the market. Names include 'road surface profiler', 'automatic road analyser', 'multi-function road monitor' and 'road assessment vehicle'. Lasers are used throughout the engineering world as accurate and remote distance sensors, and they can be used for profile measurement by aiming them down at the road from a vehicle-mounted frame and monitoring the reflection from the surface. Over distances of a few centimetres, the variation gives a measure of texture depth; over distances of metres, the data can be interpreted in terms of longitudinal profile (averaging out the texture depth element). Most commonly, a series of at least three sensors is mounted longitudinally and the measurements from each are compared. Sensors can also be mounted transversely, giving a measure of rut depth. Such surveys can be carried out at normal traffic speed.

It is possible to derive an approximate equivalent IRI value from the data, but this under-utilises the information available. Therefore, in many countries including the UK, the data are interpreted scientifically as a measure of variance (or standard deviation) of surface profile from a moving mean. The number is less directly related to ride quality than is the IRI, but it is more readily understandable in terms of pavement distress.

25.3. Skid resistance surveys

Together with visual condition and profile surveys, skid resistance surveys also form a logical ingredient in a pavement management system. While visual condition and profile give general data relating to all aspects of condition, skid resistance is a single property. However, it is a property that is directly related to the safety of users, applying to both highways and airfield runways. In the case of airfields, there are international standards and it is necessary for airport authorities to check runway skid resistance regularly, particularly under adverse weather conditions (rain or snow). Highways are governed by standards set by individual countries, regions and cities, but in many parts of the world skid resistance is specified both as an initial requirement and to be checked at intervals. It is also sometimes necessary to carry out measurements at particular locations where accidents have occurred or where safety is under review.

The sideways-force coefficient routine investigation machine (SCRIM) is a standard device for skid resistance measurement on highways and is used in many parts of the world (e.g. BSI, 2009). The principle of operation is shown in Figure 25.2. An extra wheel is set at an angle to the direction of travel and dragged forwards (while being allowed to rotate). The result is a sideway force that, when divided by the applied vertical force, gives a SCRIM coefficient. A water tank on the back of the vehicle supplies water continuously to the measurement wheel, as wet conditions are always the most severe.

Also shown in Figure 25.2 are two examples of light weight trailer-mounted friction testing devices. Well-known names include the Tire Traction Tester, Mu Meter, Griptester and Skiddometer. These are typical of the machines used at airfields and by many highway authorities for the purpose of localised investigations into particular problems. For wet skid resistance, a water tank has to be mounted in the towing vehicle. In these cases too, a friction coefficient results, although it may differ from the SCRIM coefficient because of differences in vertical force, tyre type and mode of drag. On UK highways, for example, a calibration to SCRIM is needed.

The point was made in relation to surface design in Chapter 11 that surface friction is a season- and weather-dependent quantity and that summer conditions are normally worst (other than if ice is present) because of the effect of polishing on surface aggregate. This must, of course, be borne in mind when measuring skid resistance, whatever the device used; absolute judgements about inherent surface properties should really only be made after at least three measurements, preferably during the summer (or dry) season.

Figure 25.2 Skid resistance measurement devices

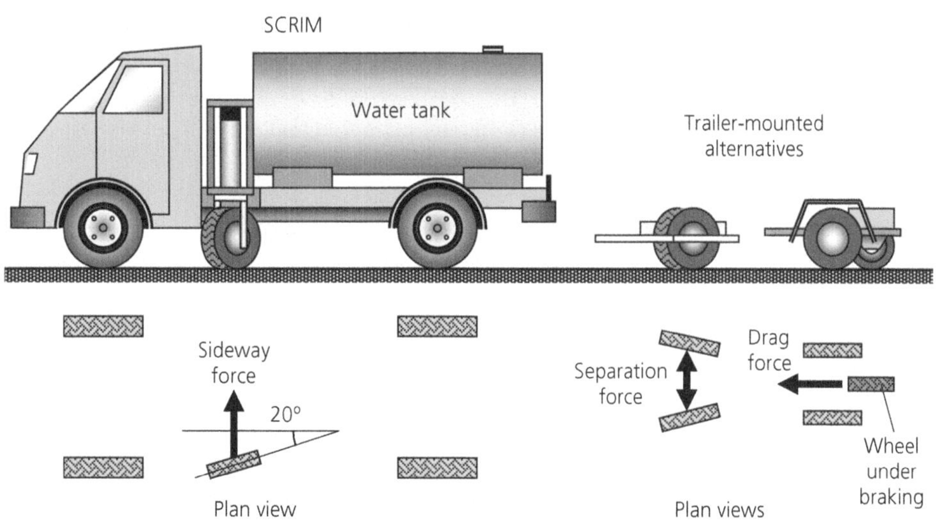

25.4. Cores and trial pits

In the end, there is no substitute for digging down and taking a look beneath the surface. Coring, using cutters, which need be no more than 100 mm in diameter, is a relatively non-destructive method of sampling a pavement, providing the hole is properly backfilled afterwards.

Trial pits are a much more labour-intensive method of sampling. They are suitable where (*a*) the bound layer thickness is low, (*b*) samples of unbound foundation material are required or (*c*) specific information is needed that demands a larger view (or a larger test specimen) than can be afforded by a core. However, modern coring machines are quite capable of obtaining cores of 300 mm diameter or more if necessary, and techniques now exist for retrieving samples of unbound material from holes as little as 150 mm in diameter (as described later), so serious thought should be given to avoiding trial pits on heavily trafficked roads if at all possible.

25.4.1 Construction and material condition information

The first and most immediate benefit to be derived is that cores and trial pits reveal what is actually there. Layers can be identified and their thicknesses measured, a visual assessment of quality (voids content, bitumen appearance, friability of hydraulically bound material (HBM), reinforcement, aggregate type) can be made and any lack of bond is evident (sometimes caused by coring – but indicative of weak bond nevertheless). In many cases, something of the history of the pavement can be deduced, as buried surface courses or additional binder courses are frequently found. If an overlay has been carried out in the past, then one can generally assume that this was done for a reason – although the nature of that reason is unlikely to be deducible from a core alone. Figure 25.3 gives examples showing the potential usefulness of cores in identifying damage mechanisms.

However, a core will not usually reveal water-related problems because water is added to cool the core cutter. Only if a layer is thoroughly saturated, or in cases where water continues to weep

Figure 25.3 Damage mechanisms revealed (images provided by Scott Wilson Ltd)

out from an interface long after coring has been completed, can it be deduced that this was almost certainly not coring water.

A small sample of underlying unbound material can readily be retrieved from the bottom of a core hole. Furthermore, if an *unbound material sampler* is used – a device that is driven into the foundation layers, usually through a core hole – then an excellent picture of foundation materials and the underlying soil can readily be obtained, although the operation adds considerably to the time taken at a single core location and it also precludes carrying out the in situ testing described below. An unbound material sampler consists of a steel tube, containing within it a plastic inner tube in two halves. The inner tube (containing the sample) is later extruded from the steel outer tube, and separated into its two halves, releasing the sample for inspection and testing.

Trial pits, of course, do allow water problems to be identified and also allow excavation deep into the foundation if necessary.

25.4.2 In situ tests

The relatively small size of most core holes means that there is a limit to the types of test that can be carried out. In fact, there is only one in situ test device that is commonly used, and that is the *dynamic cone penetrometer* (DCP) (Section 6.2.10, Figure 6.14). This gives a profile of material quality with depth, as well as the strength of the underlying soil.

It is important to appreciate the difference between the resistance to a cone and properties such as stiffness and shear strength. Two points are particularly noteworthy. The first is that, if individual aggregate particles are of comparable size to the cone (20 mm diameter), then the precise arrangement of particles can dramatically affect the penetration rate; an individual particle can temporarily block progress. The second point is that saturation of an unbound material will only slightly lower resistance to cone penetration whereas it will greatly reduce shear strength and stiffness of the aggregate under traffic loading.

While cores permit only the DCP, trial pits open up the possibility of plate tests. The traditional *plate loading test* was described in Section 6.1.2, where the point was made that the measure is a combination of stiffness and shear strength, depending on the material type being loaded. Probably more useful, the portable *dynamic plate test* (DPT), also known as the *light weight deflectometer* (LWD), is a pure stiffness test; it takes just a few seconds to carry out and is ideally suited to use within a trial pit. The equipment was described in Section 6.3.10 and Figure 6.23. The interpretation is in terms of an elastic stiffness modulus, assuming the whole foundation to be a single uniform material of infinite extent. In reality, of course, it has to be remembered that there will be contributions from all the different layers down to a depth of 1–2 m.

25.4.3 Laboratory test specimens

Many laboratory tests are designed to be carried out on cylinders of 100 mm or 150 mm diameter, which means that cores are an ideal source of such specimens. These tests include

- compressive strength of HBM (note that a height to diameter ratio of at least 1.0 is required)
- uniaxial stiffness modulus of HBM
- indirect tensile strength (ITS) of either HBM or asphalt
- indirect tensile stiffness modulus (ITSM) of asphalt

- indirect tensile fatigue test (ITFT) for asphalt
- repeated load axial test (RLAT) for asphalt deformation
- interlayer bond strength tests.

These have all been introduced in Chapters 9, 10 and 19 and descriptions will not be repeated here.

If non-cylindrical test specimens are required, then matters become rather more complicated – and therefore expensive. Beam tests, very popular among researchers and widely used in specification of concrete, require a specimen of 300 mm or more in length, obtainable from a trial pit or by sawing from a core of over 300 mm diameter.

Density and voids content can be obtained on a specimen of any convenient shape. Asphalt specimens can also be broken down into their constituents, by means of a centrifuge, with solvents used to extract the bitumen. Aggregate gradation can be checked, as can binder content. Binder quality can also be measured using a dynamic shear rheometer (DSR) (see Section 9.1.3), although it should always be borne in mind that the properties may have been affected by the method of binder extraction.

Material from an unbound material sampler yields only a small and disturbed sample. Nevertheless, pocket cone penetrometer tests, giving a measure of shear strength, are sometimes possible; determination of water content, liquid and plastic limits can also be carried out. However, if gradation of a granular layer is required, then a larger sample has to be obtained, in which case a trial pit or a very large-diameter core are the only options.

25.5. Ground-penetrating radar

Coring may be thought of as being minimally destructive, but it is certainly intrusive, as well as being time-consuming and restricted to providing information at discrete and usually widely spaced points. On the other hand, ground-penetrating radar (GPR) has the capability of providing a continuous and truly non-destructive picture of the pavement structure, and nowadays it is not even particularly expensive. It sounds too good to be true!

25.5.1 Principle of operation

Figure 25.4 shows GPR equipment and the way in which results are typically interpreted. GPR is a very powerful technique. The concept is that radio waves are transmitted from a point while a receiver at an adjacent point 'listens' for reflected waves. Reflections will take place when a wave encounters a change of material – this may be air–surface, it may equally well be asphalt–granular. Thus, if radar is pointed down into the ground, the receiver will pick up a reflected signal corresponding to each change in material properties. The time elapsed between transmission and reception is then directly related to the distance travelled down to an interface between two materials and then back up to the surface. Furthermore, as only a proportion of the wave energy is reflected while the remainder continues down into the pavement, there will be a series of reflections as each interface is encountered.

On the individual received signal trace illustrated in Figure 25.4, the strong contributions from layer interfaces (and from the surface) are clear. The profile shown on the right is generated by a processor that assigns 'black' where the received signal is above a certain amplitude only. This profile can then be interpreted, partly by image-recognition software but with human assistance also, converting from 'time' to 'depth', for which an estimate of wave speed in each medium is required. This is usually achieved by calibrating the GPR data against core records.

Figure 25.4 Ground-penetrating radar

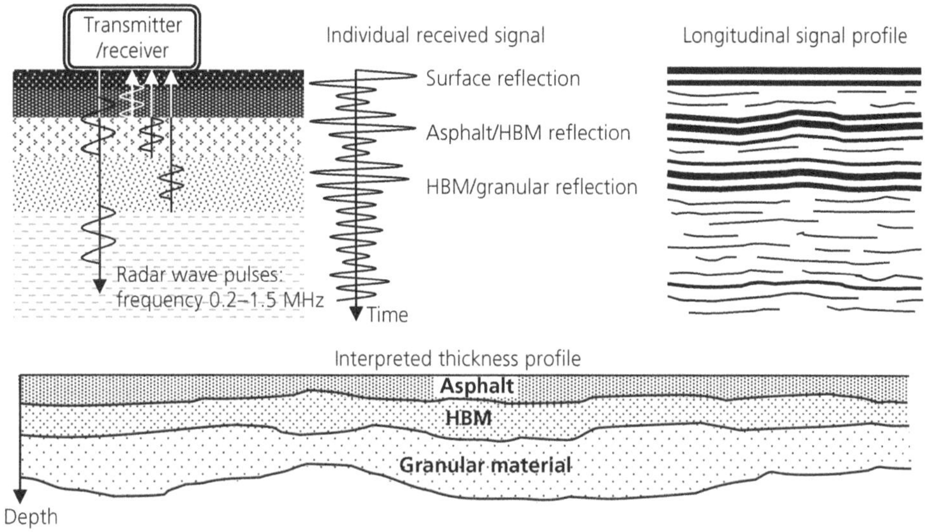

25.5.2 Information obtained

Thickness determination is the primary reason for carrying out a radar survey on a pavement and it usually yields thoroughly excellent data, accurate to around 1 cm, although the sensitivity of GPR is a function of the wavelength used. Short wavelength (i.e. high frequency ($\geq$1 GHz)) will give good resolution but poor penetration into the pavement; it is suited to identifying bound layer interfaces. Longer wavelength (lower frequency) is better suited to identifying pavement foundation issues. One has to be slightly wary of material identification and it is not uncommon for GPR interpretation to conclude that a certain material is present, for example, asphalt, when in fact it is something different, for example, HBM. When the full range of possible material types within each generic class is considered, this is hardly surprising.

GPR can also give data on the presence of moisture (because water molecules become excited at radar frequencies), on voids (because of the strength of a solid–air interface) and on steel reinforcement (because steel interferes with wave propagation). Basically, the more information that is required the more expensive the survey becomes, as additional frequencies need to be used and, for accuracy, slow travel speed is required. A normal travel speed for thickness determination is around 20 mph (30 km/h). Information from thickness determination is often extremely useful and minor errors in material identification are usually unimportant, although information on moisture, voids and reinforcement is, in the author's experience, less frequently of value. In some cases, this is because the radar interpretation appears suspect; in most cases it is because the radar information adds little to what is known already.

25.6. Deflection surveys

GPR can help the engineer 'see' beneath the surface, but it gives no direct information on the structural condition of the materials. For this, there is no alternative but to assess performance under load, which means measuring the deflection of the pavement surface under the action of an applied wheel load (or similar alternative). The stiffness of response will then give direct information, which might logically correlate to the ability of the pavement to carry multiple load applications.

25.6.1 The Benkelman beam

The Benkelman beam was developed by AC Benkelman in the 1950s (HRB, 1954, 1955) and is the oldest and simplest form of deflection test device. It has been used successfully throughout the world for many decades. Figure 25.5 illustrates the principle of operation and the interpretation procedure commonly used.

As shown in Figure 25.5, the Benkelman beam itself is a simple frame with an arm on a hinge, the rotation of which is read from a dial gauge. The equipment is placed on the ground immediately behind the twin rear tyres on one side of a goods vehicle loaded to a standard weight, the arm resting on the pavement surface between the twin tyres. When the operator is ready, the goods vehicle is slowly driven forward and a maximum reading is taken as the tyres pass the end of the arm; when it has driven forward some metres, a minimum reading is also taken. The difference relates to the deflection caused by the loaded wheel.

This is a practical, low-technology tool. Therefore, like the bump integrator, it is well suited to developing countries. One could criticise the slow vehicle travel speed, which means that any asphalt layers will appear to have a lower stiffness than would be the case under more normal traffic speed. One could also draw attention to the fact that the reference frame feet are close enough to the wheels as to be affected; the measurement is, therefore, not an absolute one. Nevertheless, as the usual method of interpretation is with reference to empirical experience, these criticisms have little validity. And even without any accepted method of absolute interpretation of the readings, the measure of relative performance between one point and the next is always useful.

25.6.2 The Lacroix deflectograph

The Lacroix deflectograph is an extension of the Benkelman beam idea. Initially developed in France, the UK has been one of the countries to make most intensive use of it. The problem with the Benkelman beam is that productivity is low – it takes a minute or two to set up, take

Figure 25.5 The Benkelman beam

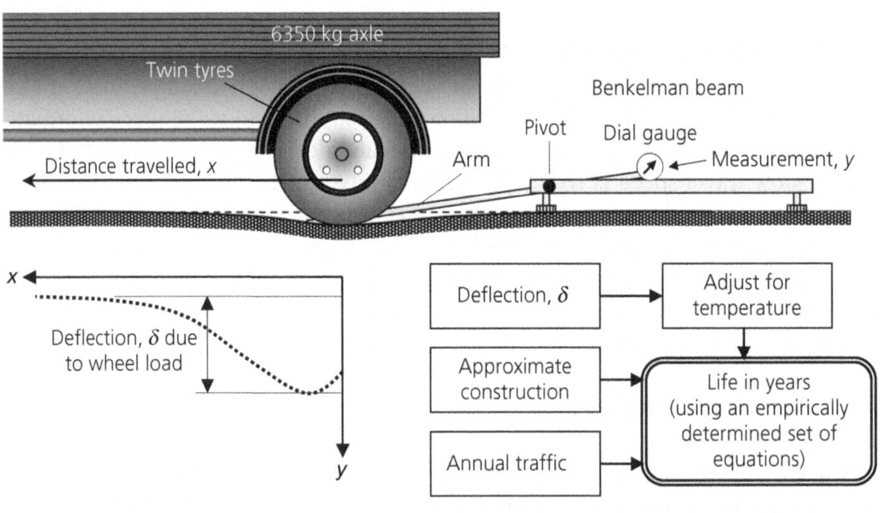

measurements and move forward to the next point. The deflectograph uses exactly the same principle (i.e. recording the deflection between the twin tyres of a rear axle), but with a quite different reference frame – one that allows the vehicle to travel continuously along the road. Figure 25.6 indicates conceptually how this is achieved. The reference frame is now in front of the measurement axle and it is repeatedly dragged forward relative to the body of the vehicle and then released. As soon as the rear tyres of the vehicle have drawn level with the tip of the measurement arm, the frame is winched forward toward the front of the vehicle again, ready for the next reading. The result is that a measurement is taken every 3–4 m and that the vehicle can travel continuously at a speed of 2–3 km/h. Readings are taken in both wheel-paths.

This is clearly a massive improvement on the Benkelman beam from the point of view of productivity, allowing a near-continuous profile of pavement stiffness to be determined. However, it is not any more accurate; in fact it is less so. The reference frame is even more greatly affected by adjacent wheels, both front and back, and the arm is shorter. However, as the method of interpretation is still according to empirically determined equations, this is not a serious criticism.

In the UK method of usage, for example, worked out by the Transport Research Laboratory over a period of many years' study, a series of charts was produced for pavements of various types and thicknesses, giving relationships between deflection and the passage of traffic (Kennedy and Lister, 1978). Thus, if the past traffic is known, then a measured deflection can be translated straight to a remaining life, by simply reading off the chart. Similar charts were produced for overlay designs, again based on monitoring the gradual deterioration (and change in deflectograph deflection) of a large number of pavements. These charts have, of course, been digitised into equations that are built into the data-processing systems currently used.

Note the temperature adjustment required when interpreting results from both the Benkelman beam and the deflectograph – and indeed any deflection test on an asphalt pavement. Infrared sensors

Figure 25.6 The Lacroix deflectograph

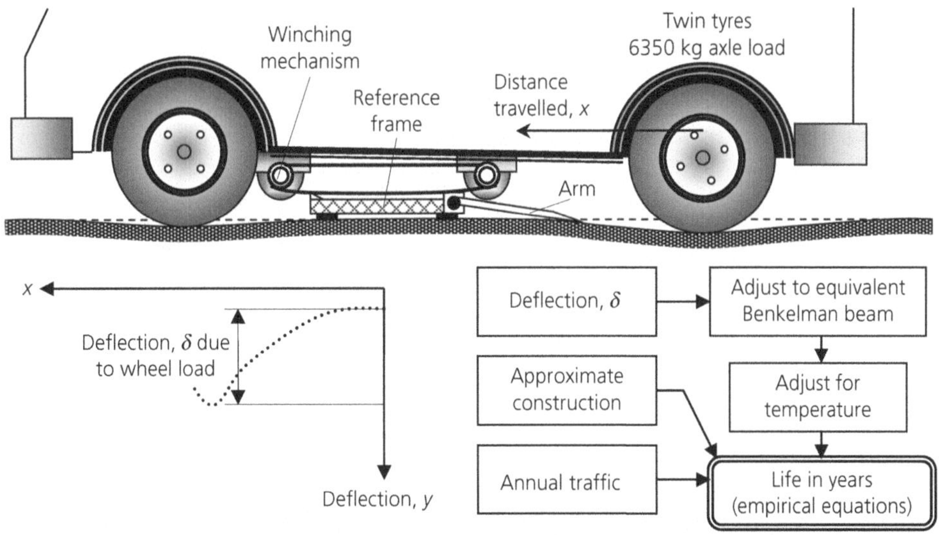

provide a means of sensing surface temperature. If the temperature at depth is required, then a small hole can be drilled into the pavement, glycerine or oil poured in, and the glycerine/oil temperature recorded once it has reached an equilibrium. Temperature adjustment of Benkelman beam or deflectograph readings is, like everything else in this analysis, based on empirical evidence.

There is no doubting the usefulness of Benkelman beam and deflectograph data, both in pavement management and also for project-level design of maintenance or rehabilitation. However, the information provided will always be crude, no matter how much empirical evidence is used. There is absolutely no way of knowing the relative contributions of the different layers to deflection. A high reading may indicate poor asphalt; it may equally indicate a soft subgrade. A standard processing package will give the same remaining life in the two cases – and the same overlay thickness – but the actual performance will not be the same at all. This is a serious drawback. It means that the Benkelman beam and the deflectograph can only be used to give a first estimate of structural condition, an estimate that may be appropriate for network-level management but which, on its own, is not sufficient for project-level design. For this, a more sophisticated deflection test is needed.

25.6.3 The falling weight deflectometer (FWD)

The FWD has taken the world of pavement structural evaluation by storm since it was first used commercially in the early 1980s. As far as productivity is concerned, at around a test per minute it is only slightly quicker than the Benkelman beam, but where the FWD scores is that it gives a very precise value of absolute deflection (accuracies of $\pm 2\,\mu m$ commonly quoted), and this opens the door to a much more sophisticated method of interpretation. Figure 25.7 illustrates the equipment and the data analysis methodology.

The machine is usually trailer-mounted. Tests are performed with the equipment stationary and with the loading plate and deflection sensors lowered onto the surface. The load pulse is then

Figure 25.7 The falling weight deflectometer (FWD)

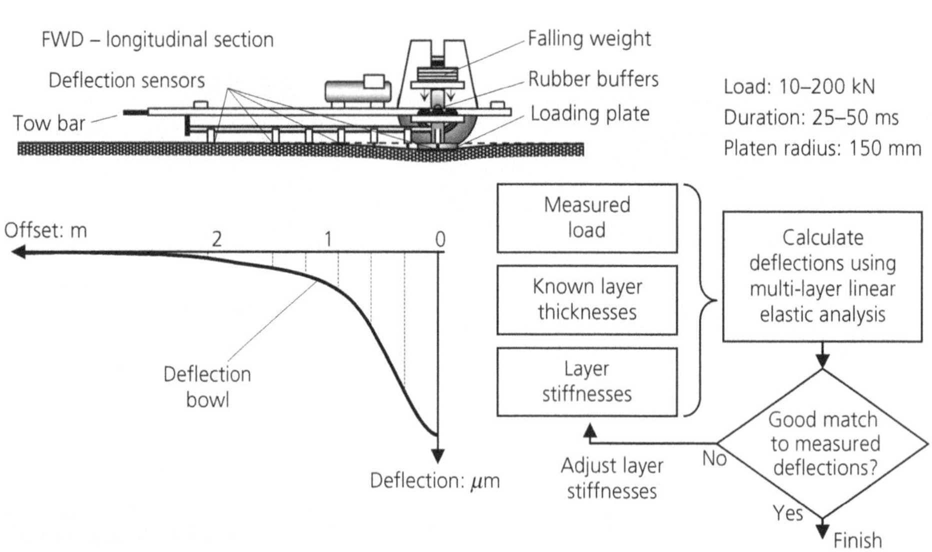

generated by the action of a falling weight onto a set of rubber buffers. Despite the stop–start nature of the testing procedure, the FWD actually has several desirable features

- the load magnitude can be selected to match a typical wheel load
- the pulse duration is similar to that from a moving vehicle
- the deflections are absolute and highly accurate (generally achieved by use of velocity transducers)
- measurements are taken not only at the load location (through a hole in the centre of the loading plate) but also at selected distances from it.

This last point is particularly important. The full set of readings describes what is termed a *deflection bowl* (or basin), the interpretation of which is presented conceptually in Figure 25.8. This means that, whereas neither the Benkelman beam nor the deflectograph are able to distinguish between two pavements with the same deflection under load but with quite different individual layer stiffnesses, the FWD can; each particular combination of layer stiffnesses gives a near-unique deflection bowl. So, with knowledge of the full deflection bowl, it should, in theory, be possible to deduce just which combination of layer stiffnesses was responsible. This procedure is termed back-analysis (or back-calculation).

There are numerous back-analysis computer programs on the market, almost all making the assumptions that

- all layers are of uniform thickness and of infinite lateral extent
- all materials are linear elastic and homogeneous
- the load consists of uniform stress on a circular area
- dynamic effects due to inertia are negligible.

Figure 25.8 Interpretation of FWD deflection bowls

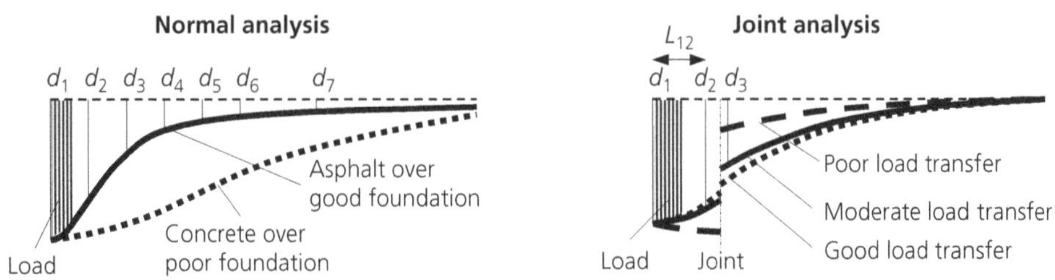

Upper pavement
- affects curvature of central part
- typical indicator: $d_1 - d_2$ or $d_1 - d_3$

Base and sub-base
- affects slope in next region
- typical indicator: $d_2 - d_4$ or $d_3 - d_5$

Subgrade
- affects deflection at distance
- typical indicator: d_6 or d_7

Load transfer efficiency
- affects 'step' across joint
- typical indicator: $d_2 - d_3$ or d_3/d_2

(note: $\neq$ 0 or 100% even if perfect, due to distance between sensors 2 and 3)

Slab support
- affects angle of loaded slab
- typical indicator: $d_1 - d_2 / L_{12}$

(L_{12} is the distance between sensors)

None of these are true of course, and this means that even the best back-analysis procedure will struggle to achieve a perfect match between the measured deflections and those calculated, and that the resulting stiffnesses must be treated with a healthy degree of scepticism. It also means that it is asking for trouble to try to subdivide the pavement into more than four layers – less if possible. Where materials differ significantly in stiffness, for example, asphalt, unbound sub-base and soft soil, back-analysis should be able to arrive at a reasonable estimate of the stiffness modulus of each, as long as no layer is too thin (and therefore too insignificant in its effect on deflection). Where layers have similar stiffness, for example, asphalt and cement-bound base, there is always a danger that back-analysis will greatly exaggerate the stiffness of one while the other is underestimated. An appreciation of the real degree of confidence (or otherwise) can only come with experience, but a general rule would be that simple, intact pavement structures can be analysed with reasonable certainty; more complex and variable cases will still generate valuable answers, but these should always be considered alongside other evidence.

One argument with considerable validity is that, whatever incorrect assumptions are made when back-analysing FWD data, the same incorrect assumptions will usually be made when evaluating remaining pavement life under design wheel loading. The calculated stress or strain in the pavement under wheel load should therefore still be approximately correct. Although this statement sounds very much like wishful thinking, studies have shown (Mamlouk and Davies, 1984) that the error involved in ignoring dynamic effects is rarely more than a few per cent for any of the back-calculated stiffness values. Other error sources (non-homogeneity, edge effects, lack of bond) will certainly be the same under traffic as under the FWD.

A final advantage over the Benkelman beam and the deflectograph is that the FWD is equally at home on concrete or asphalt, whereas the other devices have been used almost exclusively on asphalt and would give only very small deflections on concrete because of the non-absolute nature of the measurements. The FWD is particularly well suited to measuring load-transfer efficiency across a concrete joint. The loading plate is positioned on one side of the joint and deflections are measured on either side. As well as summarising the meaning of different components of a 'normal' deflection bowl, Figure 25.8 also illustrates the information that can be obtained from a joint test.

25.6.4 Rolling wheel deflectometers

The ultimate deflection test device would be one that measured a full deflection bowl, like the FWD, but which travelled at traffic speed along a highway, thus combining both quality and quantity of information. Such a device is the rolling wheel deflectometer, and several versions have been developed over the years, achieving their goal to varying degrees. There is an inevitable trade-off between measurement accuracy and travel speed. Several companies and research organisations have used lasers to measure distance from a datum – either conventional lasers requiring data to be averaged over a certain length to give a reasonable approximation to the correct deflection bowl, or Doppler lasers measuring pavement velocity in the same way as a geophone on an FWD. Surface texture and vehicle oscillation are two of the larger problems that have to be overcome, together with the fact that it is not easy to reach the area of surface immediately adjacent to the tyre (and certainly not the area that is actually loaded).

It is fair to say that the Doppler laser system now works well and it is in use in the UK for network-level surveys. However, this is surely an area where equipment will continue to develop rapidly.

25.7. Pavement evaluation planning

Resources are always limited, so it is clearly imperative that pavement evaluation is sensibly targeted; there is no point spending vast amounts on understanding exactly what problems there might be on a particular pavement if there is then no money left to carry out the necessary maintenance. This is easy to say but much harder to put into practice, not least because it is quite impossible for anyone to see the whole economic picture. All that will be attempted here, therefore, is to set out a few guiding principles, accepting that the real economic equation will vary from circumstance to circumstance. The larger question of pavement management is discussed in Chapter 28.

25.7.1 Network level

A pavement authority needs to monitor performance regularly in order to plan budget allocation well in advance, and this is true whatever the political difficulties involved in securing that budget. In fact, the greater the political constraints, the more important it is for the authority to be equipped with up-to-date information on pavement condition in order to argue the spending case from a position of strength. Some sort of regular condition evaluation is therefore essential. Table 25.1 lists the alternatives.

Traffic counts have not been mentioned previously because they are not directly connected with pavement condition; however, they are essential to highway network management. Counting can be automatic using piezoelectric strips on the road surface or radar systems attached to street furniture, or manual. The advantage of automatic counts is that they can continue day and night for long periods with little expenditure. The advantage of manual counts is that traffic can be classified accurately into different vehicle types (cars, buses, light goods, heavy goods – different axle configurations). Both are extremely useful in assigning the relative value of different spend proposals.

Axle-weight surveys have also not been mentioned so far. These can be carried out either by stopping and weighing wagons on a fixed weighbridge or by installing a weigh-in-motion (WIM) device into the pavement, many designs for which now exist. For example, the UK maintains several WIM stations around the country, purely to monitor national trends. From time to time, as changes occur, this has led to changes in the relevant design standard. Results from a fixed weighbridge can be considered accurate; those from a WIM device are less reliable and depend

Table 25.1 Network-level monitoring options

Survey type	Information	Usefulness
Traffic count	Trends; design traffic	High
Axle weight	Vehicle damage factors	Low–medium
Visual condition	Roughness (approximate) Structural condition (very approximate) Skid resistance (possibly)	Very high
Profile	Roughness Structural condition (very approximate)	High
Deflection testing	Structural condition (approximate)	Medium
SCRIM	Skid resistance	Medium

on regular calibration of the installation. Such surveys would normally only be carried out at state or country level, in order to monitor trends in goods traffic development over the years, enabling 'wear factors' (Section 4.2), sometimes known as 'damage factors', to be updated as necessary.

The only absolutely essential ingredient as far as pavement condition assessment is concerned is visual survey data. The information that results is certainly approximate, but it relates to ride quality and structural condition and also safety-related features such as skid resistance. It can be processed to give single numbers related to each – numbers that can be used to assign likely remaining life and future maintenance costs – albeit very approximately. Table 25.2 suggests a way in which indicators of ride quality, structural condition and skid resistance can be built up with contributions from different condition parameters obtained from visual surveys. It should be noted that Table 25.2 is conceptual; details will depend on the way in which each condition parameter is measured.

Visual survey data allow relatively minor maintenance to be programmed; they also give an excellent basis for estimating future costs as an average across a network, although predictions may be wildly inaccurate in any individual case.

However, major highways or airfield pavements are thick structures and there is a limit to what visual surveys can be expected to reveal. For these reasons, profile surveys, deflectograph (or rolling wheel deflectometer) measurements and SCRIM surveys should all be considered. Each will give a more accurate evaluation of individual aspects of condition – namely *ride quality, structural condition* and *skid resistance*, respectively. However, the high productivity achievable from a profile survey, together with the fact that structural condition tends to correlate very roughly with profile, propels profile surveys to the top of the list. They are second only to visual condition surveys in terms of usefulness, and it is worth considering employing them, at least on major highways, at reasonably frequent intervals (e.g. annually or biannually).

It may also be possible to justify the deflectograph (or rolling wheel deflectometer) and the SCRIM (or similar alternatives) as network-level survey tools but, if so, their use would logically be less frequent. At this level, one is moving away from the overall aim of obtaining an average network condition and a whole-network budget, towards the more specific aim of being able to decide when to carry out a more detailed survey of individual parts of the network. A combination of visual condition and profile is probably sufficient for budgeting purposes.

Table 25.2 Possible construction of condition indicators based on visual surveys

Distress type	Ride quality indicator	Structural condition indicator	Skid resistance indicator
Cracking	—	60%	—
Rutting	10%	30%	—
Edge distress	—	10%	—
Patching	20%	—	10%
Ravelling	40%	—	—
Potholes	30%	—	—
Visible texture	—	—	90%

25.7.2 Programming a project-level survey

A good pavement management system, whether entirely manual or incorporating sophisticated software, will include 'triggers' to indicate when detailed evaluation at project level is needed. These triggers may be based on individual indicators (visual, profile etc.) or a combination of them all, but the key point is that detailed evaluation should be triggered in time for cost-effective maintenance. Figure 25.9 illustrates the challenge.

The question as to what constitutes cost-effectiveness is not easily answered. Clearly it is not worth spending too early or the pavement is brought into an unnecessarily strong condition, with the possibility that future unforeseeable changes in use, changes in traffic load and so on will mean that the potential life of the pavement is never fully utilised. On the other hand, spending too late means that materials that might have been structurally sound as a base layer no longer are, requiring increased expenditure, and this is on top of the nuisance to users of allowing continuing use at a poor standard of ride quality. The actual optimum will definitely depend on the type of pavement (earlier for major highways and airfield runways; later for minor roads and industrial areas), but there will be an approximate optimum somewhere.

However, network-level surveys inevitably include a large degree of uncertainty, as illustrated in Figure 25.9. It is therefore sensible to carry out a detailed evaluation well before the optimum intervention level is reached – according to network-level survey data. For example, if the optimum intervention level is assumed to occur when the network-level 'condition indicator' falls to 60% of its original value, it would be worth programming a detailed evaluation when the survey data indicate 70–75%, because the real level may in fact already be at 60%.

Unfortunately, it is only possible to describe the principle here because of the quite different requirements of different classes of pavement and the large number of possible condition indicators in use. In summary, the following advice is offered.

Figure 25.9 Pavement condition uncertainty

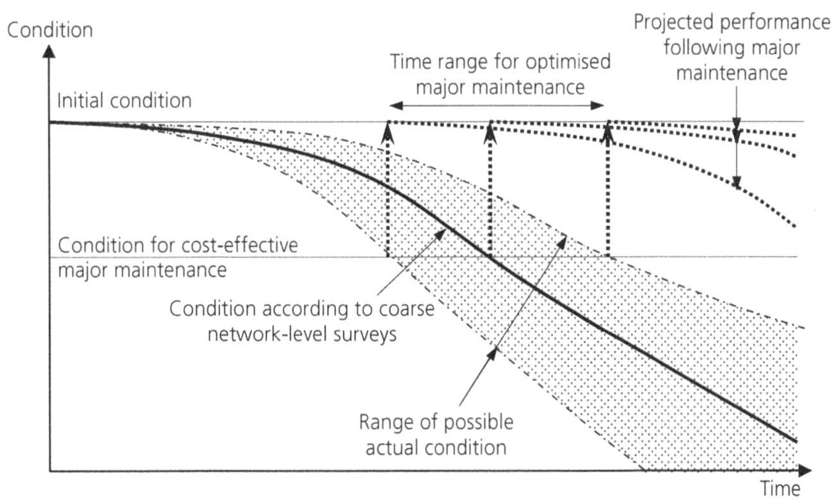

- Decide on one or more network-level condition indicators.
- Estimate what condition (according to the condition indicators) represents the optimum time for major maintenance.
- Set triggers for project-level surveys at condition indicator values significantly higher than those for optimum major maintenance.

25.7.3 Designing a project-level survey

Having established that a particular section of the network is ready for its detailed (project-level) pavement evaluation, the question arises as to what types of survey are most appropriate. There are basically two aims to this exercise, namely

- to establish the deterioration mechanism(s)
- to provide parameters for analysis and rehabilitation design.

To satisfy the first aim it is necessary to understand the root cause of any cracking, rutting or surface distress that is taking place, and the prime tools for this are likely to be cores and/or trial pits combined with a detailed visual survey. Cores through cracks can be compared with those in areas with no distress. The immediate visual evidence (surface + cores/pits) is often enough for near certainty of the distress mechanism, but laboratory tests (e.g. to establish binder hardness) may be required to prove the point.

However, to obtain the quantitative data required for pavement analysis and rehabilitation design, a much more thorough job is needed. The key design parameters are *layer thickness, stiffness modulus, fatigue strength* and, for asphalt or unbound materials, *deformation resistance*, and it is necessary to build up a near-continuous picture throughout the site. Table 25.3 lists the tools available and suggests their likely use.

Table 25.3 Project-level evaluation tools

Tool	Information obtained	Frequency
Cores, trial pits, laboratory tests	Layer thickness	Occasional
	Bound material stiffness modulus	
	Fatigue strength	
	Deformation resistance	
	Foundation strength	
Radar (GPR)	Layer thickness	Continuous
	Locations of high moisture	
Detailed visual survey	Crack density	Continuous
	Rutting locations/severity	
FWD	Layer stiffness modulus	@1–100 m
	Joint condition	
Drainage survey	Drainage efficiency	Targeted

The most logical approach is to consider the parameters needed for design and then to decide how best to obtain them.

25.7.3.1 Layer thickness

Cores/pits and GPR are the methods listed in Table 25.3 for determining layer thickness; however, construction records form another source of information. There are, therefore, three options

- rely on construction records
- use construction records supplemented by cores/pits
- use GPR, together with cores/pits and construction records.

For a pavement of relatively recent construction, records may be sufficient. However, in older pavements, particularly those that have undergone maintenance in the past, cores or pits are often essential. GPR is more of a luxury and should therefore only be used where the economic importance of the pavement warrants the expenditure, although the information obtainable from GPR is of excellent quality and allows treatment lengths to be specified with much improved confidence.

In practice, it is rare to have to rely on construction records alone because there are usually other reasons, besides thickness determination, for taking cores or digging trial pits.

25.7.3.2 Stiffness modulus

This is an essential ingredient in analytical pavement design, but may be replaced in empirical methods by a 'condition factor' of some sort. If no quantitative assessment is needed, then it is probably unnecessary to carry out measurements of FWD accuracy; sufficient reliance can be placed on the combined evidence of visual condition and cores/pits, preferably supplemented by relatively coarse deflection measurements such as those from the Benkelman beam or deflectograph. However, where quantitative analysis is to be carried out, the FWD (or similarly accurate rolling wheel devices as they become available) is required. The intensity of use will depend on the likely variability in the pavement structure and the economic importance of the pavement. Typical patterns of use might be

- @ 20 m in near-side wheel-path + occasional tests in lane centre
- @ 50 m in less trafficked lanes on a multi-lane highway
- @ 100 m in near-side wheel-path for long, overview-type surveys
- @ 20 m at ± 2–5 m offsets from runway or taxiway centreline
- @ 50 m in less trafficked parts of airfield pavements
- in a grid pattern, 20–50 m spacing, on large areas such as airfield aprons, parking areas and industrial sites
- @ 1 m over short lengths to establish an underlying crack pattern.

Due thought has to be given to the practical time demands. Traffic management will almost certainly be necessary in order to enable coring or pitting to be carried out, and the same time window should logically be used for FWD testing.

Tests in the most heavily trafficked areas (the wheel-paths) are obviously necessary. However, it is often very informative to be able to make a comparison with other much less heavily trafficked regions; hence the suggested occasional tests in lane centres, in less heavily trafficked lanes and on outer parts of airfield pavements. Tests at 1 m centres are only suggested in cases where it is suspected that a HBM base underlies an asphalt surface. The pattern of deflections can reveal the approximate spacing of underlying cracks or joints.

The reader may be surprised that the recommendations related to stiffness modulus so far only concern in situ tests rather than laboratory measurements. In fact, laboratory measurements are extremely useful, as will be explained in the following subsections, but, as tests on discrete elements, they are not necessarily representative of the effective in situ behaviour of the pavement as a whole. Laboratory stiffness measurements on asphalt specimens are strongly advised, but they should preferably not be used on their own to derive stiffness moduli for pavement design.

This is also where the possibility of a drainage survey comes in. The FWD may allow derivation of an in situ modulus at the time of test but, if a drainage survey reveals problems, then subsequent drainage improvement may allow a higher long-term modulus to be assumed in pavement rehabilitation design. Drainage is such a vital ingredient in ensuring long pavement life that a drainage survey should be considered anywhere where problems are suspected. As no traffic control is usually required, it may be carried out in targeted locations after the completion of other tests.

25.7.3.3 Fatigue strength

Quantitative information on fatigue strength is only needed where analytical design is used. For asphalt, two approaches present themselves, namely

- rely on published data for typical materials
- carry out indirect tensile fatigue tests (at least five).

In many cases it is impractical to carry out fatigue tests because of the number of different bituminous materials present on different pavement sections and in different layers. However, particularly where visual appearance suggests a reduced fatigue strength compared with normal expectations (friable, dry-looking materials), testing can be very useful. The indirect tensile test mode is suggested because core specimens are already cut to the right size; at least five tests (and preferably ten) are required in order to establish a fatigue characteristic (see Section 9.4.4). Evaluation of fatigue data will normally be comparative; that is, by comparison with the performance expected for typical new materials.

In the case of HBMs, it is reasonable to assume a standard fatigue relationship between the stress/strength ratio and life, as presented in Section 19.2, in which case the missing item of knowledge is strength. This can most readily be obtained by carrying out compressive strength tests (at least five advised) on core specimens, although indirect tensile strength testing is also possible where the appropriate equipment is available. In either case, a judgement has to be made as to the appropriate design flexural strength. This should logically be a value a little less than that derived from an average of the test data, following the reliability principles in Chapter 5. This recognises the likely real variability in material properties, whereas the test data will often exaggerate this variability because of the small sample size being tested.

25.7.3.4 Deformation resistance

The appropriate measurement for asphalt is a repeated load axial test, carried out on a core specimen. These tests should logically be carried out where rutting has occurred, and each layer of asphalt construction should be tested. In this way, it will become apparent which layer is the source of the rutting, and data will be available to predict future behaviour, including behaviour after one or more layers have been replaced. If no rutting is present, this sort of test is usually unnecessary.

For unbound materials, three levels of evaluation are possible

- use a stiffness modulus derived from FWD data and then rely on an assumed relationship between vertical strain in the subgrade (under a design wheel load) and the number of load applications to failure (see Section 7.3.1)
- supplement this by carrying out dynamic cone penetrometer (DCP) tests in core holes or trial pits
- extract samples for shear strength testing.

It is undeniably common practice to rely on calculation of vertical subgrade strain, taking stiffness moduli from FWD back-analysis; however, this is a very approximate approach. Use of the DCP is strongly recommended in any project-level investigation (buried services permitting) because it does at least provide an independent measure of the quality of unbound foundation materials. It is likely to provide supporting evidence as to whether or not reliance should be placed on subgrade strain.

The final option, carrying out shear strength tests, is an extreme one and would rarely be adopted. However, situations arise where subgrade strength is absolutely critical to performance, so, if the economic value of the pavement is high (major highway, international airport) and subgrade deformation is a real or potential problem, then shear strength tests should be considered. Samples can be obtained from trial pits or appropriate-diameter cores.

25.8. Summary

Pavement evaluation plays a key role in pavement management. There are plenty of different types of survey available, each with its particular advantages and limitations, and the trick is to use each in the most appropriate way. The choice of network-level evaluation tool (or tools) is clearly important, although it is at least as important to develop an effective system for processing the data and then for planning project-level investigations.

It is worth repeating that project-level surveys need 'designing' rather than reliance being placed on following a set procedure. Cores or trial pits are seen as almost essential for every investigation, together with a detailed visual inspection of the pavement. Other surveys depend on the situation. However, if analytical design procedures are to be used, then the FWD is a key tool, supported by appropriate laboratory tests and by use of the DCP in core holes or trial pits. Radar is less essential, and its use should depend on the likely variability in construction. Drainage surveys are important where water-related problems are suspected.

REFERENCES

BSI (2009) DD CEN/TS 15901-6:2009: Road and airfield surface characteristics. Procedure for determining the skid resistance of a pavement surface by measurement of the sideway force coefficient (SFCS): SCRIM®. BSI, London, UK.

HRB (Highway Research Board) (1954) *WASHO Road Test*. Highway Research Board, Washington, DC, USA, Special Report No. 18.

HRB (1955) *WASHO Road Test*. Highway Research Board, Washington, DC, USA, Special Report No. 22.

Kennedy CK and Lister NW (1978) *Prediction of Pavement Performance and Design of Overlays*. Transport and Road Research Laboratory, Crowthorne, UK, Report LR833.

Mamlouk MS and Davies TG (1984) Elastodynamic analysis of pavement deflections. *Transportation Engineering Journal, ASCE* **110(6)**: 536–550.

Paterson WDO (1987) *Road Deterioration and Maintenance Effects: Models for Planning and Management*. The Highway Design and Maintenance Standard Series, World Bank Publications, Washington, DC, USA.

Principles of Pavement Engineering, Third edition

Nick Thom
ISBN 978-1-83549-711-1
https://doi.org/10.1108/978-1-83549-710-420243030

Chapter 26
Diagnosis and prognosis

Once a detailed project-level survey has been carried out, it has to be properly interpreted. If the correct treatment is to be prescribed then a correct diagnosis of pavement condition must be made. The process can sensibly be compared to making a medical diagnosis – or to solving a crime. The evidence has to be assembled and then processed in a logical manner, weeding out all possible explanations for the observed symptoms other than the correct one; that at least is the theory. And this is what Sections 26.1–26.3 will attempt to do, addressing a number of likely questions and charting a path through the data to arrive at probable answers.

To make a diagnosis is clearly vital, but the question then arises as to just what the future holds. For example, it is useful to know that observed longitudinal cracking of an asphalt pavement can be traced to a debonded interface 100 mm below the surface, but it would be even more useful to know just how quickly the pavement will continue to deteriorate, including any possible deterioration to layers deeper than 100 mm. Only by knowing this – or at least being able to estimate it – can sensible decisions be made about the timing and nature of any future rehabilitation treatment. This is the ‘prognosis’.

Naturally, this is far from easy. A pavement is a complex beast, each element depending on those above and below for its own survival. Predicting the life of a new pavement is quite possible because one has the luxury of being able to ignore little practical details such as construction defects, drainage failures or substandard materials. While a project is simply a theoretical entity, the real world need only intrude insofar as forcing the designer to select a suitable design reliability. Once a pavement has been built, the annoying realities can no longer be ignored! Sections 26.4–26.8 will therefore address those realities – commencing at the easy end of the spectrum.

Note that, while this chapter is addressed at enhancing the understanding of human engineers, the topics dealt with are prime candidates for the rapidly developing field of artificial intelligence, as is also the case for medical diagnosis and prognosis.

26.1. Diagnosis: pavements with an asphalt surface
26.1.1 Why is the pavement rutting?

- Check the visual condition. If ruts are narrow with shoulders, the problem is near the surface (surface course or binder course probably); the wider the rut, the deeper the problem.
- Inspect the cores carefully. If an asphalt layer appears rich in binder, especially if that binder is soft (check with a pointed instrument), that is likely to be the cause of the problem.
- Consider carrying out repeated load axial tests to check whether materials are deformation susceptible.
- Also consider testing specimens of recovered binder to evaluate viscosity (or recovered penetration).

- Inspect any available unbound material samples (e.g. stones from the bottom of core holes). Check for visible quality. Rounded materials tend to deform much more easily than angular materials.
- Look at dynamic cone penetrometer (DCP) data, if they exist. Either a sub-base California bearing ratio (CBR) less than 20% or a subgrade CBR less than 3% would be suspiciously low.
- Check falling weight deflectometer (FWD) data, if available. An unbound material stiffness of 50 MPa or less is potentially deformable.
- Consider whether the unbound layers may be near saturation. For example, a low stiffness from the FWD, lack of obvious functioning drainage or a permeable (e.g. cracked) asphalt surface all raise suspicions.
- Check any data on the type of subgrade present. In general, clayey material can rut; sandy materials do not.

26.1.2 Why are there transverse cracks?

- What do the cracks look like? Straight and regularly spaced cracks represent reflective cracks from joints in an underlying concrete pavement. If the asphalt cover is thin, these may appear as pairs of close parallel cracks.
- If the shape and spacing are irregular, the cause is either reflection of thermally induced cracks in a hydraulically bound material (HBM) base or else low-temperature cracking of the asphalt.
- If reflective cracking is suspected, check the crack distribution. A concentration in more heavily trafficked lanes or in wheel-paths implies that traffic is playing an important part; otherwise, the effect is thermally driven.
- Close spaced (e.g. < 150 mm) cracks imply a shallow feature, affecting upper layers only; if more widely spaced, the problem may extend deeper.
- If FWD data are available, check for any loss of effective asphalt layer stiffness in the wheel-paths. This will provide evidence of the severity or otherwise of the damage.
- If short cracks are generally distributed across the pavement then these are surface-originating cracks, probably initiating from construction defects (e.g. due to rolling while too cold), the depth of which can be checked by coring.

26.1.3 Why is there longitudinal cracking in the wheel-path?

- The fact that this cracking is in the wheel-path proves that it is traffic that is doing the damage.
- Check the visual appearance. If there are multiple cracks in the wheel-path then the effect is shallow and probably implies debonding between asphalt layers. If the cracking is largely confined to a single crack, then near-surface debonding is unlikely.
- Cores through cracks are strongly advised. Check crack depth; also check for debonding. A shallow depth implies slow top-down cracking in a thick pavement. Check whether lower layers have begun to disintegrate, suggesting water ingress through the crack.
- If a FWD survey has been carried out, check the asphalt layer stiffness. If it is low or variable, this implies significant damage and/or debonding.
- Compare FWD-derived asphalt stiffnesses with those from laboratory testing of recovered samples, adjusting to a common temperature and loading rate (see Section 9.3.3). If the two measures agree, then the asphalt layers are substantially intact and well bonded. If the FWD-derived value is lower, this implies either cracking or debonding.

- Alternatively, Benkelman beam or deflectograph data will give an idea of relative condition from point to point; a high level of variability is almost always associated with significant cracking damage.
- Check for binder hardening or unusually high laboratory stiffness. Binder hardening is often associated with increased susceptibility to fatigue.
- Also check for poor binder adhesion. This will show up as unusually low stiffness, even of visually undamaged material.
- If the pavement is of high economic importance, consider carrying out indirect tensile fatigue tests on cored specimens (at least five specimens needed) and compare the fatigue characteristic with standard data.
- Check the condition of supporting layers to see if they are poorer than expected. The FWD will supply realistic foundation stiffnesses; the DCP reveals strengths.
- Using the best available evidence for the stiffness of each layer, carry out a multi-layer linear elastic pavement analysis and an associated computation of pavement life. Compare the theoretical life obtained using this approach with past traffic and current pavement condition.

26.1.4 Why is the surface ravelling?

- Ravelling (and associated potholes) can only occur if the adhesion between binder and aggregate is breaking down for some reason.
- Check whether surface course binder has hardened appreciably. If so, this may be a contributory factor.
- Also measure the surface course stiffness; the indirect tensile stiffness modulus test is suitable. The stiffness should have increased in line with binder hardening. If not, this implies a general lack of binder adhesion.
- Check gradation and binder content by an analysis of recovered specimens. Both low binder content and high filler content can contribute to lack of binder adhesion.
- If neither the binder hardness, binder content nor the filler content appears suspect, the implication is either that the particular binder–filler–aggregate combination is a poor one, probably due to chemical incompatibility, or that asphalt manufacture was deficient. Either way, the ravelling observed on the surface will also be reflected in poor resistance to cracking throughout the mixture. This type of problem is, therefore, a serious one, and implies a much-reduced structural life for the pavement.

26.1.5 Why is bitumen bleeding from the surface?

- Inspect the cores. Multiple layers of surface dressing are one obvious source of excess binder.
- If surface dressing is not the cause, determine the voids content of the surface course. Bleeding should only occur at voids contents of 2% or less.
- Also check the visual condition for rutting within the surface course. Rutting of this type is often (but not always) associated with very low voids content.

26.1.6 Why is the skid resistance low?

- Is bleeding evident? Bleeding results in near-zero texture depth and very poor wet skid resistance.
- Check for aggregate polishing using the pendulum test (Section 11.4.1). Compare with the value that would have been specified.
- Check for loss of texture depth using the sand patch test or laser texture meter (Section 11.4.3).

26.2. Diagnosis: pavements with a concrete surface

26.2.1 Why are there transverse cracks?

- Was the cracking present soon after construction? If so, the cause is likely to be shrinkage cracking during construction.
- Is the joint spacing greater than about 20 times the slab thickness? This, very approximately, marks the maximum spacing at which joints in unreinforced concrete can be expected to function properly.
- In the case of a continuously reinforced concrete pavement, transverse cracks are expected. They should form at a spacing of 1–2 m but, in a properly functioning pavement, they will remain narrow and there should be no spalling. If there are more cracks than this or if many are no longer narrow, then this implies that the pavement is not functioning as designed. Check concrete strength by carrying out compressive or indirect tensile strength tests on core specimens. Check slab thickness by means of cores and/or radar. Check foundation stiffness from FWD data. Compare with design expectations.

26.2.2 Why is there longitudinal cracking?

- Longitudinal cracking is a sure sign of overloading by vehicles.
- Check concrete strength by carrying out compressive or indirect tensile strength tests on core specimens.
- Check slab thickness from cores and/or radar.
- Check foundation stiffness from FWD data.
- Based on these measures, estimate what the pavement life should have been, either with reference to a design manual or through analytical computation.
- If the estimated life is less than the current condition suggests, then the foundation has probably deteriorated. Check for evidence of voiding under the slab, notably poor slab support as deduced from FWD deflections. Also check for subgrade softening, from FWD, DCP, unbound material samples or drainage survey data.
- If the estimated life is greater than the current condition suggests, then something has occurred that the estimate does not take into account (e.g. shrinkage cracks). If longitudinal cracking is localised then there may be particular circumstances in specific areas (e.g. defective concrete or inadequate curing).

26.2.3 Why is there faulting across joints?

- If dowels or tie-bars are present, there should be no faulting. If faulting is present, this implies disintegration of the concrete around the dowels and probably water ingress and corrosion of the bars.
- Even without dowels or tie-bars, faulting implies poor load transfer. This could be due to excessive joint spacing, poor aggregate durability or simply low-strength concrete (in which case more general cracking is also likely).
- Faulting also implies a deformable foundation, such that one side of the joint has been able to settle more than the other. This suggests a low-quality or saturated granular sub-base.

26.2.4 Why is the surface deteriorating?

- Is/was there an excess of cement mortar at the surface? Trafficking will eventually remove this excess mortar, but the result can be a decrease in ride quality. Once this loss of surface mortar has occurred, there should be no further deterioration.

- Has 'scaling' – the loss of discrete areas of surface – occurred? This is caused by frost action and implies that whatever precautions were taken against frost attack in the design (strong concrete; air entrainment) were insufficient for that location.
- Concrete surfaces can also lose their wet skid resistance over time. As in the case of asphalt surfaces, the pendulum test can be used to measure the degree to which polishing has occurred and the sand patch test can be used to measure texture depth.

26.3. Diagnosis: other pavement types

26.3.1 Grouted macadam

Grouted macadams are basically special asphalts and many of the points made already with regard to diagnosis of asphalt pavement problems also apply to grouted macadam. Grouted macadam is rightly portrayed as a jointless product – but it can certainly crack in much the same way as asphalt. However, a particular issue is the occurrence of localised distress, often around the edge of an area of grouted macadam, suggesting less than 100% grout penetration. This shows itself as localised cracking, sometimes also with minor depressions.

26.3.2 Block paving

The strength of block paving depends very much on the supporting layers. Any rutting or other deformation of the surface implies lack of strength in underlying layers, and this can be investigated in just the same way as for an asphalt-surfaced pavement. Uneven block settlement implies a sand bedding rather than mortar. Excess (>50 mm) or uneven sand thickness can lead to localised settlement; ponding of water then makes the situation worse.

If individual blocks have fractured, then the original block specification may have been at fault. Industrial applications commonly demand a thickness of 80 mm or more and a length of no more than 200 mm. Longer or thinner blocks are susceptible to fracture under normally applied tyre pressures. Otherwise, overloading (e.g. by trailer feet) has to be suspected and/or a local loss of adhesion between blocks and underlying mortar.

26.3.3 Unsurfaced pavements

Rutting can, of course, occur in an unsurfaced pavement and the points made in relation to asphalt-surfaced pavements also apply. To some extent, the surface distress possible in an asphalt is mirrored in an unbound surface also, except that internal 'cohesion' is the binder rather than bitumen. Stone loss and potholing are both likely.

Corrugations – regular undulations along a road – occur due to a resonance being set up between a vehicle's suspension system and longitudinal defects in the road surface. This is an expected occurrence on unsurfaced roads and does not necessarily imply any particular defect in design, material provision or construction. It does, however, imply a need for ongoing maintenance.

26.4. Prognosis: substantially intact asphalt-surfaced pavements

Following diagnosis, a decision has to be taken as to whether the pavement under consideration has substantially retained the properties of each component material that it had at the start of its life. Cracks may be present, but they should only be shallow or else very infrequent.

Moderate rutting may have occurred, but not enough to have actually caused loss of material integrity. Localised distress may be present, but only localised – it should not have reached the stage where it becomes typical of large sections of the pavement. Only if these conditions are satisfied can the pavement be treated as 'intact', and the following subsections applied.

26.4.1 Statistical treatment of data

As the pavement is basically intact, it can be analysed in the same manner as a new pavement and its remaining life predicted following the guidance in earlier chapters. To do this, due account must be taken of the measured properties of each layer, and a statistical view should be adopted. Predictions might sensibly be made for both an average (50th percentile) condition and a conservative (e.g. 15th percentile) condition. This is a difficult area to get right; however, the principle is exactly the same as explained in Chapter 5. The relative influence of each input parameter (e.g. asphalt stiffness, asphalt thickness, foundation stiffness) must first be estimated and, adopting the simplifications suggested in Chapter 5, the following equations may be used.

$$\text{Fatigue life} \propto E^3 k h^5 \tag{26.1}$$

$$\text{Deformation life} \propto E^3 k h^3 \tag{26.2}$$

where E is the asphalt stiffness, k is the foundation stiffness and h is asphalt thickness, and these three parameters may be considered as independent of each other.

Next, the degree of variability in each of the input parameters has to be established. This might be based on measurements, but care should be taken to subtract the variability due to the measurement itself from real variability. Thickness variability can often be estimated quite accurately from cores or ground-penetrating radar data. Failing measurements, the estimates suggested in Table 5.1 may be used. As to which percentiles should be used in remaining-life calculations, the following advice was given in Chapter 5

- 50th percentile of pavement condition – use 50th percentiles of individual parameters
- 25th percentile of pavement condition – use 33rd percentiles of individual parameters
- 15th percentile of pavement condition – use 25th percentiles of individual parameters
- 5th percentile of pavement condition – use 14th percentiles of individual parameters
- 1st percentile of pavement condition – use 6th percentiles of individual parameters.

26.4.2 Pavement life prediction

The method of choice here is certainly an analytical computation, making direct use of the percentiles from the previous subsection. However, if no analytical prediction can be made, then creative use of design manuals is always possible, for example, using 'structural coefficients' or 'condition factors'; the key point is that this must include consideration of the real properties of materials. Some design approaches (e.g. AASHTO, 1993), are flexible enough for this to be carried out without difficulty; others, particularly those presenting a catalogue of designs, are less easy to use.

Whichever predictive approach is used, the result is a set of pavement lifetimes for the various possible deterioration modes. The next step is to take account of the damage which, although it may not show up in terms of loss of stiffness, has actually taken place already. For *fatigue*, whether of asphalt or HBM, the assumption is commonly made that pavement materials follow what is known as Miner's law (Miner, 1945). This is a general law applied to fatigue damage, not only to damage of pavement materials, although it is an approximate rule rather than an exact law. It states that relative damage is cumulative. Thus, if the predicted fatigue life is, say, 30×10^6 axle loads, and the past traffic is 12×10^6, then the relative damage stands at 40% and the remaining life is predicted to be just 18×10^6 axle loads. The concept of relative damage becomes important for strengthening designs because, whatever the predicted life of the strengthened pavement, a proportion of it may have to be discounted straight away due to past damage.

In the case of *rutting*, whether of asphalt or foundation materials, it is suggested that sensible account should be taken of what has actually occurred; the current rut depth divided by the number of years since the road was last resurfaced gives a measure of rut development rate, and past trends are likely to continue into the future. This can even be used to calibrate predictions based on vertical strain in the subgrade under a standard axle load, or on repeated load axial test data from asphalt specimens, something that may prove useful when designing strengthening.

The result of all this is that a set of predicted pavement lifetimes is determined, which can then be compared with future traffic estimates, allowing informed decision-making as to the need for and optimal timing of any future maintenance. It would be foolish to pretend that such predictions are actually accurate, particularly when factors such as continuing ageing of asphalt and increasing danger of drainage failure are considered, but they will be the best possible under the circumstances, and the experienced engineer should even try to factor in the effects of continuing future change in material properties. It may be an uncomfortable truth for those who believe rigid procedures are required to achieve a mythical 'quality assurance' level, but the fact is that there is absolutely no substitute for engineering judgement here. Calculations should *never* be believed without question, especially when so many assumptions are being made.

26.4.3 Pavements with HBM bases

If the pavement is considered as substantially intact, an evaluation of the fatigue susceptibility of the HBM base is required, which – following the guidance in Chapter 5 – means that a stress ratio has to be computed at the base of the layer. The approximate equation suggested in Section 5.2.2 was

$$\text{Fatigue life} \propto k^3 h^{15} \sigma_f^{12} / E^3 \qquad (26.3)$$

Use of this equation results in the following recommendations for cracking of the HBM layer

- 50th percentile of pavement condition – use 50th percentiles of individual parameters
- 25th percentile of pavement condition – use 35th percentiles of individual parameters
- 15th percentile of pavement condition – use 28th percentiles of individual parameters
- 5th percentile of pavement condition – use 18th percentiles of individual parameters
- 1st percentile of pavement condition – use 10th percentiles of individual parameters.

Remember that, because the stiffness modulus (E) is on the bottom of Equation 26.3, the appropriate percentiles for the modulus of a strong HBM layer are the opposite of those used for other parameters; for example, a 35th percentile is replaced by a 65th and so on. There is also a new parameter to consider, namely the stress at failure (σ_f). The appropriate percentile of strength has to be determined, either by estimation or from laboratory strength tests on cores. Failing other information, it is suggested that a value 80% of the expected average strength should be assigned. Note that the thicknesses of asphalt and HBM should not be considered as independent variables here, and the percentile used should therefore be for the combined layer.

26.5. Prognosis: substantially intact concrete pavements

The same equation and percentiles apply here as those in Section 26.4.3 for HBM base pavements. As in the case of asphalt pavements, Miner's law can, in theory, be applied to fatigue life, leading to the determination of a proportion of life already 'used up'. However, in the case of concrete, this is a slightly questionable move. The problem is the slow increase in concrete strength that occurs over time, and this raises the question of what value of strength should be used in any calculation of pavement life. If the current value is used, based on tests on core specimens, then it may be argued that the calculation will overestimate the life of the pavement because greater damage would have been occurring during the earlier years while the concrete was weaker. On the other hand, if an adjustment is made to bring the strength back to what it might have been at 28 days (e.g. by taking around 80% of the current value), then this will overestimate the damage that has been occurring more recently. There is no simple answer to this dilemma; however, it seems most logical to adopt the current value (using the appropriate percentile), accepting that past damage may be underestimated but in the expectation that future damage estimates will be realistic.

On a practical level, if few cracks have occurred in the past, while the concrete was weaker, this gives considerable confidence that fewer still will occur in the future.

With regard to deterioration other than traffic-induced fatigue cracking, thermally induced cracking may be a maintenance nuisance but it does not represent structural failure. Faulting at joints is considered to be just too hard to predict; if it is occurring, then, as in the case of rutting of asphalt pavements, a calculation may be made regarding the rate of development in the past and the same rate or higher may be taken as likely to continue into the future.

26.6. Prognosis: failing asphalt-surfaced pavements

The key question to pose here is whether there is any layer (or part layer) in the pavement that is substantially damaged (i.e. cracked or partially disintegrated). This layer could be either asphalt or HBM. The evidence for this may come directly from the visual condition of cores or trial pits – or indeed from surface appearance; it may also come from the determination of a lower than expected modulus value for a layer based on falling weight deflectometer (FWD) data. However, care has to be exercised in interpreting FWD results. It must be remembered that debonding between bound layers (to be discussed in Section 26.8) can lead to very considerable loss of apparent stiffness. Core data should be able to confirm whether such debonding is present. The influence of debonding must be taken into account when evaluating material modulus. However, once this has been done, any bound layer that is found to have a genuine in situ modulus half of what would have been expected or less has to be considered as 'failed' – in the sense that it is meaningless to compute a life *before* cracking because cracking has already occurred. In fact, such a layer is almost certainly still in the process of failing and its long-term modulus would therefore be lower than that being recorded currently.

26.6.1 Full-depth crazing

Asphalt pavements can fail in many ways – and at greatly differing rates. However, in the case of full-depth crazing the prognosis is usually not good. Figure 26.1 illustrates the situation.

In theory, it may be possible to compute a life to rutting failure by considering the likely rate of deformation of the unbound layers (as outlined in Chapter 7). In practice, this sort of prediction is so sensitive to the strength parameters assumed that it is unlikely to be at all reliable. There are many such 'failed' pavements that continue to take heavy traffic loads without any dramatic problems, so it would be premature simply to assume that the pavement's life has come to an end. Even wet layers of high-quality crushed rock sub-base have considerable load-bearing capacity, particularly under the overburden pressure provided by overlying bound layers.

The point at which ride quality becomes unacceptable depends on the type of pavement and traffic speed. If the foundation is not sufficiently strong to prevent deformation, then ultimate failure

Figure 26.1 Asphalt pavement with full-depth crazing

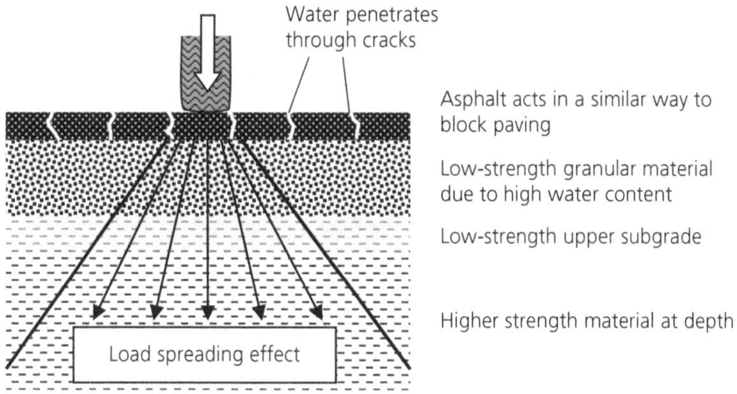

of the pavement comes when differential settlement between damaged and undamaged locations becomes excessive, which probably means when settlement in damaged locations reaches 40 mm or so. On the other hand, if the foundation is strong enough to resist deformation, failure is more likely to be defined by the degree of surface break-up of the asphalt, something that will accelerate markedly if freeze–thaw conditions are encountered. Either way, prediction definitely cannot be based simply on computations but must include a large dose of engineering judgement, based on empirical experience.

26.6.2 Lower asphalt layer disintegration

This is actually not uncommon, despite not correlating with any of the failure modes normally considered at the design stage. The cause of the failure is usually durability related, perhaps because the voids content in a road base layer was too high and/or water has managed to find its way in. Layer debonding is commonly followed by progressive failure of an overlying material as water seeps along the debonded interface.

An approximately similar situation is found when a failing pavement has been overlaid or when the surface course has been planed out and replaced with fresh material, leaving broken lower layers in place.

Evidence from cores or trial pits should be sufficient to reveal whether an underlying layer is disintegrating or not, although it has to be remembered that coring is a hit-and-miss affair. If FWD testing has been done, then the overall stiffness of the asphalt will be low, contrasting with the results of laboratory tests on core specimens from the upper layer or two.

The difficult task, if an analytical assessment is to be attempted, is to assign stiffness moduli to the intact and disintegrated layers. It may be possible to deduce something about the intact material from FWD testing or from laboratory stiffness testing; otherwise, typical values have to be assigned. For disintegrated materials, it is advised that they are treated as being no better than an unbound layer in the long term, with a modulus of 150–200 MPa at best. The computation of remaining life then considers only the upper, intact asphalt as being a bound material; the computed life should also be reduced to take account of fatigue damage using Miner's law, as described in Section 26.4.2.

26.6.3 Intermediate asphalt layer disintegration

It is not unusual to find that an intermediate layer is disintegrating, quite often the binder course, while the base remains more or less intact. In such a case, it is practically impossible to compute a meaningful life for the overlying surface course using traditional analytical design methodology because the layer is too thin. However, it is still possible to predict the life of the base. It is simply necessary to include a suitable long-term modulus for the disintegrating binder course in the calculation. When checking against a design manual prediction, it is suggested that the thickness of the disintegrated layer is reduced to an effective contribution – probably of only 15–20% of its actual thickness – reflecting its lack of contribution to reducing flexure of the base. Miner's law should also be used to take account of past damage to the base.

26.6.4 Surface layer(s) disintegration

This can occur where shallow debonding has taken place, either beneath the surface course or beneath the binder course, and the overlying materials then craze and fail. The mechanism may be aided by water or ice, but it will be traffic that actually does the damage. The key here is to realise that the apparently disastrous appearance of the pavement surface only relates to the upper layer or two. This should be deducible from the relatively close-spaced nature of the cracks compared with the overall thickness of the asphalt, from the appearance of debonding in cores and from the fact that the layer modulus as determined from the FWD is greater than would be the case if full-depth crazing were present.

Clearly, no meaningful computation of surfacing life can be made – it has already failed. However, the life of the remaining base can still be assessed, again taking a much-reduced thickness contribution from the failing layers when checking against design manual predictions.

26.6.5 Failure of a strong HBM base

In most pavements with HBM bases, a degree of failure is expected. As explained in Chapter 17, thermal stresses will always induce transverse cracking and, to control this, it is common practice nowadays to deliberately introduce cracks, often at a spacing of around 3 m. However, it is always possible that break-up of a HBM layer goes much further than this. It may have been under-designed; it may have been of substandard construction; it may just be that the durability of the material was inadequate, something which is hard to predict in advance with confidence. The long-term presence of water is another common reason for premature loss of strength, water that may have come either from the underlying unbound foundation or from above. Local degradation is often found beneath cracks in overlying asphalt.

If the asphalt is still intact, then the difficulty lies in assigning an appropriate stiffness modulus for the HBM – and values obtained from the FWD should, in general, *not* be used. As noted already in Chapter 17, it is common to assign an artificially low stiffness, often 500 MPa, to the material when designing the asphalt against reflective cracking, and there is no reason to increase this once the HBM layer starts to break up. Although the stress concentration at cracks that causes reflective cracking will reduce as cracks become more numerous, this is hardly a well-controlled process. Only if the FWD back-calculated stiffness modulus is 1000 MPa or less should direct use of this number be contemplated and, even then, allowance should be made for future loss of stiffness – engineering judgement again! Existing damage to the apparently intact asphalt can be taken into account using Miner's law.

26.7. Prognosis: failing concrete pavements

It is important to understand the deterioration mode(s) applying, which relate to the type of concrete pavement present. In general terms, transverse cracks are not unexpected; in the case of a continuously reinforced concrete pavement they will always be present. This means that, if damage is restricted to transverse cracking, then the pavement can probably be considered substantially intact. Calculations of stress in pavement-quality concrete should take into account the likelihood of poor load transfer across some of the cracks, as well as loss of slab support due to the softening effects of ingressing water, but the procedure can nevertheless follow that for intact pavements described in Section 26.5.

However, as soon as longitudinal cracking occurs then the prognosis takes a serious turn for the worse. A transverse crack can only propagate across the width of a carriageway and no further; a longitudinal crack knows no limit. Once it starts, it will tend to propagate rapidly along a wheelpath and generate further transverse cracks as it does so. In such a case, it is fairly meaningless to try to calculate a remaining life because the life before cracking is effectively zero. The rate of subsequent deterioration will be governed by the durability of the foundation under the influence of significant water ingress. A judgement has to be made as to whether the condition of the foundation layers is likely to reduce to the extent that localised settlement occurs, the effect of which at the surface will be 'faulting' – the occurrence of steps between individual intact concrete slab elements. The best clue always comes from careful observation of past performance; that is, if faulting has already commenced, then it will generally accelerate.

26.8. Prognosis: the effect of debonding

This issue is one of the hardest of all to resolve. It is common to find that one or more layer interfaces are debonded, at least to the extent that they separate under the action of coring, yet it is relatively uncommon to hear of debonding being blamed for pavement failure. This section explores the factors that need to be borne in mind.

26.8.1 The theoretical effect of debonding

One effect that is much more than theoretical is the fact that debonding between bound pavement layers significantly decreases the overall apparent stiffness of the bound layers. Calculation Sheet 26.1 presents a mathematical assessment of this effect, which is really a reduction in the second moment of area (I). A formula for the strain in the bottom of the asphalt, the parameter most commonly used in fatigue crack prediction, is also derived. Both the apparent stiffness and the strain resulting from this prediction are plotted in Figure 26.2 in relative terms.

The loss of apparent layer stiffness is very clear in Figure 26.2, although there the assumption is that the debonded interface transfers no shear stress at all, which is an extreme case. However, it is common to find that the apparent stiffness of the bound pavement layers is no more than half that expected, with no clear cause other than debonding; thus, the effect is real enough. But the most interesting finding is definitely that relating to the tensile strain at the bottom of the asphalt. According to Figure 26.2, it can be almost 60% higher than it would have been in a non-debonded pavement if the debonding occurs at just the wrong level in the pavement, about 40% of the full layer thickness, which would imply a noticeable reduction in life. Even allowing for some friction at the debonded interface, the increase in asphalt tensile strain might be expected to be at least 30%, reducing fatigue life by a factor of 3 if a fourth-power law is used.

Calculation Sheet 26.1. Debonding theory

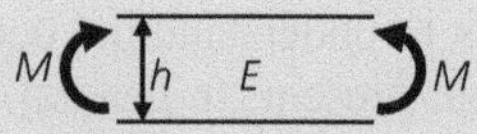

Treat bound layers as a beam in bending:

$\therefore$ Second moment of area (I) $= h^3/12$

Now assume a debonded interface at depth d:

$\therefore I =$ sum of two components $= (h-d)^3/12 + d^3/12$

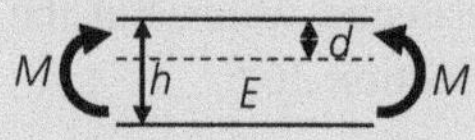

From Calculation Sheet 5.1:

Curvature (κ) $= \mathrm{d}^2y/\mathrm{d}x^2 = P\lambda^3/k$, where $\lambda = (k/4EI)^{0.25}$

$\rightarrow \kappa = P/(4^{0.75}k^{0.25}E^{0.75}I^{0.75})$

$\rightarrow \varepsilon = \kappa \times (h-d)/2$

$= 1.14P(h-d)/\{k^{0.25}E^{0.75}[(h-d)^3 + d)^3]^{0.75}\} \propto (h-d)/[(h-d)^3 + d)^3]^{0.75}$

Figure 26.2 The theoretical effect of debonding (zero interface friction)

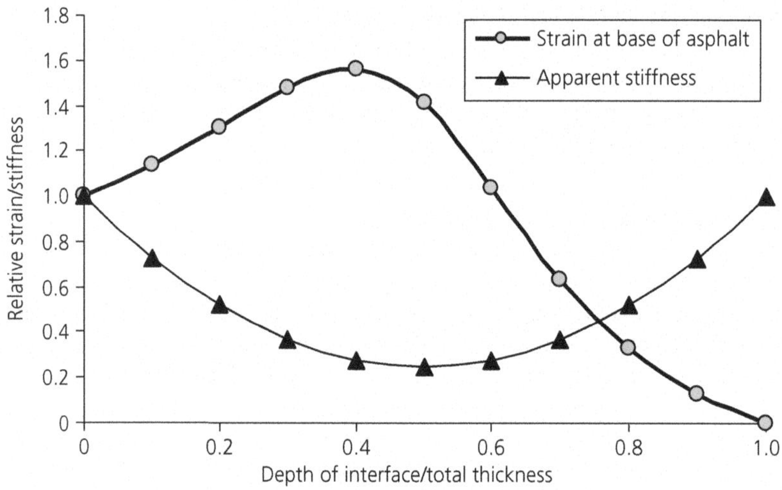

In the layer of bound material above the interface, the strain mirrors that of the lower part, with a maximum value when the interface is at a depth about 60% of the full bound layer thickness. Thus, while either the upper or lower part may fail early, the other part will not – at least in theory – which means it is hard to be certain just how much reduction in overall pavement life debonding actually brings.

One thing is certain, however, and that is that it is definitely *not correct* simply to assume that an apparent layer stiffness (e.g. from FWD data) for an asphalt with a debonded interface represents a single uniform layer. In the extreme case shown in Figure 26.2, with an interface at mid-depth, the apparent stiffness would be one-quarter of what it would have been if fully bonded, and this would lead to the calculation of a strain around 2.8 times that of the fully bonded case (following the approximate relationship in Calculation Sheet 26.1). This compares to a strain of 1.4 times that of the fully bonded case shown in Figure 26.2. The message is, therefore, be very careful. The effect on fatigue life, although real, may be much less than it initially appears.

Nevertheless, there is no doubt that the loss of effective stiffness from bound layers due to debonding leads to an increased stress and strain in the foundation, and this can be calculated in the usual way (e.g. multi-layer linear elastic computations) using the apparent asphalt stiffness, allowing prediction of pavement life against rutting failure in the foundation.

26.8.2 The real effect of debonding

Clearly, theory does not hold all the answers. While it does indeed appear to be true that debonding has much less direct negative effect on asphalt than is often thought, it is frequently accompanied by indirect effects. The first question to ask is: why did debonding occur in the first place? And the answer is most likely to be, at least in part, that the binder is either embrittled or else at a low content. This immediately warns of the likelihood that asphalt durability is low. One then has to realise that a debonded interface provides an ideal channel along which water can flow, thereby compounding the danger. If the surface is perfectly sealed this danger may be avoided, but – as soon as water enters – it will begin to attack the materials above and below the debonded interface, materials that are likely to be susceptible to water damage. Thus, it is strongly recommended that a predicted life for a debonded pavement should be treated with a healthy degree of scepticism, particularly if there is already evidence of water damage having occurred.

26.9. Summary

The principles and techniques described in this chapter lie at the heart of the decision-making process in relation to pavement maintenance. They give the possibility of predicting future performance, either by human action or by means of a well-trained artificial intelligence system. The importance of a correct statistical treatment of the data is worth highlighting. There is no fundamental meaning in a claim that a pavement has x number of years of remaining life; it is always the case that there is an average condition and a spread above and below this state. It is therefore critical, if proper decision-making is to be carried out, that the spread of possible future lives is understood. It would be criminal not to make the fullest possible use of the data.

REFERENCES

AASHTO (American Association of State Highway and Transportation Officials) (1993) *Guide for Design of Pavement Structures*. AASHTO, Washington, DC, USA.

Miner MA (1945) Cumulative damage in fatigue. *Journal of Applied Mechanics, ASME* **12(67)**: A159–A164.

Nick Thom
ISBN 978-1-83549-711-1
https://doi.org/10.1108/978-1-83549-710-420243033

Chapter 27
Strengthening design

This is where pavement engineering really does begin to look more like an art than a science. A diagnosis may often be made with reasonable confidence; after all, one is still dealing with the past, with things that have already happened. A prognosis is rather trickier, because it requires extrapolation from the present into the future, sometimes the distant future. Nevertheless, in making a prognosis one is still dealing with the same materials and layers that have been in position for many years, and so experience from the past has a high degree of relevance. But, on entering the realm of rehabilitation design, the level of confidence inevitably drops still further. The problem is not only that new materials are being introduced, but also the fact that the conditions (stress, moisture state) applying in those materials being left in place are being altered. Calculations can still be carried out – but a lower level of assurance is commonly present compared with new pavement design.

The basic tools of the trade remain unchanged. The design approaches described in earlier chapters are no less valid and the material properties detailed previously still apply. There are certainly issues regarding the properties to be assumed for existing materials, issues that require a certain amount of engineering judgement; but probably the greatest difficulty lies not in the ability to predict performance, hard though that sometimes is, but in working out just which of the rehabilitation options on offer represents the most effective strategy in a given case. Whether explicitly or otherwise, the decision requires a balance between uncertainty in future performance and the possible consequences, should expected levels of performance not be realised, and the measure of 'value' used should ideally include disruption to users, safety and environmental issues, as well as direct costs.

However, in order to stand a chance of making the right decision, one has to be as confident as possible in predicting the performance expected from the various design options. This therefore is the starting point.

27.1. Drainage improvement

It is fair to say that this is not the conventional place to start. Yet many are the experts who have been proud to state that good drainage is the most essential feature of pavement design. Whether this is strictly true or not is open to debate, but drainage is certainly extremely important and it does not lose its importance at the rehabilitation stage. The point has been made that it is normal to find that the uppermost 100–500 mm of subgrade has softened considerably in comparison with the underlying material, and the cause is certainly water. It is also common to observe that the presence of a particularly serious crack in a pavement is associated with significant subgrade softening. As a pavement deteriorates, the passage of water through the bound layers and into the foundation becomes ever easier and, the greater the time during which a pavement foundation is subjected to high levels of water, the greater the rate and extent of subgrade softening. Now, clearly, if an overlay is applied, the water source is shut off – or at least greatly restricted. But, without functioning drainage, the subgrade will not easily regain its original strength and stiffness.

So, what can be done? The first question is whether anything needs to be done at all. Whether or not advanced technology, such as closed-circuit television, is used to inspect otherwise inaccessible pipe runs, much information can be obtained from visual inspection of drains. Direct symptoms of drainage problems are

- the complete absence of any sign of a drain
- side ditches silted up
- grass growing in a drainage medium
- no sign of a direct connection between granular foundation layers and a side drain (usually investigated by digging a trial pit)
- lack of water flow though catch-pits.

None of these is sufficient for immediate panic on its own. However, the combination of a softened upper layer of subgrade and a clear drainage fault presents a likelihood that the pavement would benefit from drainage improvement of some sort. But it is still too early to panic. One must first be convinced that the softening of the subgrade really has contributed to pavement deterioration – because it doesn't always. Cracking of an asphalt pavement may appear severe, but if it is top-down and does not extend beyond around 100 mm then the subgrade is unlikely to have contributed to the problem. The same is true of rutting in an asphalt; it may also apply to reflective cracking problems over a hydraulically bound material (HBM) base. There is little point spending time and money correcting a very real fault if it is not the fault that is actually causing the observed pavement distress. It sounds an obvious point – and it is – but that does not mean it is always followed. A flowchart describing the decision-making process is shown in Figure 27.1.

Note that, even after answering 'yes' to all questions in Figure 27.1, the advice is only to 'think about' drainage improvement. The final decision simply cannot be taken at this, purely technical, stage. That will depend on factors such as economics. However, one further technical evaluation

Figure 27.1 Decision chart – drainage improvement

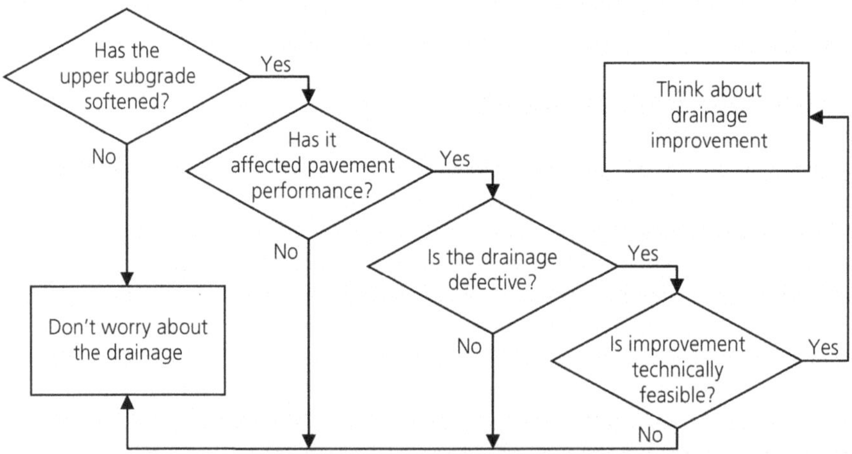

Figure 27.2 Subgrade softening seen in DCP data

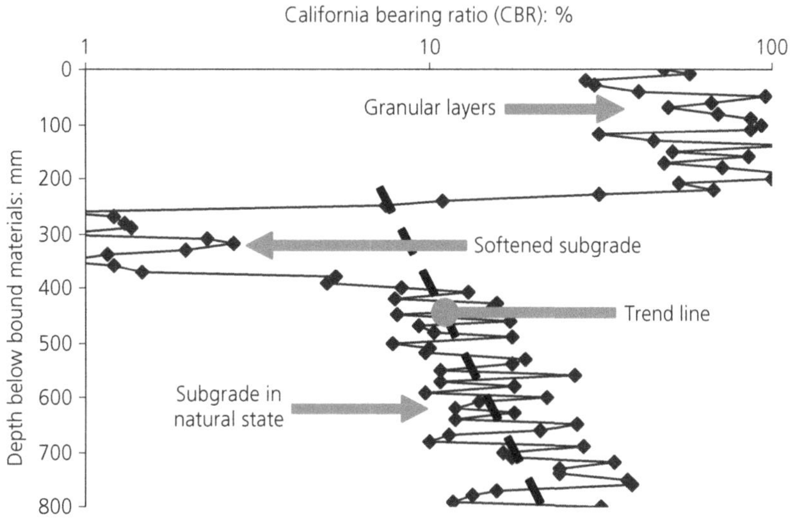

that the engineer can at least attempt is to quantify the long-term benefit likely to be derived from drainage improvement. The basic question is what the stiffness modulus and/or strength of the subgrade is likely to revert to once drainage becomes effective, and the most visually understandable means of estimating this can be seen in a dynamic cone penetrometer (DCP) trace. Figure 27.2 makes the point.

It would appear likely that, in the example shown in Figure 27.2, the subgrade originally had a CBR of 7–8% immediately beneath the unbound granular layers, and this has reduced to 1–2% due to water ingress. However, following restoration of drainage function, the soil is unlikely to return all the way to 7–8%. A clue is often to be seen in other DCP results, from less water-affected locations; but, in the absence of any actual data, it is suggested that a conservative assumption, in this case perhaps 5%, is sensible. In terms of stiffness, the influence is clearly also dependent on the thickness of the affected layer. It would be possible to carry out a multi-layer linear elastic calculation in order to estimate the influence that improvement of the affected layer would have on the overall effective stiffness modulus of the foundation; however, Calculation Sheet 27.1 presents a simplified assessment – which is acceptable bearing in mind the large uncertainty inherent in any such prediction.

Calculation Sheet 27.1 allows the increase in overall whole-subgrade stiffness due to drainage improvement to be estimated. For example, in the case shown in Figure 27.2, stiffness is reduced over a thickness h of about 0.3 m; the reduction ratio (p) is probably about 0.25, increasing to 0.7, say, after drainage improvement. The graph in Calculation Sheet 27.1 then shows that the relative whole-subgrade stiffness would increase from about 0.6 to 0.9, a 50% increase. Thus, if falling weight deflectometer (FWD) data suggest a current whole-subgrade stiffness of 30 MPa, for example, this might be expected to reach 45 MPa after drainage improvement. It's crude, but it is a great deal better than pure guesswork.

Calculation Sheet 27.1. The effect of drainage improvement

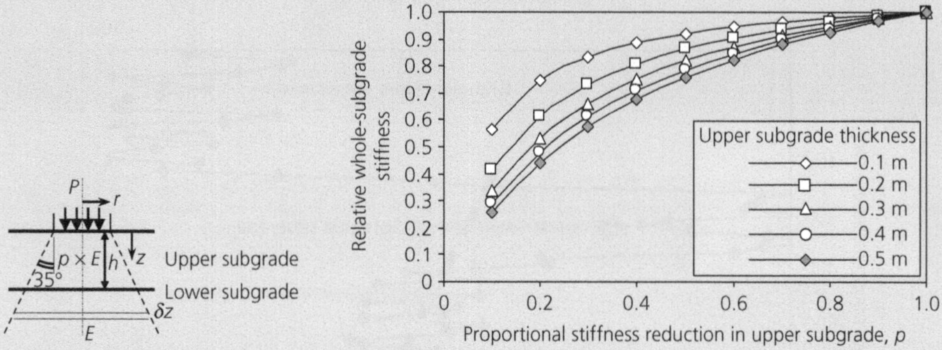

Assume 35° load spread angle ($\tan 35^\circ \approx 0.7$)

Deflection y is

$$y = \int P/[\pi(r + 0.7z)^2 E]\,dz \quad \text{(from 0 to } \infty)$$

$$\rightarrow y = P/[0.7\pi E(r + 0.7z)] \quad (E \rightarrow pE \text{ from 0 to } h)$$

$$= (P/0.7\pi E)\{1/pr + [1/(r + 0.7h)](1 - 1/p)\}$$

$$k = P/\pi r^2 y = 0.7pE/[r - (1 - p)r^2/(r + 0.7h)]$$

When $p = 1$

$$k_{p=1} = 0.7E/r$$

$$\therefore k/k_{p=1} = p/[1 - (1 - p)r/(r + 0.7h)]$$

Estimate r according to load-spreading ability of pavement structure

The plot above assumes $r = 0.75$ m

Armed with an estimate of the long-term foundation stiffness, it is now possible to use the design tools described in previous chapters and to predict the influence that drainage improvement might have on pavement life.

27.2. Overlays/inlays to flexible and composite pavements

Although the benefits of drainage improvement are too frequently neglected, they may still be insufficient to ensure adequate future life; hence the likely need for additional fresh material to be added.

The term *overlay* refers to the addition of a new layer or layers without the removal of any of the existing pavement (other than planing 10–20 mm to ensure a good key between old and new materials). An *inlay* refers to the situation where a certain depth of existing pavement is planed out and the same depth is replaced with new materials. It is of course possible (and common) to combine the two by planing out some of the existing pavement but then over-filling, giving an increased overall thickness. This section includes consideration of all such options.

The techniques for predicting the life of an overlaid or inlaid pavement are basically the same as those for a new pavement and the difficulties are similar to those already introduced in Chapter 26 when making a prognosis. The structure becomes one in which a certain thickness of new asphalt material overlies whatever is allowed to remain of the existing pavement layers. The choice as to how much (if any) of the existing pavement is removed prior to applying new material will depend on numerous practical factors as much as anything strictly technical. However, the key point for the engineer is that it must be possible to evaluate the likely pavement life provided by each option.

27.2.1 Rutting within asphalt layers

Naturally, the application of a new surface means that rutting is reset to zero. If there was little rutting in the original pavement then it is not necessary to consider the matter further, other than ensuring that new materials meet appropriate specifications and therefore have sufficient deformation resistance. If rutting is considered to be a potential problem, however, it is quite possible to carry out a calculation of the sort introduced in Section 10.4.1, assigning mixture viscosities to both old and new pavement layers. Alternatively, a qualitative judgement can be made. The most damaging shear stresses occur within the upper 100 mm of the pavement, particularly between 50 mm and 100 mm below the surface. This means that, ideally, no potentially deformable layer from the existing pavement should be left within 100 mm of the surface – and this requirement may significantly affect the design. Figure 27.3 illustrates the point.

For the inexperienced engineer it is tempting to assume that all clearly defined rutting with shoulders is occurring in the surface course – which means that it is easy to take the wrong maintenance decision. However, at the cost of relatively few tests on specimens from cores (tests such as the repeated load axial test), the truth can be discovered and the right decision made.

Figure 27.3 Conceptual illustration of rutting after rehabilitation

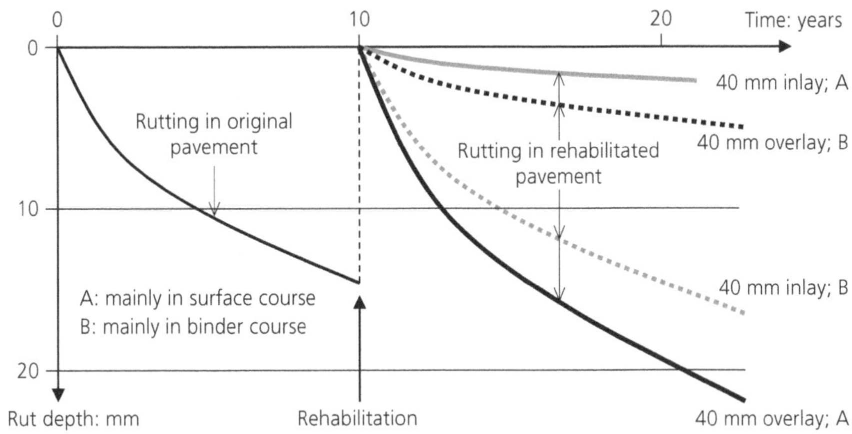

27.2.2 Rutting within foundation layers

In general, it is fair to say that if this mode of deterioration has not been present in the past then it is unlikely to occur following rehabilitation, whatever the calculations may show. This is an important point; it is not at all easy to make confident prediction of deformation in unbound materials, and the best evidence on which to base future predictions is definitely past performance. Calculations are sometimes still worthwhile, particularly if future traffic is much higher than has been the case in the past or if the design traffic load is to be increased, as may sometimes occur during airfield or industrial pavement development, but these predictions should always be 'calibrated' where feasible based on experience of past performance.

The methods for prediction are those presented in Chapters 7 and 10, remembering that appropriate long-term stiffness moduli have to be used in any computation of stress or strain under load, including adjustment of foundation stiffness if drainage improvement is to be made.

27.2.3 Pavement cracking

This is usually a significant element, often the controlling factor, in design of overlays and inlays. The same decisions have to be made as introduced in Chapter 26 as to whether existing pavement layers are to be considered intact or not, and the same general advice holds as to the most appropriate methods of analysis. This could mean, as shown in Figure 27.4, that the resulting structure comprises: fully intact and undamaged material (the inlay/overlay), over intact but partially damaged material (upper asphalt from existing pavement), over substantially damaged, and therefore non-intact, materials (lower asphalt or HBM base), over an unbound foundation. This is slightly more complicated than any of the cases covered in Chapter 26 in that the intact material is a combination of both damaged and undamaged components. This affects the application of Miner's law, which states that relative damage (i.e. proportion of life used) can never be recovered. A pragmatic approach would appear sensible, based on the ratio of thickness of new undamaged material to older damaged material. Take as an example a case where 100 mm of existing intact material is retained and it is concluded that 40% of its life has been used already. This is to be overlaid with 150 mm of new material. The effective damage might therefore be taken as 16% ($40 \times 100/250$) for the total intact asphalt layer. If, say, the calculated life with respect to fatigue cracking is 100 million vehicle passes, this therefore reduces to 84 million.

Figure 27.4 Possible overlay/inlay designs to asphalt pavements

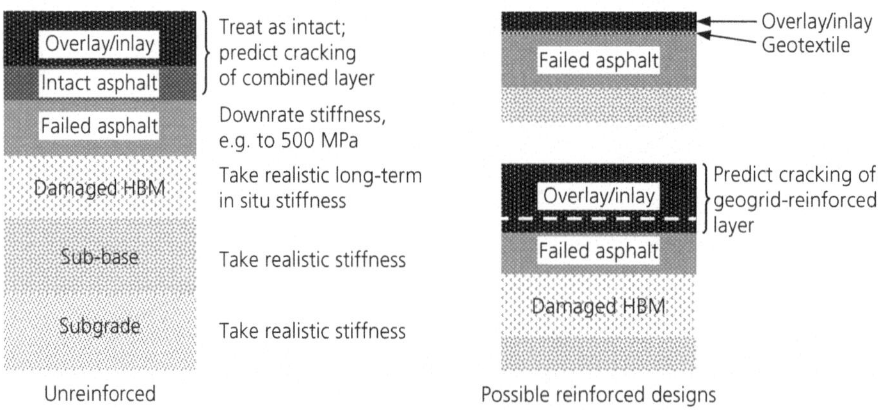

It is often sensible to check the fatigue life in more than one way. In addition to the type of calculation illustrated in the last paragraph, which would usually involve computation of the horizontal strain at the base of the lowermost intact asphalt layer, a further calculation could be carried out assuming that none of the existing pavement was still intact. This is particularly applicable to cases where it is suspected that there is a degree of full-depth cracking through the existing asphalt. The uppermost existing asphalt layers should in this case be assigned a much lower stiffness modulus, typically 500–1000 MPa as discussed in Section 17.4.2 for cracked HBM base layers, and it is the horizontal strain within the overlay that controls predicted life. No allowance for damage would then be needed. The conservative designer might take the lower of the two pavement lives found – or would at least consider carefully which prediction is the more appropriate in a particular case.

The point here is that a responsible engineer will not wish to base a design on a single approach, but will consider various means of obtaining an estimate of pavement life, including use of design manuals where they allow such a prediction.

27.2.4 Use of interlayers

An asphalt overlay/inlay to a damaged pavement is one of those that most obviously lend themselves to the use of what might be described as 'interlayer systems' – techniques for enhancing the effectiveness of an overlay. Broadly, the options are as follows, two of which were illustrated in Figure 27.4

- a geogrid or strong geotextile, providing actual enhanced strength
- a standard geotextile, acting as a separation layer
- a high-durability asphalt layer
- a granular interlayer
- an open-graded asphalt layer.

Clearly the effects of these different systems are not all the same. *Geogrids* and *strong geotextiles* actually slow down the rate of crack propagation; Section 10.6.6 suggested by a factor of up to around 4, depending on the product used. Based on the principles outlined there, Figure 27.5 includes example predictions of the benefit of a high-quality polypropylene geogrid in the case of (*a*) an overlaid fragmented asphalt pavement and (*b*) an overlaid jointed concrete subject principally to thermally induced loading (e.g. an airfield pavement); the benefit in these cases is significant. A secondary benefit, particularly of those geogrids that develop an efficient interlock with surrounding asphalt, is that cracks are prevented from opening even after penetrating full depth. Unfortunately, this benefit is hard to quantify, but it does at least mean that the slowing down of deterioration will continue throughout the life of the pavement.

Although the above list draws a distinction between 'strong' and 'standard' geotextiles, this is really just for convenience. Most geotextiles are capable of providing a degree of reinforcement, and this is particularly true for thin overlays on lightly trafficked roads. The nature of the reinforcement is not that it gives significant direct strength increase, but rather that the asphalt is encouraged to 'recover' after each load application, being slowly pulled back into its original state as tension in the geotextile fibres dissipates. Under ideal conditions (thin overlay, fragmented underlying pavement, low traffic) the increase in pavement life can be dramatic – although rather hard to quantify. It should also be remembered that geotextiles, which are placed into a layer of sprayed bitumen (usually around 1 l/m^2) act as a sealant, preventing rapid water ingress even after surface cracking recommences. Their role as separators, where the systems probably come under

Figure 27.5 Example crack propagation computations; interlayer effect

Case 1 (traffic-induced)
60 mm overlay
150 mm asphalt (cracked @ 1 m)
250 mm sub-base
Medium subgrade

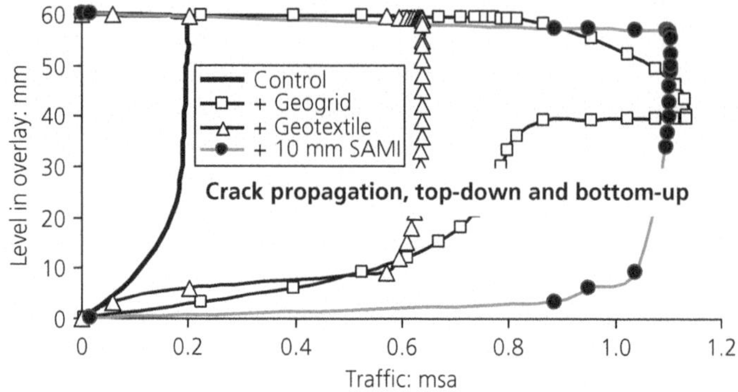

Case 2 (thermally induced)
100 mm overlay
250 mm concrete (jointed @ 5 m)
150 mm hydraulically bound sub-base
Good subgrade

Control:	1.3 years
+ Geogrid:	15.3 years
+ Geotextile:	26.3 years
+ 10 mm SAMI:	15.7 years

the heading of *stress-absorbing membrane interlayers* (SAMIs), relates principally to cases where relative horizontal movement occurs across cracks, usually due to thermal effects in a HBM base. The term SAMI is applied to a variety of interlayer systems, with or without a geotextile, the common theme being that they are tough, flexible, bitumen-rich layers that allow slow viscous slip to occur but provide a high level of crack resistance. In this case, the effect is achieved by ensuring that the geotextile is saturated with bitumen; this allows slip to occur (slowly) between the overlay and the existing pavement, thus inhibiting thermally induced reflective cracking, while maintaining the required vertical bond between layers. Figure 27.5 includes predictions for geotextiles and the effectiveness against thermal movement (case 2) is clear, while it is less relative to the geogrid under traffic loading (case 1).

The description *high-durability asphalt layer* refers to specialist products that can be placed over significantly cracked pavements, such as crazed asphalt surfaces or pavements with major transverse cracks initiated in a HBM base layer. The idea is that a crack finds it difficult to propagate through the product, which will usually comprise a thin (typically around 10 mm) layer with modified binder and, often, fibre reinforcement. These systems, which are also categorised as SAMIs, can certainly be highly effective, especially where the main danger comes from horizontal thermally induced movement in a HBM base, greatly reducing reflective cracking. Figure 27.5 includes example predictions (replacing the lowermost 10 mm of asphalt with SAMI material) and in this case they match the lives predicted for the geogrid-reinforced case very closely. Note, of course, that none of the examples in Figure 27.5 should be seen as representative, although all are based on real materials.

The final two systems, a *granular interlayer* and an *open-graded asphalt*, are both intended to form a layer through which a crack cannot propagate. They are applicable to cases where the existing pavement contains large discrete cracks, often associated with a HBM base. The effectiveness of the granular interlayer is obvious in that the material is 'ready cracked' and hence cannot

Figure 27.6 Decision chart – interlayer selection

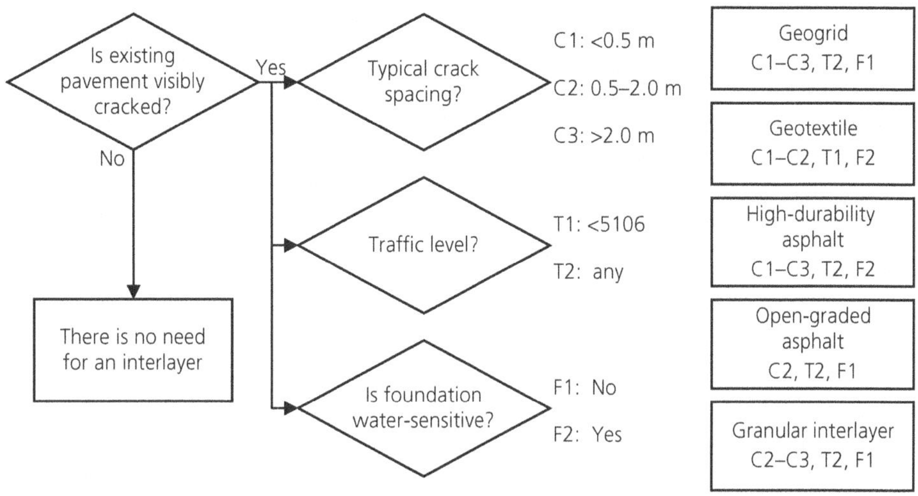

physically transmit a discrete crack. The resulting pavement comes under the heading 'inverted pavement', as described in Section 17.6.

The effect of an open-graded asphalt (logically, a cold-mix asphalt would be suitable) is similar. It is a stiffer material (perhaps 1000 MPa depending on design temperature), but forms a less complete break to reflection of underlying cracks. It would theoretically be best suited to cases where cracks in an underlying pavement are more frequent (e.g. 2 m spacing or less) and therefore reflective cracking is a slightly less dominant issue. However, this is an area where confident prediction is impossible.

Figure 27.6 is included here as an aid to decision-making. The guidance contained is of a general nature and should not be seen as prescriptive. It is always important to ascertain the type of cracking present, in particular the nature of likely movement across cracks in the future. Note that the category Fl in Figure 27.6 is not intended to disqualify the use of any interlayer systems, but simply to indicate which systems will actively protect the subgrade from water ingress. Note also that it is possible to combine geogrids and geotextiles into composite products, thus combining the two sets of benefits.

27.2.5 Repairs prior to overlay

This is an area that should not be overlooked. When a pavement is at the design stage, it is 100% uniform. As soon as it is constructed, variations in subgrade quality, material composition, compaction, layer thickness and so on immediately introduce a degree of variability. By the time major maintenance is required, this variability will be greater again, as water finds its way into some sections and not others, or as debonding occurs in some places and not others. The result is that there may be some sections that are in relatively good condition while others are suffering from advanced full-depth cracking. If an overlay or inlay is applied uniformly, then that variability will, of course, still be present. The obvious solution is to bring those sections that are showing the most advanced signs of deterioration up to a certain minimum standard, for which the overlay or inlay can then be designed. In practical terms this might mean

- replacing localised areas of badly crazed asphalt
- trenching and replacing asphalt at severe transverse cracks ('trenching' refers to the removal of a strip of asphalt either side of a crack, of total width less than 1 m; this can be replaced by fresh asphalt, with or without an interlayer system (geogrid, SAMI))
- planing out severely rutted areas, at least to the depth of the rut.

The question is: just how much is worth doing? And the answer depends, of course, on the economics of each individual case. However, as a guideline, it is probably not worth replacing more than about 10% of the total pavement area due to crazing. If crazing is more widespread, then the overlay or inlay should be designed to cope with the crazed condition. In the case of severe transverse cracks, it may be economic to treat them if they are 20 m or so apart; much more frequent than this and the condition can be considered as 'general' and the overlay or inlay designed accordingly. Planing of rutted areas, on the other hand, is a much less expensive operation and is advised in every case where they exceed 15 mm or so.

It is also strongly advised that *local strengthening* is considered. The point is that localised crazing occurs for a reason and, even after carrying out a full-depth reinstatement with new asphalt, that reason may still remain. It is most likely to be due to foundation properties, and these will not usually be significantly improved during rehabilitation. The alternative, therefore, is to use either materials of enhanced strength and durability in these local reinstatements, or to increase asphalt thickness locally. The fact that reinstatements are rarely given the same level of compaction as the original pavement (due to difficulties at edges and corners) makes such strengthening or thickening doubly desirable.

The same principle applies to repairs to severe transverse cracks. It will never be possible to completely prevent their reoccurrence, but it is certainly possible to slow down the process. A trench repair may make use of material with enhanced 'elasticity', meaning increased ability to cope with high strains without cracking. A polymer-modified binder is an obvious choice. The use of interlayer systems (see previous section) should also be considered.

27.2.6 Ultra-thin white-topping (UTWT)

This is a more radical solution (Figure 27.7). UTWT is a thin concrete overlay, which at first sight would appear to be a strange choice to make because a thin concrete slab is inherently fragile. However, when the calculations are carried out, using techniques such as finite-element analysis, it is found that the support given by an underlying asphalt pavement can certainly be adequate, although cracked asphalt is unsuitable as it will cause stress concentrations in the concrete, which will lead to fracture.

The technique is therefore restricted to structurally intact asphalt pavements – which begs the question as to why one would ever want to carry it out. However, there is a very real circumstance where UTWT has proved highly effective, notably in some US states, namely where serious rutting is taking place, which often means at heavily used highway intersections. In these cases, the continuity of support provided by the asphalt offers suitable insurance against concrete cracking, while the rigidity of the concrete prevents any further permanent deformation from occurring in the asphalt. In these particular cases, the combination provides an ideal solution. Particular points to watch are as follows.

- The concrete must be jointed and it is recommended that joints are spaced at no more than about 1 m. They can consist of simple saw-cuts, through to a depth of about half the slab thickness, which are then sealed.

Figure 27.7 Cementitious overlay options

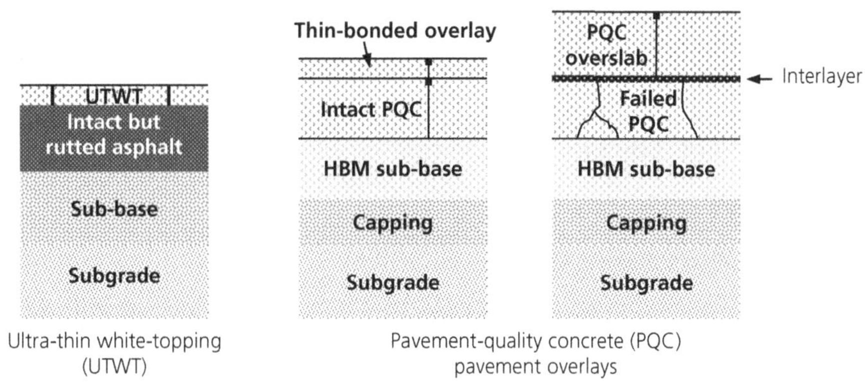

- The slab should be reinforced to give adequate toughness and, in practice, this dictates the use of fibres because the thickness is usually insufficient to allow conventional reinforcement.
- It is very important that the concrete is well bonded to the asphalt. This is best achieved by thoroughly cleaning and then washing the prepared asphalt surface.
- The asphalt surface does not have to be particularly even, although ruts greater than about 15 mm should be reduced by planing so as to avoid excessive variation in slab thickness.
- Design thickness will normally be in the range 50–100 mm.
- Placement is best achieved by slip-form paving, although hand-laying is acceptable in low-speed applications such as at intersections.
- The surface should be textured as for any other concrete pavement.

The greatest difficulty facing the engineer is how to judge the design thickness of UTWT. Current practice appears to base recommendations on proven past experience. However, the principle is that the asphalt acts as a stiffener for the concrete rather than as part of the foundation. This drastically reduces the value of tensile stress that would otherwise occur at the base of the concrete under bending. Calculation Sheet 27.2 presents an approximate, two-dimensional analysis, illustrating the way in which the system works; multi-layer linear elastic analysis would generate a much more accurate estimate of stress.

Calculation Sheet 27.2. Stress in UTWT

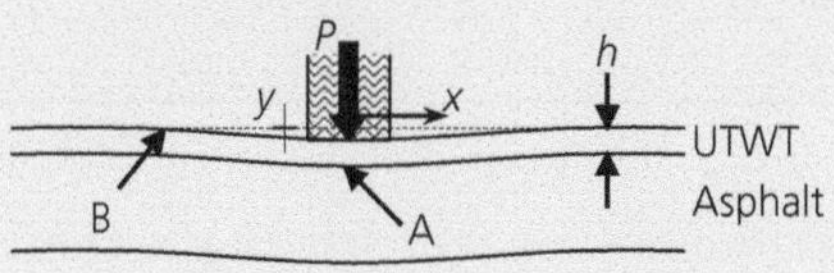

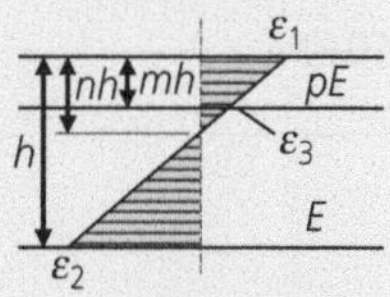

Approximate analysis using beam on elastic foundation (Calculation Sheet 5.1)

Curvature $\kappa = \mathrm{d}^2 y/\mathrm{d}x^2 = (P\lambda^3/k)\ \mathrm{e}^{-\lambda x}(\sin \lambda x - \cos \lambda x)$ where $\lambda = (k/4(EI)_{\mathrm{eff}})^{0.25}$

Maximum sagging at A (i.e. when $x = 0$)

$$\rightarrow \kappa_A = P\lambda^3/k$$

Maximum hogging at B (i.e. when $d^3y/dx^3 = 0$)

$$\rightarrow (2P\lambda^4/k)\, e^{-\lambda x} \cos \lambda x = 0 \rightarrow \lambda x = \pi/2$$

$$\kappa_B = (P\lambda^3 e^{-\pi/2})/k$$

At neutral axis depth nh, moments balance:

$$E\varepsilon_2 h^2(1-n)^2/3 = E\varepsilon_1 h^2(n-m)^2/3 + pE\varepsilon_1 h^2 n^2/3 - pE\varepsilon_1 h^2(n-m)^2/3$$

$$\varepsilon_2 = (1-n)\varepsilon_1/n$$

→ Cubic equation for n:

$$2n^3 + (-3-2m+pm)n^2 + (3+m^2-pm^2) - 1 = 0 \rightarrow n$$

Total moment:

$$M = E\varepsilon_2 h^2(1-n)^2/3 + E\varepsilon_1 h^2(n-m)^2/3 + pE\varepsilon_1 h^2 n^2/3 - pE\varepsilon_1 h^2(n-m)^2/3$$

But $\varepsilon_2 = (1-n)\varepsilon_1/n$ and $\varepsilon_1 = \kappa nh$

$$\rightarrow M = k(Eh^3/3)[(1-n)^3 + n(n-m)^2 + pn^3 - pn(n-m)^2]$$

$$\therefore (Eh^3/3)[(1-n)^3 + n(n-m)^2 + pn^3 - pn(n-m)^2] = (EI)_{eff}$$

Finally:

At A, $\varepsilon_3 = -\kappa_A(n-m)h = -P(n-m)h/[4^{0.75} k^{0.25} (EI)_{eff}^{0.75}]$ probably compressive

At B, $\varepsilon_1 = -\kappa_B hn = Pnhe^{-\pi/2}[4^{0.75} k^{0.25} (EI)_{eff}^{0.75}]$ definitely tensile

Example

$P = 40$ kN; $h = 0.3$ m; $m = 0.2$; $E = 3000$ MPa; $p = 10$; $k = 50$ MN/m^3

→ $n = 0.42$; $(EI)_{eff} = 20.1$ MN.m^2; $\varepsilon_3 = 36\,\mu\varepsilon$ (compressive); $\varepsilon_1 = 15\,\mu\varepsilon$ (tensile);

$\sigma_t = 450$ kPa

27.3. Overlays/inlays to rigid pavements

In truth, an 'inlay' to a concrete-surface pavement is not a likely choice, although level constraints could, in theory, dictate its use. However, as long as the remaining pavement has a concrete surface prior to placement of new materials, it falls into the category of treatment covered in this section.

27.3.1 Thin bonded concrete overlays

This solution, illustrated in Figure 27.7, can only be applied successfully if the existing pavement-quality concrete (PQC) slab is intact. Basically, it is a means of increasing slab thickness; hence the need for bond. The thickness can, in theory, be as little as permitted by the stone size and the needs of workability; 30 mm or so probably represents a sensible practical minimum. Bond, which is absolutely critical to success, is achieved by milling and thorough cleaning of the existing surface.

This treatment has been extensively employed in the USA and has proved highly successful. The key is undoubtedly forward planning, because it can only be carried out on intact slabs. It is a particularly logical choice where pavement usage is about to change; for example, if heavier aircraft or port equipment are to be employed. Issues are as follows.

- The material used must be compatible with the existing PQC in terms of its thermal properties. In general, aggregate with a low coefficient of thermal expansion (e.g. limestone) is preferred.
- Thickness design is exactly the same as for a new pavement, and the assumption should be made that the strength of the composite slab is the lower of those of the overlay and the existing concrete.
- Joints in the existing pavement must be replicated (and sealed) immediately above in the overlay. If expansion joints are present, they should be formed by use of filler board or similar compressible medium.
- Paving technique and surface finish are the same as for new concrete pavements.

There is no fundamental reason why thin bonded overlays should not be used over continuously reinforced concrete (CRC), despite the inevitable presence of relatively close-spaced transverse cracking. The overlay will, of course, also crack immediately above cracks in the existing concrete. The issue of prime concern is that the presence of so many cracks may place excessive strain on the integrity of the bond between old and new materials. It is also essential that any local deterioration in the existing slab is repaired in advance.

27.3.2 Overslabbing

For the designer, this is a very simple case, also illustrated in Figure 27.7. It is appropriate when the existing concrete is severely damaged, in which case the entire existing pavement is relegated to the role of a foundation and the new slab is designed as in Section 20.3. The equivalent foundation modulus can be based on FWD data. If a modulus of subgrade reaction (k) is required, empirical relationships with foundation modulus can be found in several design manuals (e.g. $k = (E/4.97)^{1.292}$ (BAA, 1993)). Alternatively, multi-layer linear elastic calculations can be carried out to try to determine an equivalence in a particular case (the idea is to compute the deflection under a suitably large circular load area, reflecting the likely spread of load under the new PQC slab; the modulus of subgrade reaction is then given by the applied stress divided by the computed deflection). Or one can simply divide the equivalent foundation modulus (in MPa) by 2 to obtain a very rough estimate for k (in MN/m^3).

Note that, although overslabbing is most commonly used over existing concrete pavements, it can actually be used over any pavement type. The only quantity that the designer has to worry about is the support stiffness.

Note also that concrete pavements, particularly CRC pavements, are tolerant of wide variation in support stiffness. The design should be based on a conservatively low value, but the required thickness is actually relatively insensitive to this quantity. This means that it is not usually economic to carry out significant local repairs to the existing pavement prior to overslabbing.

However, there is one aspect to overslabbed pavements that is of critical importance, and that is the interface between the new pavement and the old. The new slab needs to be independent of the old in terms of horizontal movement, otherwise cracks in an existing slab may be reflected in the new slab. This typically presents two options – a heavy-duty polythene membrane or an asphalt (or mastic) interlayer.

There is much debate and disagreement between experts as to which, if either, is the more desirable. However, there is no doubt that, in engineering terms, the effects of the two solutions are different. The polythene forms a clear break in bond, allowing free vertical separation and reasonably easy shear movement. An asphalt or mastic layer (usually no more than 25 mm thick) will give good bond under load but will allow relative shear movement over a longer timescale (i.e. in response to thermally induced expansion and contraction effects). This author's view is that the asphalt or mastic solution is to be preferred if the underlying concrete has any significant remaining integrity, as it fulfils the separation need sufficiently while allowing a greater composite pavement strength. Where the underlying slab has negligible contribution to make then this argument no longer applies.

Clearly, for overslabbing of asphalt-surfaced pavements no interface treatment is required.

27.3.3 Asphalt overlays

Asphalt will usually be the favoured material for strengthening simply because it can be applied much more rapidly than concrete, and for airfield runways and major highways time is often the determining factor. However, except in the case of properly functioning CRC, design involves a choice between several quite different options. Table 27.1 lists the alternatives. Figure 27.8 illustrates the processes referred to in the table.

In general, overlay thickness design will be influenced by two considerations. The first applies only where the existing PQC is in good condition and is to remain intact, and it is that the stress induced in the concrete under load should be restricted in order to avoid further cracking. This is exactly the same design criterion as is used in new concrete pavement design (see Section 20.3) except for the presence of the asphalt layer. If multi-layer linear elastic analysis is to be used, then this has to include the asphalt layer, at a conservatively low stiffness, corresponding to relatively warm

Table 27.1 Asphalt overlay options over existing concrete pavement

Pavement type	Condition	Treatment advice
CRC	Good	No special treatment needed
	Poor	Consider an interlayer
Jointed reinforced	Any	Saw-cut, crack and seat
Jointed unreinforced	Good	Either form joints in asphalt or crack and seat
	Poor	Either crack and seat or rubblise

Figure 27.8 Concrete pre-treatment options prior to asphalt overlay

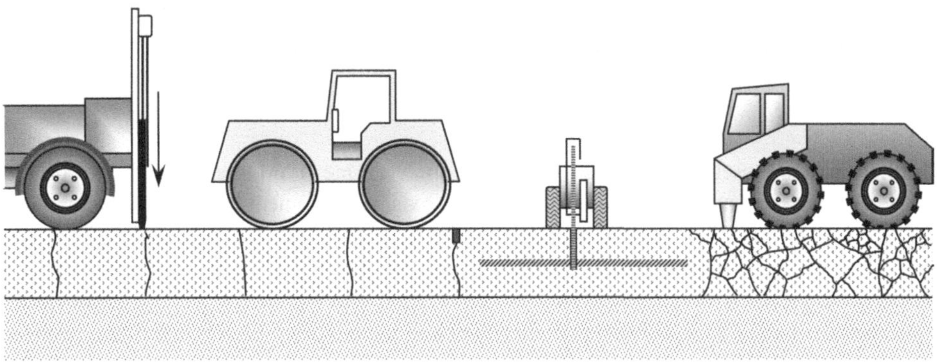

weather. Alternatively, the thickness of asphalt may be converted to an approximate equivalent PQC thickness; it was suggested to divide by 4 or 5 in Section 21.3.

The second consideration is *reflective cracking*. Only in the case of an overlay to an intact CRC can reflective cracking usually be ignored. All joints and cracks will tend to reflect; the trick is to understand the mechanisms involved. According to Chapter 17, there are two basic mechanisms. The first is thermally driven and is caused by joints and cracks opening and closing as the concrete cools and heats over each 24-hour period. The second is traffic driven and is caused by differential vertical movement across joints and cracks under each passing wheel load.

To combat the first, it is essential to minimise the length of concrete slab between cracks (or joints), ensuring only a small opening and closing movement at each. To combat the second, each crack (or joint) must have good load-transfer properties, which means narrow cracks only. The obvious answer is to break the concrete into short lengths, providing large numbers of relatively narrow cracks. This is the *cracking* operation illustrated in Figure 27.8; it is carried out using a large 'guillotine' device, dropping a heavy steel sheet onto the pavement surface, and is highly effective for unreinforced concrete.

The next operation shown in the figure, namely *seating*, ensures that any local gap between the underside of the slab and the foundation material is closed and that each individual element of concrete is properly supported.

An important issue is the optimum crack spacing. Here, three factors compete. The first is that a closer spacing means better load transfer and less thermal movement. The second is that a closer spacing means a lower overall pavement stiffness and the danger that conventional fatigue cracking might take over from reflective cracking as the dominant mode of asphalt failure. The third is that the cracking operation costs both time and money. The favoured compromise seems to be a spacing of around 1 m, usually allowing an asphalt thickness of 150 mm or less. Further advice on reflective crack prediction and thickness design is provided in Sections 17.4.1 and 17.4.2.

Unfortunately, jointed reinforced concrete presents a difficulty. Cracking and seating is no problem, but the continuity of the reinforcing steel means that, even after cracking, thermally induced movement will still be concentrated at joints, joints which are usually well over 10 m apart. This is certainly a recipe for reflective cracking. The difficulty can be overcome by *saw-cutting* through each slab at regular intervals (e.g. 3 m) to over half depth in order to cut the reinforcement – but this is an expensive operation. It should only be undertaken after careful consideration of alternatives such as the interlayer systems introduced in Section 27.2.4, or simply putting up with the reflective cracking.

Two further options present themselves for jointed concrete. One is 'do less', the other is 'do more'. The do less option is an acceptable solution when the existing pavement has relatively close-spaced joints (<6 m or so) and is in good condition. The overlay thickness is designed simply to reduce stress in the concrete to an acceptable level; no pre-treatment is carried out, but joints are *saw-cut* into the surface of the asphalt directly over those in the original slab and then sealed, as in the case of a bonded concrete overlay. The problem with this otherwise excellent solution is that future resurfacing will also have to include the same pattern of saw-cutting, and saw-cutting is not cheap.

The do more option is *rubblising*. Basically, this involves pulverising the concrete into small pieces, none more than about 150 mm in size, thus reducing it to an unbound material. The correct design stiffness for this layer is debatable and doubtless depends on the details of the rubblisation process. Some suggest up to 1000 MPa, although this may be rather optimistic; the figure advised here is no more than about 250 MPa. Asphalt thickness design then follows that for conventional asphalt pavements.

The only other case is that of CRC in poor condition – one of the most difficult maintenance headaches of all. The poor condition undoubtedly means that many of the cracks are relatively free-moving, despite the presence of reinforcing steel, as much hidden spalling has almost certainly occurred. This means that reflective cracking through any overlay is likely. It also means that confident thickness design is practically impossible. The suggestion made in Table 27.1 that an interlayer treatment should be considered is simply advice to take out an 'insurance policy'. Geogrid reinforcement is probably the most effective option because it should at least hold the asphalt together well, even if reflective cracking occurs, whereas there is little to be gained from treatments that merely allow shear movement between the asphalt and the concrete; the problem is traffic-induced vertical movement.

27.4. Cold in situ recycling

This is a massive subject and one that is likely to become increasingly important in the future. The concept has already been introduced in Section 14.4, where alternative design approaches were explained. The process involves an existing pavement being milled to a certain depth – machines with capability up to 350 mm exist – and the broken fragments are then mixed with a new binder and compacted. The result, if everything goes to plan, is a strong new pavement, although it will normally be necessary to apply a new surface course because it is unlikely that the requisite surface properties will be present from the recycled material itself.

The potential is obvious. No matter how cracked, deformed, worn, aged and so on the existing pavement is, as long as its broken remains fulfil the role of aggregate, then it is still useful. There is no need for expensive and disruptive transportation of materials (other than binder), there is no requirement for quarrying of virgin rock and there is a much-reduced energy requirement because no heating of aggregates is called for. All that is needed is the addition of an effective binder. Of course there are difficulties and the following are seen as the major issues confronting practitioners and designers.

- Binder is unlikely to be uniformly mixed into the material.
- All pavements are, to a certain extent, non-homogeneous, notably in layer thickness. In many cases, this means that the milled material will present varying proportions of asphalt, base (including HBMs), sub-base and even subgrade.
- Water content can only be controlled approximately.
- Compaction beyond about 250 mm depth will be increasingly ineffective.
- The need to work with cold, wet aggregate restricts the choice of binder.

27.4.1 Material variability

The first three points in the above list relate to variability. This is an issue that cannot be dismissed, nor can technical improvements to the process ever fully overcome it. Unfortunately, it is an issue where data are (*a*) hard to come by and (*b*) hard to interpret. In the one relevant dataset of which the author has direct knowledge, the recycled product was only marginally more variable than an adjacent fresh hot-mix asphalt when tested with an FWD, with a coefficient of variance about 1.2 times that for the hot-mix. However, when individual cores were extracted and tested, the variability was much higher, three to four times that of the hot-mix. This is logical enough; it means that the variability occurs over very short distances, presumably as more or less binder reaches individual regions and due to fluctuations in the composition of recycled material. Over a metre or so – the zone of influence of an FWD test – these fluctuations cancel each other out.

The question that has to be asked (and which can barely be answered) is whether dramatic, short-wavelength fluctuations in quality matter; after all, over a thickness of 250 mm or more, deformation susceptibility might be expected to even out. Similarly, cracking would not be able to advance far unless it propagated through both good and poor regions of the material. These are reasonable arguments and they suggest that one should not necessarily panic over apparent variability from tests on cores. On the other hand, in some respects, variability really does matter. Permeability is critically affected by small changes in gradation and voids content, which suggests that an in situ recycled material will always be relatively permeable because there will always be high-voids routes through it.

In conclusion, the following advice is offered.

- Reduce the design stiffness modulus slightly from that derived from laboratory-mixed specimens to take account of variability, perhaps by around 10%.
- Be aware that water will be able to ingress. This means that the properties of the recycled material should be assessed under wet conditions and that the issue of water reaching the foundation should be recognised.

27.4.2 Compaction

Again, direct data are hard to come by. However, it has been found that modern compactive equipment is able to achieve high density in an unbound pavement material such as a base or sub-base to a thickness of around 250 mm, and the recycled materials under consideration are effectively unbound at the time of compaction. Beyond that depth, the relative density will tend to decrease. Calculation Sheet 27.3 presents an approximate assessment of the likely effect of this in terms of the strain at the base of the layer under vehicle wheel loading, the stiffness reduction shown in the example being nothing more than an estimate. If this illustration were to be correct, the deduction would be that the designer should assume a reduced thickness for any material deeper than 250 mm, applying a multiplier of around 0.6.

Calculation Sheet 27.3. Under-compaction at base of layer

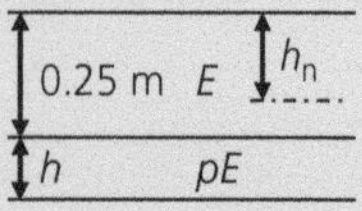

Use approach in Calculation Sheet 27.2 to determine the neutral axis depth h_n and the effective bending stiffness $(EI)_{eff}$

From Calculation Sheet 27.2:

$$\varepsilon_t = -\kappa(0.25 + h - h_n) \approx P(0.25 + h - h_n) / [4^{0.75} k^{0.25} (EI)_{eff}^{0.75}]$$

Then calculate h_{eq} to give the same ε_t with $p = 1$ (i.e. no under-compaction)

Example

$P = 40$ kN; $k = 50$ MN/m^3; $p = 0.8$ at $h = 0.05$ m, $p = 0.7$ at $h = 0.1$ m, $p = 0.6$ at $h = 0.15$ m

$\rightarrow h_{eq} \approx 50–60\%$ of h

27.4.3 Choice of binder

Mixing at ambient temperature dictates the use of either a hydraulic binder, bitumen emulsion or foamed bitumen, which means that full strength cannot be achieved immediately. Even in the case of Portland cement, adequate strength to prevent structural damage under traffic load would only be achieved after some days. The other binders all demand a long curing period, sometimes months rather than weeks, which presents a dilemma. It will almost certainly not be economic to avoid use of the pavement for such a long period but, if it is used, then structural damage has to be expected, even while curing is taking place. The unknown quantity is the extent to which such damage will self-heal during subsequent curing and also how long the curing process takes.

27.4.4 Hydraulic binders

The aim of using a hydraulic binder is to achieve a base of comparable quality to a plant-mixed HBM layer, accepting the increased variability and the lower density expected at the bottom of the layer. However, if surfacing layer construction is to be carried out shortly after recycling, followed by opening to traffic, then the issue of early-life damage raises its head, as discussed in Section 17.8.2. Once a realistic long-term stiffness modulus is determined, this can then be used in the relevant pavement design approaches in Chapter 17.

27.4.5 Cold bituminous binders

The actions of bitumen emulsion and foamed bitumen are similar, in that both binders tend to coat particles of aggregate (in this case recycled road fragments) unevenly, and both take time to gain in effectiveness. These issues were detailed in Chapter 13.

Pavement design approaches were presented in Chapter 14, where expected stiffness modulus values were suggested as a proportion of what would be expected from a hot-mix using the same aggregate. Curing (i.e. strength gain) is critically dependent on water loss from the material, which means that open-graded materials are preferred, allowing air to access the voids and hasten evaporation. This should, in theory, present no problem for in situ recycled pavements, because recycled asphalt layers are typically low in fine material, but, on the other hand, the variability inherent in an in situ process will mean that there will be some regions with lower voids, and which therefore lose water less easily. The inclusion of around 1% cement or lime is a technique commonly used to reduce the free water content and so to encourage curing. Clearly, the timing of surfacing application is another important factor in the equation and a responsible organisation will be careful to take account of the weather, not applying the surface immediately after rain has raised the water content, for example.

It is not really possible to advise on the relative advantages of emulsion and foamed bitumen for in situ recycling because continuing progress continues to be made in both technologies.

The issue of early-life damage applies to cold-mix asphalt in a similar way as to hydraulically bound materials, the difference being the inherent ability of a bituminous material to self-heal. The comments made in Section 14.3.1 apply.

Pavement design can be carried out following the principles given in Chapter 14 for cold-mix asphalt pavements. The recycled layer can be treated as a conventional asphalt, as long as the long-term stiffness is expected to be sufficiently high for it to be considered intact. If not, then the recycled layer has to be considered as an exceptionally high-quality unbound base and the surfacing above designed on this basis. The stiffness assumed for the recycled layer would normally be lower than that for an equivalent plant-mixed material because of the greater variability of in situ treated mixtures. Empirical design guidance on cold recycled pavements can be found elsewhere (Merril *et al.*, 2004).

27.5. Summary

Although it would be possible to include advice on other pavement types, such as block paving, grouted macadam or even unbound pavements, the main principles are now in place. Block paving, if it is to be strengthened, will usually be removed, leaving a standard base and sub-base; grouted macadam is basically an asphalt and can be treated as such in any strengthening analysis; an unbound pavement will readily form the foundation to any future bound pavement structure. As for all the more conventional pavement types already covered, the secret is to make appropriate assumptions on long-term properties; the actual analysis techniques are similar to those for new pavements.

So, in an ideal world, the advice contained up until this point would enable soundly based engineering decisions on future maintenance to be made, and a range of excellent rehabilitation design options to be developed. The question is: how does this fit into the real world, where resources are never sufficient and time windows are not always available? This is not solely an engineering matter, although some of the constraints will be due to engineering issues; it is a management matter, and the topic is known as 'pavement management'.

REFERENCES

BAA (British Airports Authority) (1993) *Pavement Design Guide for Heavy Aircraft Loading*. BAA, Group Technical Services Aircraft Pavements, London, UK.

Merril D, Nunn ME and Carswell I (2004) *A Guide to the Use and Specification of Cold Recycled Materials for the Maintenance of Road Pavements*. Transport Research Laboratory, Crowthorne, UK, Report TRL611.

Nick Thom
ISBN 978-1-83549-711-1
https://doi.org/10.1108/978-1-83549-710-420243034

Chapter 28
Pavement management systems

This is a subject which this book can only touch on, and is part of the much larger subjects of highway management, airfield management and so on. For example, the Highway Design and Maintenance suite of programs is very widely used in highway management. HDM-4 (Highway Development and Management version 4) is a powerful computer-based system developed under World Bank sponsored research, shortly to be upgraded to HDM-5. It is used both in support of development proposals and also in ongoing highway network management. Nevertheless, it is important that pavement engineers understand the wider impact of the computations and predictions that they make, together with the fact that there is never any one right answer (although there may be numerous wrong ones). The first matter that needs to be addressed is the fact that there are often good practical reasons why certain otherwise excellent maintenance solutions have to be rejected – or else given a high price tag. Once these constraints have been appreciated, it will then be possible to evaluate questions such as whether to continue with increasingly costly minor maintenance or whether to adopt a more radical rehabilitation solution with a much higher up-front cost.

Incidentally, although this section is presented in terms of 'cost', this is not necessarily entirely financial. Environmental and social costs can also be added into the equation using suitable equivalence factors – a highly political decision.

28.1. Practical constraints

In general, nothing in pavement engineering is impossible – but some things are rather impractical. In most cases the impracticality is obvious and it is not the intention here to labour points unnecessarily. However, the issues are worth listing.

28.1.1 Multi-lane highways

The point here is that one cannot raise (or lower) the surface level of one lane without doing the same to all other lanes on a carriageway. Thus, while a simple overlay solution may appear to be the most cost-effective in isolation, the full cost (both monetary and in delays) has to be added in (i.e. similar overlays to all lanes).

28.1.2 Highway structures

The clearance to overbridges places an obvious upper limit on raising levels. It is not difficult to carry out checks on the actual clearance, nor particularly costly, and the information is vital so that inappropriate solutions can be rejected right at the outset.

Less obviously, the carrying capacity of underbridges also has a limit and even a thin overlay adds a significant extra load. This too must be established at an early stage to allow efficient design.

Of course, even if structures limit the thickness of overlay permitted, this constraint only applies to the pavement actually over or under the structure. It is quite possible to ramp the surface up or down either side, over a length that depends on the vertical alignment constraints for that class of highway.

28.1.3 Kerbs and barriers

Both kerbs and barriers can be raised – but at a cost. In both cases, there is usually a degree of flexibility, allowing small increases in pavement level, and it is therefore important to know just how much flexibility is available in a particular case.

It is fair to say that this particular restriction is one of the prime drivers towards the use of geosynthetic interlayers, the philosophy being that their use at least maximises what can be achieved within the level constraints applying.

28.1.4 Airport runways

The situation here is similar to that for multi-lane highways in that it is generally only the central 20 m width, from a pavement that may be 60 m wide in total, which is damaged by aircraft. This often means that it is desirable to apply an overlay to this central part only, which is of course not permissible. However, it is often permissible to ramp down from a relatively thick overlay over the central part of the runway to little or zero thickness at the edge. The crossfall constraints are tight on a runway (generally 1–1.5%) and so the 'feathering out' of the overlay may only be possible at a crossfall difference of around 0.5% (at which rate a 100 mm overlay takes 20 m to reduce to zero thickness).

28.1.5 Absolute utilisation requirements

In general, closure of an airport for more than a matter of hours is not a viable economic option. Processes that can be carried out at night such that the airport is fully operational the next morning may be acceptable; processes that demand a longer period without traffic are not. The range of maintenance options on runways is therefore limited. Closure of runway ends may be possible for longer periods, because operations can still continue with a reduced runway length. Closure of taxiways and aprons may also be acceptable where alternative routes or stand areas are available.

A similar restriction may apply to highways. Authorities generally specify that a certain number of lanes have to be kept open at all times, restricting access to the work area. This means that processes which take a long time to complete (or cure fully) carry too high a price on major highways. It may also mean that some processes physically cannot be carried out because of the restricted working width.

Finally, in industry, 'time is money', which means that industrial pavements may also carry a similar requirement for continuous usability. The designer has to be aware of all such restrictions before even thinking about possible designs.

28.1.6 Absolute safety requirements

These requirements come in the following guises.

- Maintain a minimum skid resistance coefficient.
- Do not exceed a maximum rut depth.
- Avoid all loose material (on an airport runway).

Any of these requirements may dictate that a new surface is required no matter what the structural strength of the pavement may be, thereby cutting back significantly the maintenance options available.

28.2. Project-level optimisation

This section sets out the principles to be followed. The problem is that the interface between network-level management and project-level design is not an easy one to manage (an issue that has already arisen in relation to pavement evaluation planning). It is not particularly easy for an engineer to come up with a confident maintenance design; it is quite impossible to second-guess the budget constraints that apply. On the other hand, until the magnitude of maintenance requirements is known, it is impossible to manage the budget properly. It is a classic vicious circle.

In practice, the most common way in which this circle is broken is for the authority to stipulate design lives to which the pavement engineer has to work, but this is not a particularly efficient system and is always likely to result in some work being carried out either earlier or to a higher design life than is really required, while other projects languish unfunded.

28.2.1 Working with a predetermined design life

At a straightforward level, the designer may simply be asked to come up with the best solution to strengthen the pavement to take another *x* years – say 20 – of traffic, assuming that the strengthening is carried out in the immediate future. For example, applying the predictive technique outlined in Section 10.6.4 to the case of a 250 mm-thick cracked asphalt pavement with a measured modulus (converted to design temperature) of 1500 MPa over a fairly standard unbound foundation, and taking a design traffic of 50 msa and UK performance assumptions, it is computed that a 215 mm overlay is required.

But is this really an optimum solution? Perhaps only certain discrete sections comprising about 25% of the road are in this condition, while others are substantially intact and would theoretically need only 75 mm of overlay. It would be a waste to put 215 mm everywhere. The obvious solution is to look into inlay treatments and, in this case it is found that, theoretically, if 230 mm of existing pavement were planed out in the cracked areas and replaced with new asphalt then the required surface level would be just right to give a 75 mm overlay to the rest of the pavement. The saving over 75% of the road would be 140 mm of asphalt; the extra cost over 25% of the road would be 90 mm – plus planing costs. Clearly this would be a much more cost-effective solution; it would also be safer because less reliance would be placed on the performance of existing materials.

Then, of course, other practicalities have to be taken into account. If 230 mm need to be planed out, this leaves only 20 mm, and this is practically impossible; therefore the whole thickness of existing asphalt would almost certainly have to be removed, and that thickness is hardly likely to be uniform. This will demand the application of a regulating course at the bottom of the new asphalt – and so the costs mount up. Next, non-pavement costs must be considered. Are crash barriers present? If so, is 75 mm within existing tolerance? In many cases in developed countries the default requirement from a highway authority is that surface levels should not be raised at all because of the difficulties it is presumed to cause. But in this example, which typifies the situation on numerous major highways, if no raising of levels is permitted, then the 75 mm overlay required in substantially intact areas would have to become full reconstruction of the asphalt (still assuming conventional asphalt materials). Is this really sensible?

Of course, circumstances vary greatly from situation to situation, but in this case it may well be that the default requirement should be reassessed. If the requirement was a little lower, then it might, for example, be possible to propose an inlay that included a high-modulus high-fatigue-resistant material.

This example highlights the need for a balanced 'least-worst' approach to project-level design, and this presents a challenge to the decision-making structure within highway authorities, even where they are the sole party involved. For a fully balanced approach, decision-making should also take account of delays and environmental costs; and it is not too difficult to derive estimates. If the agency in charge of the highway adopts standard rates for lane-closure time, quarried volume, land-fill volume and so on, and provides this information to the engineering team, the team will be more likely to produce a genuinely optimised solution rather than one that is artificially constrained.

28.2.2 Variable design life

In the previous section, the designer was working to a set design life. However, there is no reason why one should not consider alternative design lives in order to minimise whole-life costs. Of course, the procedure now takes on a much more complex nature, and it is fair to say that a lot of estimation is needed. The following additional information is required

- an estimate of annual maintenance costs as a function of time (*a*) since the last major strengthening treatment and (*b*) since the last resurfacing
- resurfacing cost and likely frequency required
- a set of designs and associated costs for different design lives
- a projection of how these design costs would increase (due to ongoing pavement deterioration) with every year for which they were delayed
- a discount rate for future costs.

The first of these can be standardised (as average costs) by the highway (airport…) authority for different types of road (or other pavement) and traffic level. The second is even easier to standardise. The third represents the main task for the designer. The fourth can again be estimated according to a set of standardised rules. The fifth is a political choice; it was mentioned in Section 2.2 and is discussed below.

Turning to the designer's main task, what is required is to carry out the exercise described in the previous section for perhaps three different design lives (e.g. 10, 20 and40 years). Carrying on from the example in the previous subsection, if 50 msa represents a 20-year design traffic, then designs for 10, 20 and 40 years require a raising of the highway by 50, 75 and 100 mm, respectively. Assuming that the highway authority has managed to find a means of delivering a least-worst design, including the raising of kerbs and crash barriers, then the next consideration is to realise that the cost of raising these components is practically the same, whatever the lift involved. This means that it is almost certainly going to be cost-effective to extend the design life of the pavement. An extra £3 per square metre spent now might save £10 or more in 20 years' time – at today's prices – taking into account the much reduced need for major maintenance in the future.

However, one critical element in all this is the discount rate. The idea is that a certain sum of money today is assumed to increase in purchasing power with time due to continuing economic growth. The corollary is that future costs can, therefore, be 'discounted' at a certain annual rate. A debate rages as to the correctness of this assumption in our twenty-first century world, and the result is that discount rates range from zero to about 10%. One further issue is that, although economics may sometimes indicate that it is sensible to design for 100 years, no one can be at all sure that the need for any particular pavement will extend anything like this long. The recommendation here is that, no matter what the economic arguments, the uncertainty element demands the use of a discount rate. If, for example, 3% is used, then £10 in 20 years' time equates to just £5.60 today. This still represents a significant saving compared to a cost of £3 of course, and the discount rate would have to rise to 6.2% to reduce the saving to zero. Nevertheless, it does highlight the critical importance of the discount rate or, more broadly, the way in which future costs are taken into account. Inevitably, due to the way in which government, in its broadest definition, tends to be structured, there is a pull towards a short-term view. Funds are rarely allocated specifically for roads, which means there is constant pressure to save money now to allow other public-sector expenditure to proceed.

The final problem is that this only takes account of direct costs to the highway authority, and this is where the equation becomes much more speculative and political. Road users elect local and national government representatives, and so 'user costs' become both financial and political. If the roads are in excellent condition, then the public hardly notices and authorities probably get very little credit, but when they start to deteriorate the public certainly notices! They notice poor pavement condition and they notice disruption due to road works. This therefore adds pressure to maximise the design life and minimise future disruption, and has tended to lead to 30-year or even 40-year designs being favoured, despite the short-term cost.

28.3. Network-level optimisation

This is a subject that stretches far beyond the scope of this book. However, if pavement management is to be successful at network level, there has to be a sound basis of project-level optimisation in place. If an authority is able to generate simple simulations including future maintenance costs and a discount rate, as illustrated in the previous two sections, this immediately puts some of the tools in place for construction of a network-level decision-making computational tool. 'What if?' scenarios can be explored, examining the effect of delaying maintenance for example, enabling operators to distribute the available budget in the most cost-effective manner. It also allows engineers to generate the sort of data that politicians need to see in order for them to appreciate the effect of budget restrictions.

However, network-level pavement management forces non-pavement costs to be taken seriously, including costs that are not strictly financial. These have already been introduced in Chapter 2 and they should play a significant role in sensible management.

28.3.1 The cost equation

Pavement costs are clear enough and have been dealt with in previous sections. The non-pavement costs that could be taken into account are

- user costs (due to speed restrictions, vehicle wear and tear, fuel consumption and delays at maintenance works)
- accident costs
- environmental costs (air pollution, energy usage).

The most common method of dealing with user costs is to link them to the parameter known as the international roughness index (IRI), defined in Section 25.2.1. The following are examples of the forms of equation typically used for this. The first is based on a study carried out in the USA (Zaabar and Chatti, 2014), with predictions averaged over a range of vehicle types and speeds; the second is determined from vehicle operating speed data from Russia, presented by Golov *et al.* (2022).

$$\text{Vehicle operating cost (VOC)} = \text{VOC}_{\text{IRI}=1} \times [1 + 0.0083(\text{IRI} - 1)^2] \quad (28.1)$$

$$\text{Time cost (TC)} = \text{TC}_{\text{IRI}=1} \times [1 + 0.075(\text{IRI} - 1)^{1.1}] \quad (28.2)$$

There is nothing fundamental about these two equations or any of the many others that have been developed around the world; they are based on experience and estimation, and should definitely not be applied outside the limits of the data upon which they are based. The problem is that the economic equation is so different in different countries, even in different parts of one country, that it is quite impossible to give guidance other than that calibration is *always* required.

Accident costs can be linked to *pavement condition*, but only very approximately. Golov *et al.* (2022) plotted the effect of IRI on accident risk (from nine major roads in Russia) and found it to increase up to an IRI of about 5 m/km and then decrease, presumably as drivers were forced to drive slowly. This is broadly supported by a US study (Mamlouk *et al.*, 2018). The Russian study suggested an approximately 40% increase in accident risk at an IRI of 5 m/km compared to an IRI of 1 m/km (or indeed 10 m/km). Unsurprisingly, *skid resistance* also influences accident risk; for example, McCarthy *et al.* (2021), taking data from several thousand sections across three US states, found, on average, a factor of three on accident risk in wet weather between the best and worst levels of skid resistance, reducing to a factor of just two for dry conditions. Of course, risk then has to be translated into cost – and that is very much a political decision.

The issue of environmental costs is bound to grow in importance and in some countries there is now a degree of political will to address the environmentally damaging effects of congestion (social costs as well as pollutants) but, in the end, the pavement will have little effect on congestion until it reaches such a degraded state that it is well outside the range of pavement conditions dealt with in this book. And where pavements reach such a level, pollution and social issues are probably not top of the political agenda.

Energy usage, however, has a closer link to pavement type and condition. It is an area that is still subject to research, with claim and counter-claim. It is very clear that energy cost is related to overall pavement unevenness, as measured by the IRI and as included in equations such as Equation 28.1 for vehicle operating costs; it is less clear whether there is any significant difference between the energy required to travel on different pavement types, when in good condition (see Section 2.3.3).

Returning to the main issue, the key point is that network-level management should include a means of assessing the day-to-day cost of using each element of the network, and a convenient way of achieving this is to base equations on IRI or an alternative pavement condition indicator.

28.3.2 Practical management

In the end, network-level management is often simply a matter of avoiding damaging mistakes, particularly due to delaying maintenance too long. Every section of highway, airfield, cycle track, footway, industrial pavement and so on requires continuing investment in order to maintain it at a certain operational standard. In theory, it is possible to work out an approximate minimum expenditure with an optimum maintenance strategy for each individual section, but if the sum total of all these expenditures is not available each year then the task simply cannot be achieved. No matter how well optimised the network-level budget allocation, if the total spend is insufficient then the unavoidable consequence is a reduction in operational standard. Routine maintenance would have to be scaled back, which means that ride quality and safety standards would suffer, but with considerable savings to be made – of course at the expense of extra costs to the vehicle operators. These are all issues that are not difficult to explore once a sound foundation of project-level projections is made and a suitable relationship between pavement properties and 'running costs' is formulated. Unfortunately, integration of project-level predictions into network-level management requires a very clearly thought-out system, and so is generally only carried out in a relatively superficial manner.

28.4. Summary

Politicians would be lost without a supply of soundly based information from the real world and those politicians, both local and national, who deal with spend on transportation infrastructure are no exception. It is therefore very important that serious engineering input goes into the development of systems for managing pavements, alongside the required input from other disciplines (economics, sociology etc.). The trouble is that there are numerous different ways in which a pavement system can be mismanaged and funds wasted, and mismanagement is almost certain unless the practical skills of engineers are brought in.

REFERENCES

Golov E, Evtyukov S, Protsuto M, Evtyukov S and Sorokina E (2022) Influence of the road surface roughness (according to International Roughness Index) on road safety. *Transportation Research Procedia* **63**: 999–1006.

Mamlouk M, Vinayakamrthy M, Underwood BS and Kaloush KE (2018) Effects of the International Roughness Index and rut depth on crash rates. *Transportation Research Record* **2672(40)**: 418–429.

McCarthy R, Flintsch G and Izeppi E de L (2021) Impact of skid resistance on dry and wet weather crashes. *Journal of Transportation Engineering, Part B: Pavements* **147(3)**: 04021029.

Zaabar I and Chatti K (2014) Estimating vehicle operating costs caused by pavement surface conditions. *Transportation Research Record* **2455(1)**: 63–76.

Nick Thom
ISBN 978-1-83549-711-1
https://doi.org/10.1108/978-1-83549-710-420243035

Index

The manufacturer's authorised representative in the EU for product safety is Easy Access System Europe – Mustamäe tee 50, 10621 Tallinn, Estonia, easproject.com, +372 5696 8939, gpsr.requests@easproject.com

Printed and bound by CPI Group (UK) Ltd, Croydon, CR0 4YY
16/07/2026
02168889-0002